T0330997

Sustainable Management of Environmental Pollutants through Phytoremediation

Traditional remedial technologies can be cost-prohibitive and sometimes contribute to environmental contamination themselves. In order to better manage the issues of global pollution, phytoremediation, a plant-based cleanup method, has gained attention as an efficient, affordable, and environmentally sustainable alternative to traditional remedial technologies for the cleanup of a variety of hazardous pollutants. The demand for advanced technologies having potential to sustainably manage waste and pollutants in the environment will help to continue the quest for more novel treatment methods. *Sustainable Management of Environmental Pollutants through Phytoremediation* discusses all the aspects of sustainable environmental management through phytoremediation, making it a valuable resource for both academics and researchers in developing and developed countries.

- Examines technology advancements made toward the recycling and management of waste.
- Designed in a way to cover scientific principles, modeling and methods, designs, and reference data.
- Discusses the utilization of waste for renewable energy for economic growth and further social benefits.

Sustainable Management of Environmental Pollutants through Phytoremediation

Edited by
Ananya Kuanar, Alok Prasad Das, Dattatreya Kar,
and Maulin P. Shah

CRC Press
Taylor & Francis Group
Boca Raton London New York

CRC Press is an imprint of the
Taylor & Francis Group, an **informa** business

Designed cover image: Shutterstock

First edition published 2025
by CRC Press
2385 NW Executive Center Drive, Suite 320, Boca Raton FL 33431

and by CRC Press
4 Park Square, Milton Park, Abingdon, Oxon, OX144RN

CRC Press is an imprint of Taylor & Francis Group, LLC

© 2025 selection and editorial matter, Ananya Kuanar, Alok Prasad Das, Dattatreya Kar, and Maulin P. Shah individual chapters, the contributors

Library of Congress Cataloging-in-Publication Data
Names: Kuanar, Ananya, editor. | Das, Alok Prasad, editor. | Kar,
Dattatreya, editor. | Shah, Maulin P., editor.
Title: Sustainable management of environmental pollutants through
phytoremediation / edited by Ananya Kuanar, Alok Prasad Das,
Dattatreya Kar, and Maulin P. Shah.
Description: First edition. | Boca Raton, FL : CRC Press, 2025. |
Includes bibliographical references and index. |
Identifiers: LCCN 2024024502 | ISBN 9781032580517 (hardback) |
ISBN 9781032580524 (paperback) | ISBN 9781003442295 (ebook)
Subjects: LCSH: Phytoremediation.
Classification: LCC TD192.75 .S87 2025 | DDC 628.4--dc23/eng/20241023
LC record available at https://lccn.loc.gov/2024024502

ISBN: 978-1-032-58051-7 (hbk)
ISBN: 978-1-032-58052-4 (pbk)
ISBN: 978-1-003-44229-5 (ebk)

DOI: 10.1201/9781003442295

Typeset in Times
by KnowledgeWorks Global Ltd.

Contents

About the Editors

Dr. Ananya Kuanar is a plant biotechnologist with diverse research interests. Her research area is based on plant tissue culture, somaclonal variation, molecular analysis, and neural network modeling. Further, her interest is in metabolic profiling of bioactive natural products using analytical approaches and phytoremediation.

Dr. Alok Prasad Das's work is focused on fundamental research in the development of indigenous biomedical and healthcare instrumentation technology and technology modernization in the area of indigenous diagnostics. In addition to this, he is also investigating biosensors for rapid endotoxin detection in fluid systems used for the production of clinically applicable compounds and the development of simple single-step chromogenic methods for rapid detection of food pathogens and toxins. His current Google Scholar Citations is 1800 and h-index 23.

Dr. Dattatreya Kar is a plant biotechnologist with diverse research interests. His research area is metabolic profiling of bioactive natural products using analytical approaches, phytoremediation, and molecular analysis.

Dr. Maulin P. Shah has been an active researcher and scientific writer in his field for over 20 years. He received a B.Sc. degree (1999) in microbiology from Gujarat University, Godhra (Gujarat), India. He earned his Ph.D. degree (2005) in environmental microbiology from Sardar Patel University, Vallabh Vidyanagar (Gujarat), India. His research interests include biological wastewater treatment, environmental microbiology, biodegradation, bioremediation, and phytoremediation of environmental pollutants from industrial wastewaters. He has published more than 350 research papers in national and international journals of repute on various aspects of microbial biodegradation and bioremediation of environmental pollutants. He is the editor of 200 books of international repute and has edited 30 special issues specifically in industrial wastewater research, microbial remediation, and biorefinery of wastewater treatment. He is associated as an editorial board member with 30 highly reputed journals.

1 Post-Phytoremediation Strategies for Biomass Management Toward Its Sustainable Utilization and Economic Opportunities in the Context of Heavy Metals Pollution

Soumya Ranjan Patra, Sandeep Kumar Kabi, Manish Kumar, and Nabin Kumar Dhal

INTRODUCTION

Along with significant advances in phytoremediation research, there has been a rise in practical field applications. The wide range of contaminants that are removed by this environmentally friendly treatment method include petroleum hydrocarbons, antibiotics, dangerous heavy metals, emerging pollutants including polychlorinated hydrocarbons, pesticides, and many more (Mushtaq et al., 2020). The methods for handling biomass from post-phytoremediation heavy metal contamination do present some problems, though. This chapter addresses post-phytoremediation biomass as a problem. However, because most noninfectious biological waste is biodegradable, and because one organism's waste is another's raw material (Tian et al., 2021), the same principle holds true in this case. Innovative techniques must be explored to recycle used biomass. Therefore, innovative and fascinating biomass waste management strategies are the topic of discussion.

HEAVY METALS AND THEIR EFFECTS ON LIVING ORGANISMS

Most heavy metals have high atomic weights, specific gravities more than 5, and high atomic numbers. A few metalloids, transition metals, basic metals, lanthanides, and actinides are also considered to be heavy metals. The majority of heavy metals,

DOI: 10.1201/9781003442295-1

including arsenic, mercury, lead, cadmium, cobalt, tin, and nickel, are harmful to life (Järup, 2003). In trace concentrations, elements like copper, chromium, iron, zinc, selenium, and others are both nutrients and poisonous heavy metals when they build up excessively in the body. Because these heavy metals are poisonous, heavy, and constitute a health risk due to their widespread use in society, some of their impacts are listed here (Muthusaravanan et al., 2020).

- **Lead:** Lead exposure can result in neurological damage, especially in young children. It affects brain development, leading to learning disabilities, reduced IQ, behavioral problems, and hearing loss. Lead exposure also affects adults, causing high blood pressure, kidney damage, and reproductive issues (Kamran et al., 2013).
- **Mercury:** Mercury primarily affects the nervous system. It can cause cognitive and motor function impairment, memory loss, tremors, and developmental delays in children. Prolonged exposure to high levels of mercury can lead to severe neurological damage (Chen et al., 2022).
- **Arsenic:** Arsenic exposure is linked to various health issues, including skin lesions, cancer (skin, lung, bladder), cardiovascular disease, and respiratory problems. Long-term exposure can result in chronic arsenic poisoning, leading to organ damage and an increased risk of cancer.
- **Cadmium:** Cadmium exposure primarily affects the kidneys, leading to kidney damage and impaired kidney function. It can also cause lung damage, resulting in respiratory problems. Long-term exposure to cadmium is associated with an increased risk of lung cancer.
- **Chromium:** Hexavalent chromium, a highly toxic form of chromium, can cause severe respiratory problems, including lung cancer (Mancuso, 1997). It can also lead to skin irritation, ulcers, and dermatitis. Ingesting large amounts of trivalent chromium can result in gastrointestinal issues.

The effects of heavy metals on living organisms vary depending on the specific metal, duration of exposure, concentration, and individual susceptibility.

CAUSES OF HEAVY METAL CONTAMINATION

Metallic compounds are now more widespread both on land and in water due to a sharp increase in the use of heavy metals (Kapoor & Singh, 2021). Anthropogenic activity is the primary source of heavy metal pollution (Cristaldi et al., 2017). This activity results in the extraction of metals from their natural sources through excretion, trash dumps, landfills, runoffs, cattle manure, and chicken dung, as well as smelting and usage in foundries and other businesses (Järup, 2003; Pandey & Bajpai, 2019). The use of insecticides, pesticides, fertilizers, and other agricultural practices that contain heavy metals is a secondary source of contamination. Additionally, natural factors including volcanic activity, metal corrosion, metal evaporation from soil and water and sediment resuspension, soil erosion, and geological weathering can increase heavy metal pollution.

DIFFERENT REMEDIATION PROCESSES

There are different methods for remediation of heavy metals from soil: physical, chemical, and biological.

- **Physical methods:** In this approach, treatments involve physical activities like *soil replacement* (off-site disposal and replacement of contaminated soil with noncontaminated soil through excavation), *soil isolation* (*in situ* techniques like surface capping (covering of contaminated site with a layer of water proof material) and encapsulation (often known as the "barrier wall," "cutoff wall," or "liner" approach, is a remedial alternative to surface capping)) (Song et al., 2016), *vitrification* (the application of high-temperature treatment, which produces vitreous material, in order to restrict the mobility of heavy metal[loid]s inside soil), and *electrokinetic remediation* (by creating an electric field gradient with the right intensity on both sides of an electrolytic tank containing saturated contaminated soil, it is possible to remove heavy metals and their metabolites. The contamination is diminished by separating the heavy metal[loid]s in the soil via electrophoresis, electric seepage, or electro-migration.) (Khalid et al., 2017).
- **Chemical methods:** Soils are chemically treated to neutralize the toxic effects of heavy metals by the use of various chemicals and chelating agents. This approach includes *immobilization techniques* (by adding immobilizing agents to the polluted soils, the mobility, bioavailability, and bioaccessibility of heavy metal[loid]s in soil are reduced.), *encapsulation* (the immobilization of poisonous metal solutions by encasing them in controllable solid blocks), and *soil washing* (mobilizing the heavy metals from soil by using different reagents and extractants like chelating agents EDTA and EDDS) (Khalid et al., 2017).
- **Biological methods:** The most harmless and environmentally friendly methods to reduce the toxicity naturally; these include *phytoremediation* (hyper-accumulating plants are used to remove pollutants by chelate-assisted phytostabilization, phytovolatilization, and phytoextraction) and *microbial-assisted phytoremediation* (the process of using microbes to encourage the soil's heavy metal[loid] content to be absorbed, precipitated, oxidized, and reduced. These bacteria in the soil can shield plants from the harmful effects of heavy metal[loid]s or even encourage hyperaccumulator plants to take up more metal; Khalid et al., 2017; Figure 1.1).

EMERGING RISKS AND GAPS IN POST-PHYTOREMEDIATION BIOMASS MANAGEMENT

Using plants to remove, degrade, or stabilize pollutants from soil, water, or the air is known as phytoremediation (Pilon-Smits, 2005). There are a number of dangers and gaps that may develop in the handling of the resultant biomass after the

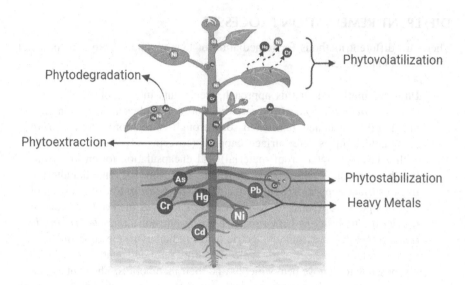

FIGURE 1.1   Schematic representation of phytoremediation strategies.

phytoremediation process is finished. The following are a few of the new dangers and holes in post-phytoremediation biomass management:

1. **Contaminant concentration:** While phytoremediation can effectively reduce contaminant levels, the biomass generated may still contain residual contaminants. Proper management is required to ensure that the biomass does not pose a risk of recontaminating the environment or entering the food chain (Shen et al., 2022).

2. **Disposal challenges:** The disposal of phytoremediation biomass can be challenging. Depending on the level of contamination, it's possible that the biomass will be labeled as hazardous waste, necessitating special handling and disposal procedures. Identifying appropriate disposal options that minimize environmental impact is crucial (Abhilash & Yunus, 2011).

3. **Bioavailability of contaminants:** Phytoremediation can sequester contaminants within plant tissues or immobilize them in the soil. However, certain conditions, such as changes in soil pH or microbial activity, can potentially release these contaminants back into the environment. Understanding the long-term stability and bioavailability of contaminants in phytoremediation biomass is important to prevent recontamination.

4. **Potential secondary effects:** The use of certain plant species for phytoremediation may introduce new risks or gaps in biomass management. For example, invasive plant species used for remediation can pose a threat if their biomass is not properly managed after the process. Careful consideration of the selection and management of phytoremediation plant species is necessary to prevent unintended ecological consequences.

5. **Economic viability:** The economic viability of post-phytoremediation biomass management can be a challenge. Large-scale phytoremediation

projects can generate substantial amounts of biomass, requiring significant resources for its management and disposal. Developing cost-effective strategies for handling and utilizing the biomass, such as energy production or material recovery, is crucial to ensure the sustainability of phytoremediation projects (Muthusaravanan et al., 2020).

Addressing these emerging risks and gaps in post-phytoremediation biomass management requires multidisciplinary collaboration among scientists, environmentalists, regulators, and industry stakeholders. Continued research and innovation in biomass management technologies, along with the development of comprehensive guidelines and policies, are essential to maximize the environmental benefits of phytoremediation while minimizing potential risks.

SUSTAINABLE UTILIZATION OF POST-PHYTOREMEDIATION BIOMASS

The fate of post-phytoremediation biomass depends on several factors, including the level of contamination, the type of contaminants present, and the regulations and guidelines of the particular region. Following are some possible options for managing post-phytoremediation biomass.

COMPOSTING

Organic materials, such as post-phytoremediation biomass, are naturally decomposed through the process of composting to create a nutrient-rich soil supplement known as compost. To break down the biomass, the composting process needs the proper proportion of organic materials, moisture, oxygen, and microbes. (Lin et al., 2022).

Here's a general overview of the composting process for biomass:

a. **Shredding:** The post-phytoremediation biomass is typically shredded or chopped into smaller pieces to increase the surface area and promote faster decomposition.

b. **Mixing:** The shredded biomass is mixed with other organic materials like yard waste, food scraps, or agricultural residues to provide a balanced carbon-to-nitrogen ratio and ensure proper decomposition. This mixture can also include bulking agents like wood chips or straw to improve aeration.

c. **Piling or composting bin:** The biomass mixture is piled or placed in a composting bin or container. The size of the pile or bin affects the composting process, with larger piles generally retaining heat better.

d. **Moisture management:** The biomass should have adequate moisture content for composting, typically around 40–60%. Water may need to be added if the mixture is too dry or turned if it's too wet to maintain the ideal moisture level.

e. **Turning:** Regular turning or mixing of the compost pile helps to aerate the biomass, ensuring oxygen supply for the microorganisms involved in the decomposition process. This turning can be done manually or with the help of composting equipment.

f. **Decomposition and maturation:** Over time, the biomass undergoes decomposition by microorganisms, including bacteria, fungi, and other composting organisms. The process generates heat, and the temperature inside the compost pile rises. This heat helps break down the organic matter, including any contaminants present, into more stable and less harmful forms.

g. **Curing and maturing:** After the active composting phase, the compost pile is allowed to cure and mature. This allows any remaining organic matter to further break down, resulting in a stable, dark, and crumbly compost.

COMPACTION

Compaction is a process that involves reducing the volume or bulk density of biomass through mechanical means. It is often used to facilitate transportation, storage, or disposal of large quantities of biomass (Muthusaravanan et al., 2020).

The compaction process typically involves the following steps:

a. **Preparation:** The biomass is collected and, if necessary, preprocessed to remove large debris or foreign objects that may hinder the compaction process.

b. **Compaction equipment:** Specialized compaction equipment, such as balers or compactors, is used to compress the biomass. These machines apply pressure to reduce the volume and increase the density of the biomass.

c. **Compression:** The biomass is loaded into the compaction equipment, which compresses it into compacted blocks, bales, or pellets. The specific method and size of the compacted biomass depend on the equipment used and the desired end product.

d. **Handling and storage:** The compacted biomass is then ready for transportation, storage, or further processing. It can be more efficiently transported to a composting facility, a landfill, or a biomass energy plant, depending on the intended use.

It's important to note that composting and compaction are distinct processes, and the choice between them depends on the desired outcome and the specific requirements of the biomass management plan. Composting focuses on organic matter decomposition and nutrient recycling, while compaction primarily aims to reduce volume for logistical purposes.

COMBUSTION AND GASIFICATION

Both combustion and gasification can be used for the thermal conversion of post-phytoremediation biomass into energy. However, the suitability of each method is dependent on various elements, including the biomass's composition, its impurities, and the intended application of the energy (Vocciante et al., 2019). Here's an overview of how combustion and gasification can be applied to post-phytoremediation biomass:

Combustion

To generate heat or power, combustion entails burning biomass in the presence of oxygen. If the pollutants contained in the post-phytoremediation biomass are not extremely volatile, poisonous, or ecologically sensitive, burning may be a good alternative. The following are ways that the combustion process can be beneficial:

a. Organic pollutants can be successfully destroyed by high-temperature burning, which can also lower the quantities of these contaminants and lessen their environmental impact.
b. Heat and power generation: Steam can be created by using the heat emitted during combustion to turn turbines and create energy. The heat can also be applied to heating spaces or industrial processes.
c. Emission control technology can be added to combustion systems to lower air pollution. Particulate matter, nitrogen oxides (NOx), sulfur oxides (SOx), and other pollutants can be minimized by using technology such as particulate filters, scrubbers, and catalysts.

Gasification

By using a thermochemical process termed gasification, biomass is transformed into a gas mixture known as syngas. Carbon monoxide (CO), hydrogen (H_2), carbon dioxide (CO_2), and other trace gases make up the majority of the syngas (Edgar et al., 2021). Post-phytoremediation biomass can be gasified for a number of advantages:

a. **Syngas production:** Post-phytoremediation biomass can be gasified into syngas, which can be used as a flexible energy source. The syngas can be used to produce electricity, as fuel for motors or turbines, or as a raw material for the synthesis of chemicals and biofuels.
b. **Contaminant separation:** The gasification process can enable the separation of contaminants from the biomass. Depending on the contaminants involved, they can be captured as ash or in other forms during the gasification process, reducing their environmental impact.
c. **Emissions control:** Gasification systems can incorporate gas-cleaning technologies to remove impurities from the syngas. These technologies, including tar removal, gas scrubbing, and catalytic conversion, help in producing a cleaner syngas suitable for various applications.

The choice of combustion or gasification for post-phytoremediation biomass should be based on a careful analysis of the unique biomass features, pollutant profiles, environmental factors, and the intended energy usage. For the purpose of processing and converting post-phytoremediation biomass in the most suitable and ecologically responsible manner, site-specific studies and expert assistance are advised.

BIOLEACHING THROUGH MICROBES: BIOSORBENT PREPARATION

In the bioleaching process, precious metals are removed from ores and other materials using microorganisms. Microbes, including bacteria and fungi, are essential for dissolving the minerals and liberating the desired metals. Contrarily, biosorption describes the process of removing metals and other contaminants from solutions using biomass, either living or nonliving. Both procedures provide ecologically suitable substitutes for conventional metal recovery and mining techniques.

Here's an overview of the preparation steps for both bioleaching and biosorbent materials:

Bioleaching

Utilizing microorganisms to remove metals from ores and concentrates is known as bioleaching. The bacteria are grown in a controlled setting before being added to the ore. They interact with the ore minerals, solubilizing the metals and releasing them into the solution.

1. **Microorganism selection:** Choose microorganisms that are capable of metabolizing the target minerals. Common bioleaching microorganisms include acidophilic bacteria such as *Acidithiobacillus ferrooxidans* and *Acidithiobacillus trioxidanes*.
2. **Culture medium preparation:** Prepare a suitable growth medium that provides the necessary nutrients for the microorganisms to thrive. The medium should mimic the conditions found in the target ore environment, such as low pH and high metal content.
3. **Inoculation:** Inoculate the selected microorganisms into the culture medium. Allow them to grow and multiply under controlled conditions.
4. **Ore preparation:** Crush and grind the ore material to increase its surface area and facilitate microbial interaction.
5. **Bioleaching setup:** Create a bioleaching reactor or tank where the ore and microbial culture are combined. Maintain optimal conditions for microbial growth, including temperature, pH, and aeration.
6. **Metal recovery:** As the microorganisms interact with the ore, they oxidize the minerals and solubilize the target metals. The metals are then collected from the solution using precipitation or other separation techniques.

Biosorbent Preparation

Biosorbents are materials that can adsorb and accumulate metals from solution. These can be living organisms, such as algae or bacteria, or nonliving materials like agricultural waste or modified polymers.

1. **Biosorbent selection:** Choose an appropriate biosorbent material based on its affinity for the target metals. This could include algae, bacteria, fungi, activated carbon, or other organic materials.
2. **Biosorbent pretreatment:** Clean and prepare the biosorbent material to remove impurities and ensure its effectiveness in metal uptake.
3. **Surface activation (if needed):** In some cases, the biosorbent's surface can be modified or activated to enhance its metal-binding capacity. This might involve chemical treatments or modifications.

4. **Equilibrium studies:** Conduct equilibrium studies to determine the optimum conditions for metal binding, including pH, contact time, temperature, and concentration.
5. **Batch or column studies:** Perform batch or column experiments to assess the biosorption capacity of the material under realistic conditions.
6. **Regeneration (if applicable):** Depending on the purpose, the biosorbent may be renewed and utilized again. This procedure desorbs the metals from the biosorbent so that it can be used repeatedly.

Both bioleaching and biosorption are adaptable and eco-friendly methods for removing and extracting metals. The decision between them is influenced by various elements, including the type of ore or waste, the target metal, and the particular process objectives.

BIO-OIL

Bio-oil can be produced from post-phytoremediation biomass through a process called pyrolysis. Pyrolysis involves heating biomass in the absence of oxygen to break down the organic materials into different components, including bio-oil, biochar, and gases (Khan et al., 2023). Here's a general procedure for preparing bio-oil from post-phytoremediation biomass (Suer & Andersson-Sköld, 2011):

1. **Biomass collection and preparation:** Collect the biomass that has been used for phytoremediation. This biomass could be plants that have absorbed contaminants from the soil. Remove any large debris, rocks, or non-biomass materials from the collected biomass.
2. **Biomass drying:** To improve the efficiency of the pyrolysis process, the biomass needs to be dried to reduce its moisture content. High moisture content can lead to incomplete pyrolysis and lower bio-oil yields.
3. **Size reduction:** The biomass should be chopped, shredded, or ground into smaller pieces. This increases the surface area of the biomass, promoting better heat transfer during pyrolysis.
4. **Pyrolysis process:** The pyrolysis process involves heating the biomass in the absence of oxygen. This can be achieved using various types of pyrolysis reactors, such as fixed-bed reactors, fluidized bed reactors, or rotary kilns. The temperature range for bio-oil production typically lies between 400°C and 600°C.
5. **Vapor condensation:** As the biomass is heated, it releases vapors containing bio-oil and other volatile compounds. These vapors need to be condensed into liquid form. This is usually done by passing the vapors through a cooling system, such as a condenser, where they are cooled and condensed into bio-oil.
6. **Bio-oil separation and upgrading:** The condensed bio-oil contains a mixture of organic compounds, including water, acids, phenols, and hydrocarbons. It is often necessary to separate and upgrade the bio-oil by removing impurities and undesirable components. This can involve processes like filtration, centrifugation, and chemical treatment.

7. **Storage and analysis:** The purified bio-oil can be stored in appropriate containers for further analysis and testing. Analytical techniques, such as gas chromatography-mass spectrometry (GC-MS), can be used to determine the composition of the bio-oil and ensure its quality.

It's crucial to keep in mind that the particular pyrolysis process settings and parameters will rely on elements like the kind of biomass, the contaminants it has absorbed, the desired grade of the bio-oil, and the kind of pyrolysis reactor employed. In order to further refine the bio-oil and make it appropriate for different purposes, such as the generation of biofuel or chemical feedstock, post-processing might be required (Shen et al., 2022).

The production of bio-oil through pyrolysis is a complex process that requires careful optimization and consideration of various factors to achieve the desired yield and quality of bio-oil.

BIOGAS

Biogas production from post-phytoremediation biomass can be a beneficial way to utilize and manage the biomass generated during the phytoremediation process. After plants have absorbed or accumulated contaminants from soil, water, or air (He et al., 2020), they can be harvested, and the biomass can be processed for biogas production.

Here are the steps involved in biogas production from post-phytoremediation biomass:

1. **Biomass collection:** Harvest the plants used in the phytoremediation process once they have reached maturity and have absorbed the contaminants. The biomass can include various plant parts such as leaves, stems, and roots.
2. **Biomass preparation:** Sort and clean the harvested biomass to remove any debris or unwanted materials. The biomass may also need to be chopped or shredded to increase its surface area, which facilitates the biogas production process.
3. **Anaerobic digestion:** Anaerobic digestion is the main process used to convert biomass into biogas. In this step, the biomass is added to an anaerobic digester, which is a sealed container where microorganisms break down the organic matter in the absence of oxygen. The digester can be a large tank or a specialized system designed for biomass digestion.
4. **Microbial activity:** Inside the anaerobic digester, bacteria and other microorganisms decompose the organic matter in the biomass through a series of biological reactions. This process produces biogas, which is primarily composed of methane (CH_4) and carbon dioxide (CO_2).
5. **Biogas collection:** The biogas produced during anaerobic digestion is collected and stored for further use. The digester system includes mechanisms to capture the biogas, which can then be directed to storage tanks or used immediately as an energy source.

6. **Biogas utilization:** The collected biogas can be used for various purposes. It can be burned as fuel to generate heat and electricity or processed further to remove impurities and increase the methane content for use as a renewable natural gas (RNG) or vehicle fuel.

It's important to note that the composition and quality of the biomass can affect the biogas production process. Certain contaminants absorbed by the plants during phytoremediation may have inhibitory effects on the anaerobic digestion process. Therefore, it is crucial to assess the biomass quality and potential contaminants before utilizing it for biogas production.

BIOFORTIFICATION

Biofortification of post-phytoremediation biomass involves enriching the harvested plant material with essential nutrients, aiming to improve its nutritional value. This can be especially valuable when using plants that have undergone phytoremediation to clean up contaminated soils. Here's a general procedure for biofortifying post-phytoremediation biomass:

1. **Selection of suitable plants:** Choose plant species that are not only effective in phytoremediation but are also suitable candidates for biofortification. Look for plants with the ability to accumulate and tolerate essential nutrients.
2. **Phytoremediation process:** Cultivate the selected plants in contaminated soils to facilitate phytoremediation. Allow the plants to accumulate and sequester heavy metals or other contaminants from the soil.
3. **Harvesting biomass:** Once the phytoremediation process is complete, harvest the plant biomass. Ensure that the harvested biomass is free from contaminants and thoroughly cleaned.
4. **Nutrient application:** Apply nutrient sources to the harvested biomass to enhance its nutritional content. Common nutrient sources include fertilizers, foliar sprays, or nutrient-enriched solutions.
5. **Biofortification techniques:** There are different methods to introduce nutrients into the biomass:
   a. **Fertilizer application:** Apply fertilizers containing the desired nutrients to the soil or growing medium where the biomass will be cultivated. The plants will take up these nutrients during growth.
   b. **Foliar spray:** Spray a nutrient solution directly onto the leaves of the harvested biomass. Nutrients can be absorbed through the leaves and transported to other plant parts.
   c. **Hydroponic cultivation:** If suitable, grow the harvested biomass hydroponically in a nutrient-rich solution to ensure direct uptake of essential nutrients.
6. **Monitoring and adjustments:** Regularly monitor the nutrient content of the biomass during the biofortification process. Adjust nutrient application rates as needed to achieve the desired nutritional enhancement.

7. **Harvest and processing:** After the biofortification process, harvest the biomass again. This biomass is now enriched with essential nutrients.
8. **Processing and utilization:** Process the biofortified biomass into various products like food, supplements, or animal feed. Depending on the target application, the biomass can be further processed, dried, or incorporated into different products.
9. **Nutrient retention and quality assurance:** Test and verify the nutritional content of the final biofortified products to ensure that the desired nutrient levels have been achieved and maintained.
10. **Distribution and consumption:** Distribute the biofortified products to appropriate markets or consumers, making sure to communicate the enhanced nutritional value.

It's important to note that biofortification requires careful planning, monitoring, and adherence to safety guidelines. The exact procedure may vary based on the type of plant, the target nutrients, and the intended application of the biofortified biomass.

Nanoparticle Synthesis

Utilizing plant material that has been used to remove or remediate heavy metals or other environmental contaminants from the environment is a necessary step in the process of creating nanoparticles from post-phytoremediation biomass. The generated biomass has the potential to be an important source for the manufacture of nanoparticles. The following is a general process for creating nanoparticles from biomass after phytoremediation:

1. **Collection and processing of biomass:** Harvest the plant material that has been used for phytoremediation. Ensure that the collected biomass is thoroughly cleaned and dried to remove any contaminants and moisture.
2. **Biomass pretreatment:** Depending on the type of biomass, it might need to be further processed before nanoparticle synthesis. This could involve grinding, milling, or any other method to increase the surface area and accessibility of the biomass.
3. **Nanoparticle extraction:** There are various methods to extract nanoparticles from the biomass. One common approach is to use extracts of the biomass obtained through processes like maceration or sonication. These extracts often contain natural compounds that can act as reducing and stabilizing agents for nanoparticle synthesis.
4. **Nanoparticle synthesis:** The exact synthesis method depends on the type of nanoparticles you want to create. Here are a few common techniques:
   a. **Green synthesis:** Utilize the bioactive compounds present in the biomass extracts as reducing and stabilizing agents. Mix the biomass extract with metal precursors and heat or stir the mixture to induce nanoparticle formation.
   b. **Microbial reduction:** Introduce microbial cultures to the biomass extracts containing metal ions. Microorganisms can reduce metal ions and promote nanoparticle formation.

    c. **Hydrothermal synthesis:** Subject the biomass extracts to high-temperature and high-pressure conditions. This can lead to the formation of nanoparticles.

5. **Characterization:** It's essential to characterize the nanoparticles after production to confirm their properties. The size, shape, crystal structure, and optical properties of the nanoparticles can be determined using methods including X-ray diffraction (XRD), transmission electron microscopy (TEM), and UV-visible spectroscopy.

6. **Stabilization and storage:** To prevent aggregation and ensure stability, the synthesized nanoparticles might need to be coated or capped with biomolecules or stabilizing agents. This step helps preserve the nanoparticles' properties over time.

7. **Application:** The synthesized nanoparticles can find applications in various fields, such as medicine, catalysis, and environmental remediation.

It's important to note that the exact procedure can vary based on factors like the type of post-phytoremediation biomass, the desired nanoparticles, and the specific contaminants present in the environment. Additionally, working with nanoparticles requires proper safety precautions due to their potential health and environmental impacts.

Solid/Composite Wood Products

Perennial plants utilized in phytoremediation have been taken from the treated site and used to create items out of solid and composite wood. The birch tree's tolerance to a metal was directly influenced by the physicochemical properties of the soil where it was grown; the birch clone that was taken from a metal-contaminated site had great resistance to related metals but low resistance to other metals. As a result, *in situ* operations for birch tree phytoremediation of heavy metals should be planned. When it comes to the phytoremediation of heavy metals, trees offer numerous benefits over small plants. Compared to small plants, trees produce more biomass and have a larger capacity for metal accumulation, which results in higher metal accumulation per kilogram of biomass per square area. According to Jobling and Stevens (1980), however, the physical and hydrological characteristics of trees make them unsuitable for remediating mine debris heaps. The following steps were taken to transform wood logs into timber:

1. Mature tree from treated site was identified and harvested.
2. The harvested logs are stored in the site temporarily to evaporate the free water content and to cut the small side branches.
3. The harvested logs are transported to the sawmill using trucks equipped with lift gears.
4. In the sawmill, the logs were processed into wooden boards using round and band saw machines, and the process involved rough cutting of logs (breaking) and resawing to improve the precision and accuracy of cutting.

Phyto-Mining

Phyto-mining of post-phytoremediation biomass involves utilizing plants that have been grown to remediate contaminated soils to also extract valuable metals or minerals from the soil. This process can help recover economically valuable resources while cleaning up the environment (Brooks et al., 1998). Here's a general procedure for phyto-mining of post-phytoremediation biomass (Sheoran et al., 2009):

1. **Selection of suitable plants:** Choose hyperaccumulator plant species that have the ability to accumulate high concentrations of metals or minerals in their tissues. These plants are essential for effective phyto-mining.
2. **Phytoremediation process:** Cultivate the selected hyperaccumulator plants in contaminated soils to facilitate the phytoremediation process. Allow the plants to accumulate and sequester metals or minerals from the soil.
3. **Harvesting biomass:** Once the phytoremediation process is complete, harvest the hyperaccumulator plant biomass. Ensure that the harvested biomass is free from contaminants and thoroughly cleaned.
4. **Metal extraction:** Extract the valuable metals or minerals from the harvested biomass. Different techniques can be used for extraction, including:
   a. **Acid leaching:** Treat the biomass with an acidic solution that dissolves and releases the target metals.
   b. **Biological leaching:** Use microorganisms to break down the biomass and release metals through natural processes.
   c. **Thermal decomposition:** Subject the biomass to high temperatures to volatilize and collect metals in vapor form.
   d. **Chemical extraction:** Utilize specific chemical reagents to selectively extract metals from the biomass.
5. **Metal recovery:** Collect and process the extracted metals or minerals to obtain usable metal concentrates. Depending on the metals and minerals involved, further refining and processing might be necessary.
6. **Environmental monitoring:** Monitor the soil and environment to ensure that the phyto-mining process has not caused significant ecological harm. Evaluate the potential for any residual contamination.
7. **Economic assessment:** Evaluate the economic feasibility of the phyto-mining process by comparing the costs of cultivation, harvesting, and extraction with the value of the recovered metals.
8. **Processing and utilization:** Process the recovered metals or minerals into usable forms for industrial or commercial applications.
9. **Restoration and replanting:** After phyto-mining, consider replanting the site with suitable vegetation to restore the ecosystem and prevent soil erosion.
10. **Regulatory compliance:** Make sure that all legal and environmental criteria are upheld during the entire phyto-mining process.

The viability of the complex process of phyto-mining, which depends on factors like the type of metals or minerals, the hyperaccumulator plants, and the local environmental conditions, must be understood. Careful planning, monitoring, and safety precautions are necessary for phyto-mining to be successful and have the least detrimental consequences on the ecosystem.

CONCLUSIONS

To ensure a project's long-term success and sustainability following the application of phytoremediation, a post-phytoremediation management plan must be created. The following are some important recommendations and post-phytoremediation management considerations:

1. **Monitoring and maintenance:** Consistent monitoring is necessary to evaluate phytoremediation's efficacy and spot any possible problems or recontamination. Testing of the soil, water, and plant health should all be done as part of the monitoring process. To ensure the continuous development and health of the chosen plants, maintenance practices including watering, trimming, and eliminating invasive species should be continued.
2. **Risk assessment:** It is essential to carry out a thorough risk assessment to ascertain whether any lingering toxins pose a threat to both human health and the environment.
3. **Stakeholder engagement:** Engaging stakeholders throughout the post-phytoremediation phase is important to maintain transparency and gain support for the project. Regular communication with local communities, regulatory authorities, and other relevant parties should be established to address concerns, provide updates on progress, and address any potential issues.
4. **Long-term land use planning:** Depending on the site's intended future use, land use planning should be considered in the post-phytoremediation phase. This may involve deciding whether the area will be repurposed for industrial, residential, recreational, or agricultural purposes. The selection of appropriate plant species for land reclamation or land cover purposes should be based on the site's future land use objectives.
5. **Ecological restoration**: Phytoremediation can significantly improve the ecological condition of contaminated sites. However, additional ecological restoration measures may be necessary to enhance biodiversity and restore the natural habitat. This could involve reintroducing native plant species, creating wildlife habitats, or implementing other ecological restoration techniques.
6. **Documentation and knowledge sharing:** Comprehensive documentation of the phytoremediation project, including data collection, methodologies, and outcomes, is essential. This information can serve as a valuable resource for future projects and contribute to the broader knowledge base of phytoremediation. Sharing the lessons learned and best practices with the scientific community and relevant stakeholders can further advance the field and encourage the adoption of phytoremediation in similar contexts.

Finally, post-phytoremediation management ought to prioritize ongoing observation, risk evaluation, stakeholder involvement, long-term land use planning, ecological restoration, and documentation. By taking care of these issues, the site can be properly managed following phytoremediation, resulting in a successful restoration of the environment and preservation of human health.

REFERENCES

Abhilash, P. C., & Yunus, M. (2011). Can we use biomass produced from phytoremediation? *Biomass and Bioenergy*, 35(3), 1371–1372. https://doi.org/10.1016/j.biombioe.2010. 12.013

Brooks, R. R., Chambers, M. F., Nicks, L. J., & Robinson, B. H. (1998). *Phytomining*, 3(9). https://doi.org/10.1016/S1360-1385(98)01283-7

Chen, W., Guan, Y., Chen, Q., Ren, J., Xie, Y., & Yin, J. (2022). The mark of Mercury(II) in living animals and plants through using a BODIPY-based near-infrared fluorescent probe. *Dyes and Pigments*, 200, 110134. https://doi.org/10.1016/J.DYEPIG.2022.110134

Cristaldi, A., Conti, G. O., Jho, E. H., Zuccarello, P., Grasso, A., Copat, C., & Ferrante, M. (2017). Phytoremediation of contaminated soils by heavy metals and PAHs. A brief review. *Environmental Technology & Innovation*, 8, 309–326. https://doi.org/10.1016/J.ETI.2017.08.002

Edgar, V. N., Fabián, F. L., Julián Mario, P. C., & Ileana, V. R. (2021). Coupling plant biomass derived from phytoremediation of potential toxic-metal-polluted soils to bioenergy production and high-value by-products-a review. *Applied Sciences (Switzerland)*, 11(7), 2982. https://doi.org/10.3390/app11072982

He, J., Strezov, V., Zhou, X., Kumar, R., & Kan, T. (2020). Pyrolysis of heavy metal contaminated biomass pre-treated with ferric salts: Product characterisation and heavy metal deportment. *Bioresource Technology*, 313, 123641. https://doi.org/10.1016/j.biortech.2020.123641

Järup, L. (2003). Hazards of heavy metal contamination. *British Medical Bulletin*, 68(1), 167–182. https://doi.org/10.1093/BMB/LDG032

Jobling, J., & Stevens, F. R. W. (1980). *Establishment of Trees on Regraded Colliery Spoil Heaps*. Edinburgh: Forestry Commission.

Kamran, S., Shafaqat, A., Samra, H., Sana, A., Samar, F., Muhammad, B. S., Saima, A. B., & Hafiz, M. T. (2013). Heavy Metals Contamination and what are the Impacts on Living Organisms. *Greener Journal of Environmental Management and Public Safety*, 2(4), 172–179. https://doi.org/10.15580/GJEMPS.2013.4.060413652

Kapoor, D., & Singh, M. P. (2021). Heavy metal contamination in water and its possible sources. *Heavy Metals in the Environment: Impact, Assessment, and Remediation*, 2021, 179–189. https://doi.org/10.1016/B978-0-12-821656-9.00010-9

Khalid, S., Shahid, M., Niazi, N. K., Murtaza, B., Bibi, I., & Dumat, C. (2017). A comparison of technologies for remediation of heavy metal contaminated soils. *Journal of Geochemical Exploration*, 182, 247–268. https://doi.org/10.1016/j.gexplo.2016.11.021

Khan, A. H. A., Kiyani, A., Santiago-Herrera, M., Ibáñez, J., Yousaf, S., Iqbal, M., Martel-Martín, S., & Barros, R. (2023). Sustainability of phytoremediation: Post-harvest stratagems and economic opportunities for the produced metals contaminated biomass. Journal of Environmental Management, 326, 116700. https://doi.org/10.1016/j.jenvman.2022.116700

Lin, C., Cheruiyot, N. K., Bui, X. T., & Ngo, H. H. (2022). Composting and its application in bioremediation of organic contaminants. Bioengineered, 13(1), 1073–1089. https://doi.org/10.1080/21655979.2021.2017624

Mancuso, T. F. (1997). Chromium as an industrial carcinogen: Part II. Chromium in human tissues. *American Journal of Industrial Medicine*, 31(2), 140–147. https://doi.org/10.1002/(SICI)1097-0274(19970204)31:2<140::AID-AJIM2>3.0.CO;2-3

Mushtaq, M. U., Iqbal, A., Nawaz, I., Mirza, C. R., Yousaf, S., Farooq, G., Ali, M. A., Khan, A. H. A., & Iqbal, M. (2020). Enhanced uptake of Cd, Cr, and Cu in Catharanthus roseus (L.) G. Don by Bacillus cereus: Application of moss and compost to reduce metal availability. *Environmental Science and Pollution Research*, 27(32), 39807–39818. https://doi.org/10.1007/s11356-020-08839-5

Muthusaravanan, S., Sivarajasekar, N., Vivek, J. S., Vasudha Priyadharshini, S., Paramasivan, T., Dhakal, N., & Naushad, M. (2020). Research Updates on Heavy Metal Phytoremediation: Enhancements, Efficient Post-harvesting Strategies and Economic Opportunities. In: Naushad, M., Lichtfouse, E. (eds) *Green Materials for Wastewater Treatment. Environmental Chemistry for a Sustainable World* (vol 38, pp. 191–222). Cham: Springer. https://doi.org/10.1007/978-3-030-17724-9_9

Pandey, V. C., & Bajpai, O. (2019). Phytoremediation: From Theory Toward Practice. In: Pandey, V. C., Bauddh, K. (eds) Phytomanagement of Polluted Sites: Market Opportunities in Sustainable Phytoremediation (pp. 1–49). Elsevier. https://doi.org/10.1016/B978-0-12-813912-7.00001-6

Pilon-Smits, E. (2005). Phytoremediation. *Annual Review of Plant Biology*, 56, 15–39. https://doi.org/10.1146/ANNUREV.ARPLANT.56.032604.144214

Shen, X., Dai, M., Yang, J., Sun, L., Tan, X., Peng, C., Ali, I., & Naz, I. (2022). A critical review on the phytoremediation of heavy metals from environment: Performance and challenges. *Chemosphere*, 291, 132979. https://doi.org/10.1016/J.CHEMOSPHERE.2021.132979

Sheoran, V., Sheoran, A. S., & Poonia, P. (2009). Phytomining: A review. *Minerals Engineering*, 22(12), 1007–1019. https://doi.org/10.1016/J.MINENG.2009.04.001

Song, U., Kim, D. W., Waldman, B., & Lee, E. J. (2016). From phytoaccumulation to post-harvest use of water fern for landfill management. *Journal of Environmental Management*, 182, 13–20. https://doi.org/10.1016/J.JENVMAN.2016.07.052

Suer, P., & Andersson-Sköld, Y. (2011). Biofuel or excavation? - Life cycle assessment (LCA) of soil remediation options. *Biomass and Bioenergy*, 35(2), 969–981. https://doi.org/10.1016/J.BIOMBIOE.2010.11.022

Tian, X., Stranks, S. D., & You, F. (2021). Life cycle assessment of recycling strategies for perovskite photovoltaic modules. *Nature Sustainability*, 4(9), 821–829. https://doi.org/10.1038/s41893-021-00737-z

Vocciante, M., Caretta, A., Bua, L., Bagatin, R., Franchi, E., Petruzzelli, G., & Ferro, S. (2019). Enhancements in phytoremediation technology: Environmental assessment including different options of biomass disposal and comparison with a consolidated approach. *Journal of Environmental Management*, 237, 560–568. https://doi.org/10.1016/j.jenvman.2019.02.104

2 Recent Advances in Phytoremediation of Chromium-Contaminated Soil

Types, Potential, and Limitations

M. Sivasakthi, S. Sathiyamurthi, S. Nalini,
S. Praveen kumar, M. Santhosh kumar,
R. Ragavaraj, and J. Prabakaran

INTRODUCTION

Heavy metals (HM) are pollutants that have been found to have significant and detrimental impacts on terrestrial and aquatic ecosystems (Pushkar et al., 2021). Moreover, they have been observed to disrupt the balance and integrity of ecological systems to a considerable extent (Zulfiqar et al., 2022). According to Wei et al. (2022), the principal source of pollution stems from the unregulated expansion of urban areas and industrial activities, which fail to acknowledge the significance of maintaining a sustainable and unpolluted ecosystem and thus have caused a substantial surge in HM contamination, leading to disturbances in the natural environment (Qianqian et al., 2022). According to Xu and coworkers (2018), the World Health Organization (WHO) documented about 1.7 million deaths worldwide as a result of people being exposed to hazardous compounds such as heavy metals.

The increasing incidence of HM pollution in the ecosystem raises the possibility of human exposure to HMs (Zulfiqar et al., 2019). The potential harm posed by heavy metals to living beings stems from their capacity for decomposition (Qianqian et al., 2022). Heavy metals frequently absorb and move throughout ecosystems at various tropic levels. According to Banerjee and coworkers (2019), it is possible for untreated waste to include HMs that can move through irrigated water and groundwater and then be absorbed by crops. The potential effects for living species can be fatal when they are exposed to heavy metals via air, water, or food sources. The quick response to the breakdown of HMs is a matter of great concern (Yaashikaa et al., 2019).

Chromium, denoted as Cr in the periodic table, is abundant in the earth's mantle, ranking as the 17th most prevalent element. Its harmful effect in crops is mostly

DOI: 10.1201/9781003442295-2

influenced by its different valency states. Chromium (Cr) finds widespread application across different sectors, including mining, tanneries, steel production, chemical manufacturing, and Cr plating (Pushkar et al., 2021). The prevalence of chromium as an ecological hazard has increased due to its escalating utilization in industrial facilities (Wei et al., 2022). Chromium is a widely distributed ecological pollutant that poses substantial risks to the ecological system (Ao et al., 2022). It is a metallic compound possessing an atomic number of 24, which falls into the VI-B group of the periodic table. In mineral form it exhibits a lustrous appearance, possesses a robust steel-grey coloration, and demonstrates a notable resistance to melting, as noted by Owlad et al. (2009). Figure 2.1 displays the yearly global extraction of chromium in thousands of metric tonnes. According to Chug et al. (2016), the durability of various oxidation states of Cr, ranging from I to VI, was shown to be highest in trivalent and hexavalent Cr. Hexavalent Cr is a more hazardous metal than trivalent Cr due to its cancer-causing, mutagenic, and oxidative properties (Wei et al., 2022). The cytostatic and carcinogenic properties of Cr^{6+} are 100 times more potent than those of Cr^{3+} (Mamais et al., 2016). Moreover, Cr^{6+} exhibits high solubility and readily undergoes absorption, garnering increased attention (Xiao et al., 2017), and it has no discernible metabolic function in floras (Srivastava et al., 2021). The biogeochemical functioning

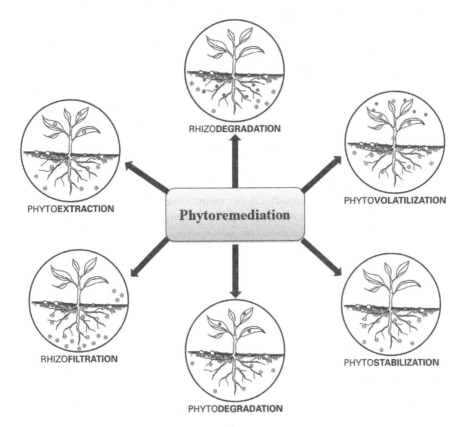

FIGURE 2.1   Types of phytoremediation.

of chromium is governed by various soil properties, including ST, pH, OM, EC, S, Fe, and Mn oxides, MA, and SMC. Additionally, plant physiological parameters like RSA, RT, RE, and plant type also play a significant role (Ao et al., 2022).

Previous research has extensively examined the mechanisms of chromium pollutants on agricultural production, lipid peroxidation, ROS generation, and potential remedial strategies (Ao et al., 2022). This chapter gives a summary of recent research on the transport, deposition, pollution, and removal of chromium. Observation of the harmful effects of chromium on essential biochemical processes in plants has revealed a consequential reduction in both growth and production. The sequestration and reduction of Cr^{6+} with the incorporation of organic substances, such as biochar, manure, and compost, has also been identified. This chapter also highlights modern techniques, such as biological remediation, which also covers phytoremediation, microbe-based remediation, and nano remediation methods that reduce hazards to human health and the surrounding ecosystem. To delineate forthcoming research objectives and requirements, this study incorporates research deficiencies about the biogeochemical processes of chromium in soil–plant systems, as well as constraints associated with the utilization of *in situ* remediation materials for Cr-contaminated soil.

SOURCE AND CONCENTRATION OF Cr

The metal Cr was discovered in 1797 after being identified as a constituent of crystalline crocoite ($PbCrO_4$). This mineral was particularly valued for its strong coloring ability, attributed to chromium's presence. The term "chromium" originates from the Greek language "chroma," denoting the concept of color. Chromium occurs naturally in the environment in rocks and soil as well as in liquid and vapor forms (Shanker et al., 2005). The principal sources of chromium consist of iron oxides with a high degree of crystallinity, as well as fragments that are tightly bonded and derived from primary rocks (Quantin et al., 2008). Ma and Hooda (2010) have identified many Cr compounds that occur naturally, such as "crocoite ($PbCrO_4$), vauquelinite ($CuPb_2(CrO_4)$ PO_4OH), tarapacaite (K_2CrO_4), chromite ($FeCr_2O_4$), and bentorite ($Ca_6(CrAl)_2(SO_4)_3$)." According to Hsu and coworkers (2015), chromium has the potential to undergo coprecipitation with minerals such as aluminum (Al), manganese (Mn), and iron (Fe), and to form complexes with SOM complexes. Various types of parent rocks have a diverse range of chromium contents. There is a notable disparity in Cr concentrations between sedimentary and igneous rocks compared to ultramafic and mafic rocks (Kabata-Pendias, 2000). The former two types of stones generally exhibit lower chromium levels, ranging from 15 to 120 mg/kg. The latter two types demonstrate more significant Cr concentrations, with values ranging from 1600 to 3400 mg/kg for ultramafic rocks and 170 to 200 mg/kg for mafic rocks. Chromium content in the earth's crust ranges from 0.1 to 0.3 mg/kg, as reported by Nriagu and Nieboer (1988).

SOURCES OF CHROMIUM

Annually, a substantial amount of chromium is removed from the earth due to its numerous commercial and agricultural uses. The worldwide mining of Cr has experienced a significant surge since the 1950s. The year 2000 saw the production or mining of over 137,000 metric tonnes of chromium (Lilli et al., 2015). According to Lukina and coworkers (2016), China, India, Kazakhstan, and South Africa are recognized as the primary global consumers of chromium. China is the foremost global producer of ferrochromium and stainless steel while concurrently maintaining its status as the top user of Cr (Gao et al., 2022). The primary global reserves of Cr are concentrated in South Africa and Kazakhstan, according to Gao et al. (2022). The production of excessive amounts of Cr is associated with leather processes and many human activities, including industries such as paper and pulp manufacturing, electroplating, metallurgy, ceramics polishing, alloying, pressure treatment, lumber storage, and resistant brick construction. Chromium contents in tanning wastewater varied from 217 to 2375 mg/L (Chen et al., 2018). Industrial cooling towers because chromium pollution at their highest levels and road dust at their lowest levels (Santos and Rodriguez, 2012). Chromium-containing solid and liquid waste discharge is an essential component of the ecosystem, and chromium is also anthropogenically released into sediments and soils by air transport (Gao and Xia, 2011). Biosolids and phosphorus nutrients have been found to contain significant levels of chromium. Cr deposition in an environment encompassing water, soil, and air is initiated by geochemical reactions and spontaneous events, such as eruptions of volcanoes or the dispersion of lava dust (Lilli et al., 2015).

CHROMIUM KINETICS IN SOIL

Chromium values in soil are typically 40 mg/kg. Chromium has diverse potential oxidation states, whereas the Cr^{+3} oxidation state is notably stable. The Cr^{3+} and Cr^{6+} oxidation states are commonly observed in chromium compounds, while the Cr^{+}, Cr^{4+}, and Cr^{5+} oxidation states are comparatively infrequent. The primary mineral that contains this metal is chromite, with the chemical formula $CrFeCr_2O_4$, which typically consists of around 70% pure Cr_2O_3 (Lakshmi and Sundaramoorthy, 2010). Most soils typically contain naturally occurring Cr in the form of relatively stable Cr(III), which requires gradual liberation by introducing acidic discharges (Chandra et al., 2010). The process of soil manganese oxides facilitates the conversion of $Cr3^+$ to Cr^{6+}. However, it is essential to note that only a limited quantity of chromium is susceptible to oxidation. Chromium exhibits complete incorporation into the soil matrix while also displaying excellent interactions with organic molecules present in Fe and Mn oxides and hydroxides (Mishra et al., 2009). According to Balamurugan and coworkers (2014), chromium exhibits a high affinity for binding to organic molecules present in oxides and hydroxides of Fe and Mn.

PERMISSIBLE LEVEL OF Cr

The CCME found that 64 mg/kg of Cr is the highest amount that should be in soil to protect the ecosystem, according to Hoffman et al. (2015). Several countries, such as Australia, Austria, Serbia, and Turkey, have established a maximum permissible chromium concentration in agricultural soils at 100 ppm. Canada has set the limit at 64 ppm, while China has put it at 200 ppm. The Czech Republic has a range of 100 to 200 ppm, England and Wales have put it at 400 ppm, and the European Community has a 50 to 150 ppm range. Poland has set the limit at 15 ppm, and the United States has put it at 1500 ppm (Ding et al., 2014). There is 5 mg/L of Cr(III) in irrigation water, 8 mg/L in fresh water, and 50 mg/L in seawater. Conversely, chromium (VI) concentrations in these respective water sources are reported as 8, 1, and 1 µg/L (Zayed and Terry, 2003). According to Lilli et al. (2015), various monitoring organizations have recommended a maximum allowable concentration of 50 mg/L of Cr(VI) in potable water. WHO (1996) says that 100 mg/kg is the maximum amount of chromium that can be found in soil.

SPECIATION OF CHROMIUM AND AVAILABILITY

Learning about the speciation of heavy metals is of utmost importance to effectively identify potential risks and implement appropriate remedies (Shahid et al., 2012). Cr^{3+} is commonly found in the form of charged ions, whereas Cr^{5+} exists as cations such as Cr_2O_7, CrO_4, and hydroxy chromate ions (Shadreck and Mugadza, 2013). During the process of reduction and oxidation, volatile compounds such as Cr^{2+}, Cr^{4+}, and Cr^{5+} are generated as intermediate phases. According to Cornelis and coworkers (2005), significant variations of Cr can experience ongoing transformations in various forms due to different physical and chemical processes.

Chromium (VI) is a substance that exhibits various chemical structures and exerts contrasting effects. The substance exhibits increased mobility toxicity toward biological entities and can induce mutagenic, carcinogenic, and teratogenic effects. Prado et al. (2012) state that interacting with other components increases its hazard level. Chromium (VI) pollutants in soil have been detected at deficient concentrations, specifically less than or equal to 1 mg/kg (Amin and Kassem, 2012). According to Martí et al. (2013), chromium has a rapid precipitation rate at a neutral pH when in the oxidation state of III. Moreover, it possesses minimal toxicity, enabling it to maintain its transportability within the environment. Evaluating metal speciation is crucial in enhancing ecological risk estimates related to chromium pollution (Shahid et al., 2012).

According to Ding and coworkers (2014), simultaneous oxidation/reduction reactions of Cr^{6+} and Cr^{3+} take place in soil, and these events are thermodynamically induced. The impact of chemical processes on Cr transformations in soil, including precipitation, reduction, oxidation, and hydrolysis, is significant as it disrupts the chemical equilibrium specific to different species (Zayed and Terry, 2003). The activity of Cr is influenced by innumerable factors, such as microbiological habitats, metal concentrations, cation exchange capacity, levels of competing cations, and the pH and redox state of the soil (Taghipour and Jalali, 2016).

PHYTOREMEDIATION (PR) OF Cr

The existence of pollutants containing heavy metals in soil can potentially cause detrimental impacts, and consequently the field of phytoremediation has emerged as a critical area of scientific investigation aimed at mitigating this problem (Escudero Oñate et al., 2017). According to Sharma and coworkers (2023), several techniques such as phytoextraction, phytostabilization, and rhizofiltration have been employed for the purpose of remediating soil polluted with chromium. The efficacy of phytoremediation can be augmented by integrating it with other techniques, as suggested by Feki et al. (2021). The procedure includes plant cells absorbing, transporting, and detoxifying heavy metals (Kafle et al., 2022). This process may be impacted by elements like soil characteristics, plant types, and pollutant types. According to Sandrin and Maier (2003), the presence of hyper-accumulator plant types and their ability to facilitate the uptake of Cr can enhance the efficiency of phytoextraction processes for HM removal. Plants can stabilize heavy metals more quickly in the soil, and plants can hold Cr in their roots. The plant roots take up Cr, which is an important part of phytoremediation (Sandrin and Maier, 2003). Chromium ions can be transported via a root film via passive diffusion and dynamic transportation. Sandrin and Maier (2003) explained about how ABC transporters and metal protein carriers help make these events happen. Plant cells can engage in the detoxification or sequestration of chromium as a means of mitigating potential detrimental consequences. The process of detoxification entails the transformation of chromium into less deleterious compounds facilitated by several proteins, namely peroxidases, SOD, and catalases. The process of sequestration entails the containment of chromium within vessels or its association with proteins or atoms, hence impeding its interaction with cellular constituents (Shahid et al., 2017). According to Grace Pavithra and coworkers (2019), certain group of plants possess the ability to sequester chromium within cell walls using the particle exchange proteins and chelating compounds.

ADVANTAGES OF PHYTOREMEDIATION

- The technology is suitable for shallow contaminants and will be demonstrated through full-scale and pilot studies (Schnoor et al., 1995).
- The results of phytoremediation will provide a clear picture about its sustainability as it is a scientific technique that can be applied in practical settings, demonstrating its potential for remediating HMs in agro-ecosystems.
- Studies on phytoremediation show high-quality and effective responses in aquatic media, particularly for organic contamination. Full-scale studies have been conducted on swine wastewater and groundwater treatment using *Eichhornia crassipes* and *Populus* sp. However, Chien et al. (2016) discovered that the efficacy of full-scale phytoremediation in natural environments is lower when compared to laboratory-scale investigations.
- However, the advantages of this treatment method raise concerns and demand for future strategies.

DISADVANTAGES OF PHYTOREMEDIATION

- The disadvantages of phytoremediation include pollutant concentration, toxicity, availability, choice of plant, and stress tolerance, as well as potential divergence in plant species due to disability, failure, or an unknown response.
- The cultivation of phytoremediator plants (hyperaccumulators) has been successful in addressing the accumulation of pollutants in the edible parts.

- Hyperaccumulators typically selectively accumulate only specific metal elements (XI et al., 2008).
- Chelate-enhanced phytoremediation is a slow process, and its effectiveness is seasonal (Chintakovid et al., 2008).
- However, it has a significant footprint in handling and disposing of contaminated plants. This green technology mobilizes radionuclides through translocation in plants, but not all compounds (Ghosh and Singh, 2005).
- Dissolved contaminants in groundwater are not suitable for aquatic phytoremediation.

TYPES OF PHYTOREMEDIATION

PHYTOEXTRACTION (PE) OR PHYTOACCUMULATION

PE is a scientific technique devised by researchers for the purpose of eliminating detrimental substances such as toxic compounds or HMs from soil and/or water sources (Yan et al., 2020). Remediating extracted soil polluted with metallic substances, radioactive elements, and organic contaminants is paramount. The methodology entails the absorption of contaminants by plant roots, followed by their subsequent diffusion to the higher leaves of the plant. According to Irshad and coworkers (2021), the concentration of soil contamination can be reduced by harvesting and removing stored toxins. PE represents an attractive alternative to conventional remediation techniques, such as excavation, disposal, incineration, and chemical treatments, due to its low costs, reduced invasiveness, and energy efficiency (Feki et al., 2021). The utilization of this approach leads to waste reduction, decreased energy consumption, and the potential for *in situ* application, hence minimizing the need for soil excavation and transfer. The effectiveness of a particular approach is contingent upon carefully selecting plant species (Sarwar et al., 2017).

The capacity of various kinds of floras to collect pollutants from soil or water exhibits variability, with certain species demonstrating greater efficacy in PE of certain toxins (Chaudhary et al., 2018). According to Suman and coworkers (2018), the process of selecting different types of plants for PE is governed by various parameters, including pollutant concentration, soil properties, and climate conditions. According to Yanitch et al. (2020), to achieve effective PE, it is crucial to select plants with rapid growth rate, high biomass productivity, and tolerance toward soil contaminants. According to de Souza et al. (2019), it is imperative for plants to possess a notable capacity for both the uptake and translocation of pollutants while also demonstrating ease in the removal of stored pollutants. The plant species *Brassica juncea* has demonstrated significant efficacy in the remediation of HM-polluted soils.

The past few decades have seen much attention given to plants with quick growth, high biomass, and ability to thrive in soils with elevated levels of HM. According to Usman and coworkers (2019), this botanical species exhibits promising capabilities for enhancing soil health to a considerable extent. The remediation of Cr-contaminated sites through PE can be achieved by employing soils, as demonstrated by Patra et al. (2018). The method entails absorbing of nutrients through the roots and subsequent movement to the aerial portions of the plant, and content of Cr in soil can be decreased by implementing a strategy that comprises the collection and removal of accumulated Cr from contaminated locations. The effectiveness of this approach is mostly contingent on appropriate plant species (Sharma et al., 2023).

CHROMIUM ACCUMULATED IN THE PLANT ROOTS

Heavy metals, such as chromium, have limited mobility within plant roots, resulting in significantly higher concentrations in roots compared to shoots, with levels potentially reaching up to 100 times greater in roots. According to Liu and coworkers (2008), Cr concentration distribution in various organs of *Pisum sativum* follows a pattern from roots through stems, leaves, and seeds. Most chromium in roots, specifically 83.2%, is present in the cell wall fraction. On the other hand, when it comes to leaf chromium accumulation (AC), approximately 57.5% is located within cellular vacuoles and cytoplasm. The increased gathering of chromium in the root is likely attributed to the development of insoluble substances containing Cr (Shanker et al., 2005).

The potential reduction in the transportation of hexavalent chromium Cr^{6+} to stem and foliage may have been undergone by the internal conversion of Cr^{6+} to trivalent Cr^{3+} for subsequent attachment to cell walls. Plants store chromium within various root compartments, such as cell walls, vacuoles, and organelles. The cellular walls serve as the initial barrier to prevent the reception of Cr into the plant. They achieve this by either immobilizing Cr in the roots by ion exchange or by creating insoluble complexes, as discussed by Haokip and Gupta (2021). According to Caldelas and coworkers (2012), certain group of plants have the ability to produce organic acids through their root systems. These organic acids have the capability to form complexes with chromium ions, hence causing their immobilization within the cell. Vacuoles are organelles that possess the ability to accumulate diverse chemicals, such as Cr (Cervantes et al., 2001). Active transport systems move it into vacuoles, which make it less poisonous and less likely to move to other tissues. The AC of chromium in plant roots plays a crucial role in the process of phytostabilization. According to Wakeel and coworkers (2020), this mechanism effectively decreases the Cr availability in the soil, hence impeding its uptake by plant roots.

CHROMIUM ACCUMULATED IN THE PLANT SHOOTS

Metal transporter gene families, such as CDF, ABC, HMA, NRAMP, ZIP, ZRT, and IRT-like protein, have a substantial effect on the transportation of HMs within biological systems (Shahid et al., 2017). These movers make it easier for metals to be taken in, stored, moved, and tolerated. Hexavalent chromium is transported to the shoots of plants by active mechanisms, specifically phosphorus and sulfate transporters. These enzymes utilize iron (Fe) and sulfur (S) channels to facilitate the upward movement of hexavalent Cr. This process leads to competition among these metals for placement inside the plant. Sulfur (S) accumulators, like the Brassicaceae family, possess the capacity to collect significant quantities of chromium through the uptake and redistribution of sulfur facilitated by S-carriers (Cervantes et al., 2001). Plants that possess the ability to hyperaccumulate iron have the capacity to absorb significant quantities of chromium and transfer it to their aboveground tissues via chromophore transport systems. According to Kim and coworkers (2006), MSN1 transcription in transgenic *Nicotiana tabacum* improved chromium absorption capacity and tolerance to this particular element. Moreover, it has been observed that Cr has the ability to enhance the expression of NtST1 and sulfur transportation-1 in *N. tabacum*, and S-transporter in plants plays a pivotal role in the absorption of sulfur and Cr (Murphy et al., 2016).

Eukaryotic organisms possess diverse sulfate transporters, each exhibiting distinct affinities. According to Pootakham et al. (2010), the bacterium *Chlamydomonas reinhardtii* is in the plasma membrane. Among these transporters, SLT1, SLT2, and SLT3 are responsible for facilitating the transport of sodium ions (Na^+) and sulfate ions (SO_4^{-2}), while the Sac1-like transporters are responsible for transportation of H^+ and SO_4^{-2}. Additionally, the SULTR1, SULTR2, and SULTR3 transporters are also involved in the transfer of in instances where there is a deficiency of sulfur (S). The processes associated with chromium resistance and detoxifying may provide divergent outcomes. According to Marieschi et al. (2015), the activation of sulfate transporters hinders the uptake of chromium. Additionally, increasing the amounts of sulfate within cells through the upregulation of sulfate absorption and removal mechanisms may be utilized to make molecules that detoxify Cr.

Amin et al. (2019) conducted research on six different biofuel plant species, including *Glycine max, Avena sativa, Sesamum indicum, Cyamopsis tetragonoloba, Abelmoschus esculentus,* and *Guizotia abyssinica*. The study discovered that *C. tetragonoloba* was among the most efficient species of plant for removing Cr from polluted soil, particularly at levels less than 75 mg Cr/kg. Roots of *A. sativa* and *A. esculentus* gathered the most Cr at 50 mg Cr/kg, whereas shoots of *C. tetragonoloba* accumulated the most at 287.42 mg/plant. *C. tetragonoloba* is the most successful plant species for extracting Cr from contaminated soil. Augustynowicz and coworkers (2020) reported the potential for phytoremediation of *Callitriche cophocarpa* Sendtn. The results showed that oxidizable Cr^{3+} at 68.2%, residual Cr^{4+} at 28.8%, reducible Cr^{2+} at 1.6, and exchangeable Cr at 1.4% are the components of hyperaccumulation. Plants at polluted sites had 33–83 times more chromium in their leaves, stems, and roots than control plots.

Chen et al. (2020) discovered that mung beans (*Vigna radiata*) can store Cr^{6+} within 7 days, with the whole plant storing 5041 mg/kg. Also, 80% of the Cr in mung bean plants is changed into parts that are not easily absorbed by the animal body.

Chintani et al. (2021) evaluated the remediation efficiency of C. *zizanioides* in Cr- and Ni-polluted soil. They used three different treatments: a placebo, Cr, and Ni at 50, 150, and 300 ppm. The findings demonstrated Cr uptake was higher than its release, while the rate of both Ni uptake and release was low. Cr accumulated at a rate of 167.8 mg/kg, and Ni accumulated at a rate of 66.3 mg/kg. The study also found that Ni-treated plants could move Ni to parts of the plant that grow in the air. This was shown by the high TF values of these plants.

Hasan et al. (2021) studied the capacity of local plants to remediate tannery wastewater polluting the Dhaleshwari River. Because it has high transfer factors and bioconcentration factors, *Eichhornia crassipe* translocates the Cr in root at 68.7 mg/kg, stem at 5.5 mg/kg, and leaf at 8.4 mg/kg. *Eichhornia crassipe* is the best choice for PR of Cr.

Levizou et al. (2019) observed that Greek oregano has the ability to accumulate chromium in its aerial and root portions. The content of Cr in the plant reached a maximum of 4.3 g/kg DM when cultivated in soil containing 0.15 to 0.2 g/kg of Cr^{6+}. Mondal and Nayek (2020) conducted a study on the PE of chromium by *Eichhornia* sp. and *Pistia* sp. The species *Eichhornia* had greater chromium deposition when exposed to a solution containing 30 mg/L of Cr, whereas *Pistia* sp. revealed the maximum accumulation at a concentration of 10 mg/L. Patra and coworkers (2020) evaluated the phytoaccumulation potential of Cr in *Sesbania sesban* L. and *Brachiaria mutica* L. Findings indicated that *S. sesban* had the highest bioaccumulation potential of Cr at 71.06% and high tolerance, while *B. mutica* had lower values at 42.45%.

Peng et al. (2021) screened chromium-reducing strains and hyperaccumulator tall fescue to remediate soil contaminated with chromium. The treatment reduced CH_3COOH extractable-Cr by 12.82–20.00% compared to CK, while residual Cr increased by 9.41–22.37%. Biomass, RL, and SL of *Festuca arundinacea* increased by 80.77–139.74%, 60.85–68.04%, and 7.06-27.10%, respectively.

Perotti et al. (2020) examined the uptake capability of the roots of *Brassica napus* to remove Cr^{6+}. They focused on the processes required and the harmful effects of solutions used after removal. The results showed that roots can handle up to 10 mg/kg, but that higher levels were bad for HR growth. Synthetic solutions were used to test how well the extraction worked, and bioassays were used to test how toxic the area was after removal. Ranieri et al., 2021 examined the impact of prolonged Cr contamination on Moso bamboo (MB) and its adaptability to the Mediterranean climate. The finding showed that the rate of MB removal of Cr varied depending on the soil contamination level, ranging from 100 to 300 ppm.

LIMITATIONS OF PHYTOEXTRACTION OR PHYTOACCUMULATION

- High metal concentrations in harvestable part of plants posing a threat to food chain pollution and potentially causing plant death. Disposal of large metal quantities in harvested plants can be challenging (CPEO, 2021). Most plant hyperaccumulators are slow growers, reducing metal accumulation efficiency (Morkunas et al., 2018). In PE tasks, metal-chelating chemicals occasionally boost the phytoaccumulation efficiency of fast-growing crops, including sorghum, maize, and lucerne (Chibuike et al., 2014).
- These chemicals restrict the process of metal sorption and precipitation by creating metal chelate complexes, enhancing the availability of metals in biological systems. Nevertheless, using chelates in PE gives rise to apprehensions regarding the potential seepage of metals into groundwater and the possible detrimental effects on plants and soil microorganisms, particularly when employed in excessive quantities (Chibuike et al., 2014)
- According to Gupta and coworkers (2016), plants have the ability to absorb metals present in soil solutions. However, it should be noted that many metals exist in both soluble and insoluble forms, rendering them inaccessible for uptake by plants. The efficacy of treatment is constrained by the depth of pollutants, whereby the treatment zone is determined by the depth of plant roots. Nevertheless, the utilization of trees in lieu of smaller plants

can effectively mitigate more severe forms of pollution due to their ability to extend their roots to greater depths inside the soil (CPEO, 2021).

* The efficacy of PE is subject to seasonal variations and is typically constrained to regions with low pollution levels, particularly in shallow soil, groundwater, and streams. Key considerations for successful soil remediation include factors such as the solubility and accessibility of metals for plant uptake, as well as the suitability of the area for PE. Hence, it is imperative to consider these aspects in conjunction with the compatibility of plants to achieve successful PR (DalCorso et al., 2019).

ADVANTAGES OF PHYTOEXTRACTION OR PHYTOACCUMULATION

* In comparison to traditional methods, the cost is relatively low.
* Contaminants are permanently removed from the soil.
* Up to 95% of the waste that must be disposed of is reduced.
* Contaminants can be recycled in some cases.

DISADVANTAGES OF PHYTOEXTRACTION OR PHYTOACCUMULATION

* The accessibility of metals found inside the rhizosphere.
* The speed with which metals are taken up by roots.
* The percentage of metal that is "fixed" within the roots of the plant.
* The ability of cells to tolerate harmful elements.

PHYTOSTIMULATION (PS)

Phytostimulation is a technique employed in the field of PR to augment the capacity of plants to effectively mitigate pollutants. The application of PS has the ability to augment the plant growth and physiological processes that can accumulate chromium in contaminated soil. PS can be achieved by utilizing microbes, organic amendments, and nutrients to enhance plant development and promote plant growth (Zulfiqar et al., 2023).

ORGANIC AMENDMENTS

Organic amendments like compost, biochar, and humic acids can help plants grow, and PR methods can be used to treat soil contamination due to Cr (Branzini and Zubillaga, 2012). According to Zulfiqar and coworkers (2023), implementing these measures has augmented nutrient availability and soil structure and facilitated the proliferation of beneficial microorganisms. Consequently, these improvements have been observed to impact plant growth and the uptake of chromium positively. According to Xu and Fang (2015), application of compost can promote the vegetative development and Cr uptake of Indian mustard in soil polluted with chromium. Similarly, biochar has been found to improve the chromium uptake and growth of sunflowers in Cr-contaminated soil.

INORGANIC AMENDMENTS

In the context of PR of chromium-contaminated soil, it is possible to employ inorganic amendments, including minerals, metals, and metal oxides, to achieve PS. Zulfiqar et al. (2023) found that the utilization of materials such as IONPs, calcium carbonate, and zeolites augmented growth and facilitated chromium accumulation by diverse processes. Xiao and coworkers (2017) have demonstrated IONPs' potential to enhance growth and boost the uptake of chromium. Besides the utilization of organic manures and fertilizers, the application of inorganic material has also been recognized as a viable approach to implementing PR techniques (Ahmed et al., 2016). Ahmed and coworkers (2021) observed the ability to augment the assimilation of Cr within plant tissues through the process of adsorption onto their surface. Multiple studies have demonstrated the efficacy of IONPs in facilitating the remediation of Cr-polluted soil, particularly in the case of *Lolium perenne* plants. Furthermore, the use of IONPs has been found to boost the overall effectiveness of PR strategies. According to Aziz et al. (2001), adding calcium carbonate, a widely occurring mineral, can potentially improve plant growth and chromium absorption in soil. The addition of calcium carbonate ($CaCO_3$) to soil has been found to have several beneficial effects, including the potential to raise soil pH, enhance nutrient accessibility, and mitigate chromium toxicity in plants. Furthermore, it has been observed that the extension of Cr-reducing enzyme activity in plant roots is facilitated by this process, leading to an enhanced uptake of Cr (Ahmed et al., 2021).

Xiao and coworkers (2017) found they could promote plant development and facilitate the uptake of chromium through the use of zeolites. Cr can be adsorbed onto the surface of zeolites, rendering it accessible for plant uptake. The study directed by Zulfiqar and coworkers (2023) demonstrates the positive impact of zeolites on the architecture of sunflowers and the bioaccumulation of Cr in chromium-contaminated soil.

NUTRIENT MANAGEMENT

The implementation of nutrient management techniques can have a positive impact on plant development and on the effectiveness of PR in soil that is polluted with metals (Antil, 2012). The process entails optimizing soil nutrient availability and supplementing deficient minerals, which can promote plant development and chromium uptake. It has been demonstrated that nutrients such as NPK can improve the uptake of chromium in soil that is polluted with this element. Dheeba and coworkers (2015) reported that the application of nitrogen fertilizer was found to have a positive influence on the growth and absorption of chromium by Indian mustard and sunflower plants. Likewise, the application of P nutrition is also observed to boost the uptake and growth of rye grass in Cr-contaminated soil. The elevated plant growth and nutrient uptake by the application of nutrients has resulted in an increasing uptake and accumulation of chromium in plant tissues.

MICROBES

Bacteria and microbes are essential components in the process of PS for the purpose of PR of soil polluted with chromium. According to Marques et al. (2011), several mechanisms, such the nutrient cycle, hormones, and solubilization of nutrients, contributed to increasing plant growth and the uptake of chromium. AM fungi represent a category of symbiotic fungi that establish interactions with the roots, thereby facilitating the flow of carbohydrates from plants to fungi in return for fungal assistance in acquiring essential elements, such as phosphorus, which are crucial for the plant's survival. AMF can promote Cr uptake by increasing nutrient availability, and it can synthesize phytohormones such as IAA, which elicit a stimulatory effect on plant development. The study conducted by Ma and coworkers (2001) demonstrated that the application of AM fungi has the potential to promote the growth of sunflowers in soil polluted with chromium. According to Fahad et al. (2015), the presence of PGP bacteria can facilitate both plant growth and uptake of chromium through a range of processes. The PGP bacteria have the ability to synthesize growth hormones, improve the solubility of minerals and nutrients, and secrete enzymes for absorption of chromium by plants. Many studies have demonstrated that AMF and PGPB have the potential to enhance the process of phytoremediation in chromium-contaminated soil (Karthik et al., 2016; Zulfiqar et al., 2023).

Endophytic bacteria, residing within the internal tissues of plants, possess the capacity to boost plant growth and facilitate the uptake of chromium by means of various mechanisms. In addition, the production of enzymes by these organisms facilitates the absorption of chromium and leads to the conversion of less harmful Cr forms (Murthy et al., 2022). According to Qadir and coworkers (2020), endophytic bacteria and rhizobacteria can improve the process of phytoremediation in Cr-contaminated soil. The presence of endophytic bacteria showed a positive impact on the growth of sunflowers and their ability to ingest chromium in soil that is contaminated with chromium. Rhizobacteria, residing in the rhizosphere, exhibit the ability to synthesize growth hormones, facilitate the solubilization of minerals, and enhance the uptake of Cr by plants through growth hormones (Chen et al., 2016).

Abbas and coworkers (2020) investigated the capacity of biochar from sugarcane bagasse and acidified dung for translocation of Cr moves in maize plants. The results showed that when biochar and acidified dung were used together, the pH of the soil changed a lot, and crop growth improved. The amount of available Cr was reduced by 36% when biochar and acidified dung were added. Agar and coworkers (2020) reported the protective role of copper and $C_6H_8O_6$ in sunflower seeds against Cr stress. The findings suggest that MT2-1 could be a valuable gene resource for Cr-related plant breeding initiatives.

Adiloğlu and Göker (2021) studied the potential of maize to uptake hexavalent chromium. The amount of chromium in the root varied from 18.57 to 34.95 mmol/kg with the highest level of chelate. Wu et al. (2018) found that addition of P to the soil helps *Leersia hexandra* Swartz roots take up more chromium (VI). With more P, both the bioconcentration factor and the transport factor went up. The maximum BCF was 3.6 folds more than the control. This shows that more P makes it easier for *L. hexandra* to move Cr from its roots (23.9%) to its above-ground parts. Mohammadi and coworkers (2018) studied how the chelating agents like EDTA, iron sulfate (Fe^{2+}), and zerovalent nano iron (Fe^0 nanoparticles) affect the growth and physiology of sunflower plants when they are under stress from Cr^{6+}. The research findings indicated that rising EDTA improved Cr uptake and decreased morphological and physiological indicators, except for MDA and H_2O_2 contents. Treatments with Fe^0 nanoparticles and Fe^{2+} decreased the amount of Cr, increased the amount of dry matter, and positively correlated with the amount of Cr in the roots and the amount of MDA and H_2O_2 in the plants.

Sharma et al. (2020) found that adding tartaric acid (TA) to Cr^{6+}-stressed *Hordeum vulgare* L. plants made them much healthier. When 1.5 mM TA was added to 0.5 mM Cr^{6+}, the amount of Cr taken up by the roots was 208.7% higher than when 0.5 mM Cr^{6+} was used alone. In *H. vulgare* plants, TA also increased bioconcentration factors in the shoots and roots, which proved that it acts as an antagonist against Cr^{6+}. The quantity of Cr applied was 150 mg/kg. Mohammadi and coworkers (2020) examined the efficacy of sunflower plants growing in soil polluted with Cr^{6+}, as well as the potential role of nano-zerovalent iron (nZFe^0). The researchers discovered that raising the concentration of Fe^0 nanoparticles lowered Cr absorption while improving plant morphological and physiological characteristics. Following Cr exposure, there was a favorable association between seedling growth, BAF, and TF. According to the findings, Fe^0 nanoparticles can improve sunflower plant performance in the presence of Cr toxicity by lowering Cr absorption in shoot and root by about 79.9 µg/g and 38.1µg/g, respectively and enhancing detoxifying enzyme activity (SOD, CAT, POX, and APX).

de los Angeles Beltrán-Nambo et al. (2021) studied the capacity of *Gigaspora gigantea* and *Rhizophagus irregularis* to protect maize from Cr-contaminated soil. The findings indicated that *G. gigantea* increased the amount of nitrogen in the plant and moved more Cr to the plant leaves at 97 mg/kg, while *R. irregularis* caused more arbuscules and spheres to form and kept more Cr in the roots at 197 mg/kg. Farid et al. (2019) studied the ability of 5-aminolevulinic acid (ALA) and citric acid (CA) to help sunflowers to remediate chromium-contaminated soil. Sunflower plants were growing in soil with various quantities of Cr, and when they were young, different amounts of ALA and CA were added. The toxicity of Cr caused a big drop in plant growth, soluble proteins, and colors that help plants make food. The amount of Cr taken up by plants decreased Cr toxicity by 20 g/L. Zanganeh et al. (2022) implemented cyanobacterial species *Oscillatoria* and *Portulaca oleracea* L. to study bioaugmentation and bioaugmentation-assisted PR methods for soil polluted with Cr^{3+} and Cr^{6+}. In the bioaugmentation-assisted PR test, the accessible amount of these metals dropped by up to 90%. Sahoo et al. (2021) found that the *Azotobacter vinelandii* SRI Az3 can help *Oryza sativa* (var. IR64) to handle Cr stress. The strain's effectiveness was evaluated at different amounts of Cr stress, and it helped rice plants handle stress up to 200 M. The result showed that levels of antioxidant enzymes were much higher and increased the Cr uptake by 35%.

LIMITATIONS

The sites where phytostimulation will be used should have minimal amounts of pollution in shallow parts. The pollutants at high concentrations may prove hazardous for agricultural crops (Schnoor, 1997).

ADVANTAGES OF PHYTOSTIMULATION

- *In situ* treatment with little disruption.
- There will be no elimination of tainted substances.
- The pollutant can be completely mineralized.
- Cost-effectiveness of construction and upkeep.

DISADVANTAGES OF PHYTOSTIMULATION

- Substantial root region growth is necessary, which takes time.
- The depth of roots is restricted.
- Plant decaying tissue can be employed as a C provider rather than contamination to reduce the quantity of pollutant biological degradation.

PHYTOVOLATILIZATION (PV)

Phytovolatilization, alternatively referred to as PE, denotes the mechanism through which contaminants are assimilated and then sequestered within the anatomical structures of plants. The aforementioned procedure encompasses the absorption of chromium (Cr), its subsequent transportation to the aerial portions, and its subsequent release into the atmosphere. The complete understanding of the mechanisms underlying chromium PV remains elusive; nonetheless, existing research indicates a potential involvement of the reduce Cr within plant tissues, leading to its subsequent release as volatile molecules (Irshad et al., 2021). PV, which is a process used in PR, entails the generation of chromium (IV) intermediates that undergo a reaction with oxygen present in the atmosphere, resulting in the synthesis of volatile Cr molecules. This method presents numerous advantages compared to alternative approaches. For instance, it eliminates the need for the extraction and disposal of polluted plant matter while also enabling the treatment of extensive regions of polluted soil. Certain plants like Indian mustard and sunflower have the ability to store significant quantities of chromium in their tissues and then emit it into the atmosphere as Cr^{3+} (Naveed et al., 2021).

PHYTOSTABILIZATION (PS)

Phytostabilization is a technique employed to immobilize metals by using plants, hence diminishing their bioavailability and impeding the capacity of organisms to absorb them (Bolan et al., 2011). This approach proves to be highly advantageous in situations when complete eradication is impractical or economically unviable, as Cr is commonly present in polluted soils as a result of industrial operations, mining, and inappropriate disposal of waste materials (Bolan et al., 2011). The effectiveness of PR strategies relies on the plants to sequester HM by adsorption, precipitation, and complexation. According to Rizwan and coworkers (2021), *Brassica juncea*, *Helianthus annuus*, and *Salix* sp. are frequently employed for the purpose of chromium PS due to their diverse capacities. The process of PS involves the accumulation and immobilization of Cr in plant tissues. The findings indicate that optimal results are achieved in soils with a slightly acidic to neutral pH range and a substantial amount of organic matter. The capacity of PS may be impacted by pH, OM, and the circumstances under which plants are grown (Kanwar et al., 2020).

Radziemska and coworkers (2018) examined the effect of trace element immobilizing soil amendments on soil contaminated with Cr and metal by *Brassica juncea L*. The results revealed that significant differences in element contents were observed in Indian mustard parts, with halloysite and dolomite causing the highest average above-ground biomass. Halloysite also increased Cr concentrations in roots at 51%, indicating the effectiveness of using Indian mustard in PS techniques. Radziemska et al. (2020) examined the effect of three different alkaline amendments (limestone, dolomite, and chalcedonite) on *Festuca rubra* L. on uptake of Cr^{6+} in contaminated soil. The result showed reduction in total Cr from the soil at 150 mg/kg.

Chen et al. (2020) investigated the effect of co-pyrolysis with rice straw on the availability of Cr in soil. The study discovered that the amount of leachable, total, and accessible Cr in the soil was greatly affected by temperatures and the amount of rice straw in the soil. The findings suggested that there is a reduction of leachable chromium at 95%, total Cr^{6+} at 86%, and bioavailable Cr at 70%.

Kalam and coworkers (2019) studied the long-term effect of the phytoremediation potential of *Dalbergia sissoo* Roxb., using heavy metal-rich wastewater from tanneries. The study showed increases in the Cr accumulation in leaves at 21–23% and barks 24.7–26.3%. All three sites had BCF of Cr that was greater than 1, which shows that the tree is very good at PR (Table 2.1).

Ahsan et al. (2018) studied the capacity of endophytes on the stabilization of Cr-polluted soil. They used three possible endophytic bacterial strains in CR-containing soil in which *Leptochloa fusca* and *Brachiaria mutica* plants were growing. The results showed that adding these endophytes increased plant accumulation of Cr in roots by 54% and 43%, shoots by 48% and 38%, and leaves by 43% and 56%.

Vishnupradeep et al. (2022) investigated the effect of plant growth-promoting bacteria (PGPB) on the growth of *Zea mays* in Cr-polluted soil under drought stress conditions. The two PGPB strains, *Providencia* sp. (TCR05) and *Proteus mirabilis* (TCR20), improved plant development, colors, proteins, phenol compounds, and percent amount of water while reducing peroxidation of lipids, proline, which is and super oxide dismutase activity. The study shows that TCR05 and TCR20 could increase Cr bioaccumulation by 76% and 46% (Table 2.2).

Ullah and coworkers (2021) investigated the ability of two types of grass, *Brachiaria mutica* and *Leptochloa fusca*, to help stabilize the soil. They did this by mixing different amounts of Cr into the soil. The study found that *L. fusca* had more antioxidant enzyme activity (up to 50 mg/kg) than *B. mutica* as Cr levels increased, but this activity decreased (100 mg/kg) when Cr levels were at their highest. *L. fusca* had the highest root (93.7 g/plant) and shoot (24.7 g/plant) Cr buildup, while *B. mutica* had the lowest root (BCF 2.0), shoot (BCF 0.08), shoot (TF 0.06), and MTI (87%). *L. fusca* had more Cr in its roots (18.4 g/plant) and its leaves. Patra et al. (2017) conducted the pot culture experiments and showed that chromium accumulation in lemongrass plants (*Sesbania sesban L.* and *Brachiaria mutica L.*) increased in concentration, while protein and chlorophyll content decreased. Additionally, increased Cr accumulation (71.06% and 42.45%) and antioxidant enzyme levels

TABLE 2.1

Impact of Several Plant Species in the Presence of Chromium-Induced Stress

Plant Species	Effects	Accumulation Cr %	Removal Cr	Reference
Pyrolysis rice straw	Immobilization and reduced Cr (VI)		Leachable Cr – 95% Total Cr(VI) 86% and bioavailable Cr 70%	Chen et al., 2020
Dalbergia sissoo Roxb.	Hyperaccumulation	Leaves 21–23% Bark 24.7–26.3%		Kalam et al., 2019
Brassica sp.	Decreased antioxidant enzyme activities and accumulation	Grain 4.12% Shoot 2.27% Root 2.17%	APX 42.5% GP 45.1% CAT 45.4% GST 47.8% GR 47.1% RG 48.2%	Naveed et al., 2021
Brassica juncea L.	Increases of Cr concentrations in the roots	Root 51%		Radziemska et al., 2018
Festuca rubra L.	Reducing total Cr from the soil		150 mg/kg	Radziemska et al., 2020
Conocarpus erectus	Increasing accumulation of chromium in root and shoot			Tauqeer et al., 2019
Brachiaria mutica and *Leptochloa fusca*	Antioxidant enzymatic activity increased and reduced at maximum Cr level	Root, shoot 87% Root, shoot 56%		Ullah et al., 2021
Lavandula dentata	Reduced the bioavailable chromium		44.6 mg/kg – 7.5 mg/kg	Ye et al., 2020
Cyamopsis tetragonoloba, Glycine max, Avena sativa, Abelmoschus esculentus, Sesamum indicum, and *Guizotia abyssinica*	Hyperaccumulation	287.42 mg/plant		Amin et al., 2019
Callitriche cophocarpa Sendtn.	Hyperaccumulation	Oxidizable Cr(III) 68.2% Residual Cr(IV) 28.8% Reducible Cr(II) 1.6% Exchangeable Cr 1.4%		Augustynowicz et al., 2020

(Continued)

TABLE 2.1 *(Continued)*

Impact of Several Plant Species in the Presence of Chromium-Induced Stress

Plant Species	Effects	Accumulation Cr %	Removal Cr	Reference
Vigna radiata	Hyperaccumulation	80%	5041 mg/kg	Chen et al., 2020
Chrysopogon zizanioides	Increased Cr uptake		66.3 mg/kg	Chintani et al., 2021
Eichhornia crassipes	High translocation and bioconcentration of Cr	Root 68.7 mg/kg Stem 5.5 mg/kg Leaf 8.4 mg/kg		Hasan et al., 2021
Eichhornia sp. and *Pistia* sp.	Hyperaccumulation		30 mg/L 10 mg/L	Mondal and Nayek 2020
Sesbania sesban L. and *Brachiaria mutica* L.	Higher bioaccumulation	71.06% 42.45%		Patra et al., 2020
Brassica napus	Removed Cr from soil		98%	Perotti et al., 2020
Phyllostachys pubescens	Removes Cr from soil		100 mg/kg to 300 mg/kg	Ranieri et al., 2020
Bidens pilosa L., *Chenopodium album* L., *Malvastrum coromandelianum* L., *Garcke, Oxalis corniculata* L., *Parthenium hysterophorus* L. *(an invasive species), Polypogon monspeliensis* L., *and Rumex dentatus* L.	Higher bioaccumulation	50 mg/kg		Samreen et al., 2021
Vigna mungo var.	High Cr uptake in roots	300 µM		Rath and Das, 2021
Hydrocotyle umbellata	Higher accumulation	98.11%		Bokhari et al., 2022
Citrus aurantium L.	Increase in plant antioxidative enzymes—superoxide dismutase (SOD), catalase (CAT), peroxidase (POD), and ascorbate peroxidase (APX)—and increased lipid peroxidation	97.67%		Shiyab et al., 2019

TABLE 2.2

Impacts of Different Plant Species and Amendments under Chromium Stress

Plant species	Amendments	Effects	Accumulation Cr	Reduced Cr	Reference
Zea mays	Sugarcane bagasse derived biochar, acidified manure	Reduces the antioxidant, decreases Cr⁺, minimizes Cr mobility		Cr(III) (44%) and Cr(VI) (22%)	Abbas et al., 2020
Zea mays L.	Chelator (EDTA)	Hyperaccumulation	34.95 mmol/kg		Adiloğlu and Göker, 2021
Leersia hexandra	Phosphorus	Hyperaccumulation	23.9%		Wu et al., 2018
Helianthus annuus	EDTA, iron sulfate (Fe²⁺), and zerovalent nano iron	Reduced Cr concentration			Mohammadi et al., 2018
Helianthus annuus	5-aminolevulinic acid (ALA) and citric acid	Cr toxicity decreased and increased uptake		20 g/L	Farid et al., 2019
Hordeum vulgare	Tartaric acid	Increases antioxidative potential and chromium accumulation	208.7%		Sharma et al., 2020
Helianthus annuus	Nano-zerovalent iron	Increased activity of detoxifying enzymes and reducing Cr uptake		Shoot 0.799 mg/g Root 0.381mg/g	Mohammadi et al., 2020
Helianthus annuus	Copper and ascorbic acid	Hyperaccumulation			Agar et al., 2020
Hordeum vulgare	EDTA	Decreased the Cr		61.7%	Sharma et al., 2020

suggest enhanced damage control activity. The study highlights the potential of chromium in lemongrass cultivation.

Tauqeer and coworkers (2019) investigated the ability of *Conocarpus erectus* species to stabilize certain Cr metal-contaminated soil. The research revealed that *C. erectus* has a good chance of stabilizing Cr in the soil because it has bioconcentration factors for Cr that are less than 1.

Naveed et al. (2021) examined the effect of compost, biochar, and composted biochar on the growth, physiology, biochemistry, and health risks of *Brassica* in Cr-contaminated soil. The findings revealed that Cr stress slowed growth and the amendments made the plants grow better, lowering the amount of Cr in the grain, shoots, roots, and soil by 4.12%, 2.27%, and 2.17%, respectively. Co-composting with biochar also increased the effectiveness of PS. Overall, these changes made the bad effects of Cr toxicity less severe.

APPLICABILITY

PS is a method used to maintain metal concentration in large-scale areas with low contamination, especially when other methods are not feasible due to soil toxicity. Plants must tolerate high contaminants, produce high root biomass, immobilize contaminants, and hold them in the roots to effectively remediate these areas (Miller, 1996). Based on empirical observations, field research suggests that implementing PS techniques can significantly decrease lead (Pb) levels in sand/Perlite combinations. Furthermore, this approach demonstrates the ability to stabilize radionuclides at low concentrations and mitigate the leaching of metals by facilitating their conversion from a soluble oxidation state to an insoluble oxidation state. Blaylock et al. (1995) have demonstrated that plants possess the ability to decrease the levels of accessible and hazardous chromium (VI) to a less toxic form, chromium (III).

LIMITATIONS

PS helps in places that have shallow pollution and minimal pollution. Plants that absorb pollutants in their roots and root zone are usually efficient at thicknesses of up to 24 inches. Metals that are easily transferred into plant tissues could restrict the usefulness of the PS because of potential food chain effects (Blaylock et al., 1995).

PHYTOSTABILIZATION ADVANTAGES

- There is no need to dispose of hazardous materials or biomass.
- Highly effective when immediate inhibition is required to conserve both surface and groundwater.

DISADVANTAGES

- Contaminants are still present in the soil.
- Extensive fertilization and soil amendments are used.
- Supervision is essential.

PHYTODEGRADATION (PD) OR PHYTOTRANSFORMATION

Phytodegradation (PD) refers to the process through which plants and associated microbes facilitate the conversion or dissolution of contaminants into less detrimental forms (Sandrin and Maier, 2003). PD is a process requiring the application of plants and their symbiotic microbes to facilitate the degradation of chromium substances into less harmful forms, particularly chromium (III) (Said and Lewis, 1991). Phytoreduction and enzymatic reduction are among the direct methods by which PD can occur, while modifications to the microbial population in the soil are examples of indirect mechanisms (Sandrin and Maier, 2003). Table 2.3 presents a

TABLE 2.3

Impacts of Different Plant Species and Microbes under Chromium Stress

Plant Species	Microorganism	Effects	Accumulation Cr	Reduced Cr	Reference
Leptochloa fusca and Brachiaria mutica	Enterobacter sp. HU38, Microbacterium arborescens HU33, and Pantoea stewartii ASI11	Entophytic augmentation increased uptake, translocation, and accumulation of Cr in the roots and shoots	Roots 54% and 43% Shoots 48% and 38% Leaves 43% and 56%		Ahsan et al., 2018
Zea mays	Providencia sp. (TCR05) and Proteus mirabilis (TCR20)	Bioaccumulation of Cr in plants	76% and 46%		Vishnupradeep et al., 2022
Gigaspora gigantea	Rhizophagus irregularis	Translocation of Cr to the aerial part of the plant	97 and 194 mg/kg		de los Angeles Beltrán-Nambo et al., 2021
Portulaca oleracea L.	Oscillatoria sp.	Reduction in the bioavailable fraction of Cr(III), Cr(VI)		90%	Zanganeh et al., 2022
Oryza sativa	Azotobacter vinelandii	Increases antioxidant enzymes and accumulation	35%		Sahoo et al., 2021
Sesbania sesban	Bacillus xiamenensis PM14	Increased Cr uptake	47.33%		Din et al., 2020
Helianthus annuus L. and Solanum lycopersicum L.	Pseudomonas sp. (strain CPSB21)	Increase in superoxide dismutase and catalase activity and reduction in malonialdehyde and increase in Cr uptake	Root 60 µg/g Shoot 20 µg/g and Root 30 µg/g Shoot 10 µg/g		Gupta et al., 2018
Vetiveria zizanoides	Bacillus cereus (T1B3)	Reduce Cr		82%	Nayak et al., 2018
Rhizophagus irregularis	Arbuscular mycorrhizal fungi, Brachiaria mutica	Reduced Cr content	Increase uptake 75% at 30 mg Cr, 78% at 60 mg Cr	Reduced to 62% at 90 mg Cr	Kullu et al., 2020

(Continued)

TABLE 2.3 *(Continued)*
Impacts of Different Plant Species and Microbes under Chromium Stress

Plant Species	Microorganism	Effects	Accumulation Cr	Reduced Cr	Reference
Cicer arietinum	*Stenotrophomonas maltophilia,* *Bacillus thuringiensis, B. cereus*, and *B. subtilis*	Cr-tolerant and higher accumulation Cr	Root 6.25–60.41% Shoot 11.3–59.6%		Shreya et al., 2020
Festuca arundinacea	*Bacillus* sp. AK-1 and *Lysinibacillus* sp. AK-5	Hyperaccumulation	Extractable Cr 12.82–20.00% Residual Cr 9.41-22.37%.		Peng et al., 2021

comprehensive summary of various microorganisms that exhibit the capability to alleviate the occurrence of chromium.

Din and coworkers (2020) showed that the application of *Bacillus xiamenensis* PM14 resulted in a significant enhancement of *Sesbania sesban* overall chromium absorption by 47.33%. Additionally, the infected plants exhibited reduced levels of proline, malondialdehyde content, and electrolyte leakage when subjected to a chromium stress of 200 mg/kg, as compared to the plants that were not inoculated (Table 2.4).

Gupta and coworkers (2018) investigated Cr^{6+}-resistant plant growth-promoting *Pseudomonas* sp. (strain CPSB21) affected the growth of plants in farming soils that had been contaminated by tannery effluent. Sunflowers and tomatoes were used as test vegetables in a pot experiment. The study found that putting CPSB21 into the soil made Cr^{6+} less poisonous, improved plant growth, and made it easier for plants to absorb nutrients. The strain also improved the activity of SOD and catalase, decreased the amount of malondialdehyde, and made sunflower plants better at taking in Cr^{6+} (root by 60 μg/g, shoot by 20 μg/g) but not tomato plants.

Kalola and Desai (2020) found that *Halomonas* sp. DK4 from chrome electroplating sludge can effectively remove 81% of Cr(VI) from aqueous solutions in optimized MSM medium within 48 hours, and 59% in the presence of 15% NaCl concentration within 72 hours.

Kullu et al. (2020), investigated the effect of *Rhizophagus irregularis* on chlorophyll's brightness and Cr bioaccumulation in the grass *B. mutica*. They found that the AMF association decreased the amount of chromium by 62% at 90 mg Cr. The AMF union also kept plants from getting sick from too much Cr by making more antioxidants.

TABLE 2.4

Impacts of Different Microbes under Chromium Stress

Microbial Group	Species	pH	Temperature (°C)	Removal Efficiency (%)	Reference
Fungi	*Saccharomyces cerevisiae*	5	25	90	De Rossi et al., 2018
	Aspergillus sydowii	5	28	99.6	Lotlikar et al., 2018
	Arthrinium malaysianum	3	30	24.9	Majumder et al., 2017
	Penicillium oxalicum SL2		30	67	Long et al., 2018
	Aspergillus niger (CICC41115)	7	37	100	Gu et al., 2015
	Saccharomyces cerevisiae	3.5	25	100	Mahmoud and Mohamed, 2017
	Aspergillus sp. FK1	5		85	Srivastava and Thakur, 2006
Bacteria	*Acinetobacter* sp. B9	7	30	67	Bhattacharya and Gupta, 2013
	Enterobacter cloacae strain CTWI-06	7	37	94	Pattnaik et al., 2020
	Escherichia coli VITSUKMW3	7.5	30	40	Samuel et al., 2012
	Staphylococcus aureus strain K1	8	35	99	Tariq et al., 2019
	Bacillus subtilis PAW3	6	35	100	Wani et al., 2018
	Cellulosimicrobium funkei strain AR6	7	35	80.43	Karthik et al., 2016
	Acinetobacter sp. AB1	10	30	100	Essahale et al., 2012
	Streptomyces sp. MC1	7.4	30	52	Polti et al. 2011
	Bacillus subtilis MNU16	7	30	75	Upadhyay et al., 2017
	Pseudomonas sp. JF122	6.5	30	100	Zhou and Chen 2016
	Acinetobacter guillouiae SFC 500 – 1A	10	28	62	Ontañon et al., 2015
	B. mycoides 2000AsB1	7	30	100	Wang et al., 2016
	Streptomyces werraensis LD 22	7	41	51.7	Latha et al., 2015
	Arthrobacter sp. Sphe3	8	30	100	Ziagova et al., 2014
	Nesterenkonia sp.	88	35	100	Amoozegar et al., 2007
	Pseudomonas sp.	7	30	97	Desai et al., 2008
	Bacillus sp.			100	Molokwane et al., 2008
	Bacterial stain			100	Chai et al., 2009a
	Streptomyces	7	30	96	Polti et al., 2009
	Acinetobacter	10	30	100	Essahale et al., 2012

Nayak et al. (2018) evaluated rhizospheric bacteria that could help the chromium-loving plant *Vetiveria zizanoides* grow and store more metals. *Bacillus cereus* (T1B3) strain, which was found in mine waste, was able to remove 82% of Cr^{6+}.

Shreya et al. (2020) employed *Cicer arietinum* trees to test bacteria that can handle chromium. They found *S. maltophilia, B. thuringiensis, B. cereus,* and *B. subtilis* strains. The research discovered increases in Cr tolerance and increases in Cr uptake in roots by 6.25–60.41% and shoots by 11.3–59.6%. This shows that the inoculant may protect plants from damaging effects.

Suresh et al. (2021) discovered that *B. thuringiensis* (V45) and *Staphylococcus capitis* (S21) possess the capability to alleviate the lethal effects of Cr(VI) in sediment found in the tannery industry. The strains V45 and S21 have the ability to withstand chromium concentrations of up to 1000 µg/mL. Additionally, strain S21 has the capacity to endure Cr^{6+} concentrations of up to 340 µg/mL. Reducing Cr^{6+} at an initial concentration of 50 µg/mL can result in respective reductions of 86.42% and 97.34%.

APPLICABILITY

PD refers to a plant-based technological approach that facilitates the conversion of organic chemicals into nontoxic substances. This process has been observed to effectively transform many substances, including chlorinated solvents, munitions wastes, and pesticides (US EPA, 1998). Additionally, it has the potential to be employed for the purpose of eliminating pollutants from petrochemical sites, fuel spills, landfill leachates, and agricultural chemicals (Schnoor, 1997). In order to achieve successful implementation, it is imperative that the compounds undergo a transformation process that renders them harmless or considerably less hazardous compared to their original parental substances. PD can be employed in conjunction with other cleanup techniques or as an adjunctive treatment.

PHYTOREDUCTION (PR) OR PHYTOTRANSFORMATION

Phytoreduction refers to the mechanism by which plants take up and convert Cr(VI) into a less harmful form, Cr(III), by the action of enzymes present within their cellular structure (Shukla et al., 2007). This methodology requires the use of electrons obtained from compounds that have been reduced, such as $C_{10}H_{17}N_3O_6S$ and $C_6H_7O_6$. In their study, Rai et al. (1992) observed that several types of plants, including *B. juncea, H. annuus,* and *M. sativa,* exhibited the capacity to reduce the levels of Cr^{6+} in soil.

The rhizospheric microorganisms, specifically *Bacillus subtilis* and *Pseudomonas aeruginosa,* have the capability to engage in PR. This method involves the utilization of chromate reductase (CR) to effectively limit Cr^{6+} inside the root zone. By doing so, these microorganisms facilitate the reduction process and encourage the establishment of a microbial community within the root zone (Wang et al., 1989).

ENZYMATIC REDUCTION (ER)

ER is a natural process used by microbes and plants to convert HMs into harmless and less mobile or nonmobile forms (McLean and Beveridge, 2001). The technique employs enzymes, notably CR and nitroreductase (NR), which use electrons acquired from the reduced compounds to reduce Cr ions. The procedure comprises the utilization of enzymes, specifically CR and NR, which utilize electrons to decrease the level of chromium ions. Rhizospheric microbes, like *P. putida* and *A.oxydans*, utilize CrO_4 reducers to facilitate the decrease of chromate. Moreover, certain bacteria, including *Clostridium* spp. and *Escherichia coli*, possess the ability to reduce chromium due to nitroreductase (Kumar et al., 1995)

Wani and coworkers (2019) observed that the plant growth-boosting strain PAW3 exhibited remarkable efficacy in lowering Cr^{6+} levels and impacting the growth of cowpea plants. PAW3 was able to reduce the levels of Cr^{6+} across a wide range of conditions, including pH ranging from 5 to 7, concentrations of Cr^{6+} (100–200 µg/mL), and temperatures (20–35°C). This reduction process converted Cr^{6+} to Cr^{3+}, with the supernatant containing approximately 30 ± 1 µg/mL of Cr^{3+} and the debris containing approximately 70 ± 2.7 µg/mL. The PCR amplification confirmed the presence of Cr^{6+} reductase gene in PAW3, in contrast to other strains which did not exhibit the expression of this particular gene.

Rath and Das (2021) studied the Cr^{6+} levels on the effect of protein content on *V. mungo* var. B3-8-8 under chromium-stressed soil. Protein downregulation (29.2 kDa and 32.6 kDa) was observed at higher Cr concentrations. The root showed higher Cr uptake (300 µM) than the shoot, which could be attributed to inefficient transport or detoxification (DT). At the lower concentration of Cr, *V. mungo* was found to maintain haemostasis by enzyme activity involved in the cellular DT mechanism.

Sharma et al. (2020) found that application of EDTA to *H. vulgare* seedlings reduced the lipid peroxidase by Cr^{6+}. This led to an enhancement of the seedlings' antioxidative defense mechanism and pigment composition. The inclusion of EDTA in the presence of Cr^{6+} resulted in a reduction in both root and shoot bioconcentration factors, suggesting a decrease in the absorption of chromium by 61.7% in the seedlings from the surrounding medium.

LIMITATIONS

Usually, the efficacy of this strategy necessitates the passage of more than one growing season. The depth of the soil should not exceed three feet, while the proximity of groundwater to the surface should be no more than 10 feet. Contaminants can enter the food chain by insect or animal consumption of plant material. To enhance the plant uptake, it may be necessary to use soil additives, such as chelating compounds. These additives have the ability to disrupt the bindings that attach pollutants to soil particles (Miller, 1996).

Both economically and environmentally friendly.

DISADVANTAGES OF PHYTOREDUCTION

- To be effective, a minimum of one season of growth is required.
- The soil has to be no more than three feet deep, and groundwater must be within 10 feet of the surface.
- The pollutants can return to the food web when consumed the plant material.

PHYTOFILTRATION (PF) OR RHIZOFILTRATION

Phytofiltration refers to the mechanism by which pollutants are adsorbed into the root system of plants. This method is mainly helped by the synthesis of molecules inside the roots, which can boost the affinity of contaminants, such as metal ions (Singh et al., 2015). PF, also referred to as rhizofiltration, is an established technique in which the roots of plants facilitate the degradation of pollutants present in many water sources, including wastewater, surface water, and groundwater. Sánchez-Galván et al. (2022) described the method and its applicability to both aqueous and terrestrial plants. Furthermore, it has been successfully employed in various settings, such as ponds, tanks, and basins, enabling the effective remediation of pollutants within the natural setting.

Bokhari et al. (2022) evaluated the PF abilities of *Hydrocotyle umbellata* in removing heavy metals from Nullah Lai wastewater, as well as the associated environmental risks. The researchers conducted monitoring of heavy metal concentrations at various time intervals and observed a phytofiltration effectiveness of 98.11% for chromium. The proliferation of *H. umbellata* resulted in the buildup of HMs in both the roots and shoots, with chromium exhibiting translocation factor values of less than 1.

Shiyab et al. (2019) performed a greenhouse experiment to determine the phytotoxic impact of Cr on sour orange seed germination and seedling growth in Cr-contaminated soil. The research findings indicate that when concentrations of chromium exceed 100 parts per million (ppm), there is an observed elevation in the levels of various antioxidative enzymes in plants, including SOD, CAT, POD, and APX. Additionally, this elevated concentration of Cr has been found to lead to an increase in lipid peroxidation of 97.67%.

Plants with fast-growing root systems, known as rhizofiltration plants, have demonstrated the ability to efficiently extract hazardous metals such as Cu, Cd, Cr, Ni, Pb, and Zn from aqueous solutions over a prolonged duration. According to the

Environmental Protection Agency (US EPA) in 1998, it is possible to eliminate low-level radioactive contamination from liquid streams.

LIMITATIONS

Rhizofiltration works well with low concentrations and large amounts of water, but Salt et al. (1995) says that plants that are good at moving metals to shoots shouldn't be used because they leave behind more polluted plant waste.

ADVANTAGES OF PHYTOFILTRATION

The capacity to effectively employ both land-based and water-based plants for both *in situ* and *ex situ* purposes, hence obviating the necessity of transferring pollutants into plant shoots.

DISADVANTAGES PHYTOFILTRATION

- The adjustment of pH should be performed periodically.
- It may be necessary to cultivate plants initially in a controlled environment such as a greenhouse or nursery.
- Plants are regularly gathered and afterward discarded.
- The tank design should be thoroughly considered.

REFERENCES

Abbas, Aown, Muhammad Azeem, Muhammad Naveed, Abdul Latif, Saqib Bashir, Amjad Ali, Muhammad Bilal, and Liaqat Ali. "Synergistic use of biochar and acidified manure for improving growth of maize in chromium contaminated soil." *International Journal of phytoremediation* 22, no. 1 (2020): 52–61.

Adiloğlu, Sevinç, and Merve Göker. "Phytoremediation: Elimination of hexavalent chromium heavy metal using corn (*Zea mays* L.)." *Cereal Research Communications* 49 (2021): 65–72.

Agar, Guleray, Mahmut Sinan Taspinar, Ertan Yildirim, Murat Aydin, and Merve Yuce. "Effects of ascorbic acid and copper treatments on metallothionein gene expression and antioxidant enzyme activities in *Helianthus annuus* L. Exposed to chromium stress." *Journal of Plant Growth Regulation* 39 (2020): 897–904.

Ahmed, Bilal, Asad Syed, Asfa Rizvi, Mohammad Shahid, Ali H. Bahkali, Mohammad Saghir Khan, and Javed Musarrat. "Impact of metal-oxide nanoparticles on growth, physiology and yield of tomato (*Solanum lycopersicum* L.) modulated by *Azotobacter salinestris* strain ASM." *Environmental Pollution* 269 (2021): 116218.

Ahmed, F., M. S. Hossain, A. T. Abdullah, M. A. Akbor, and M. A. Ahsan. "Public health risk assessment of chromium intake from vegetable grown in the wastewater irrigated site in Bangladesh." *Pollution* 4, no 2 (2016): 425–432. doi: 10.7508/pj.2016.04.005.

Ahsan, Muhammad T., Muhammad Najam-ul-Haq, Abdul Saeed, Tanveer Mustafa, and Muhammad Afzal. "Augmentation with potential endophytes enhances phytostabilization of Cr in contaminated soil." *Environmental Science and Pollution Research* 25 (2018): 7021–7032.

Amin, Alaa S., and Mohammed A. Kassem. "Chromium speciation in environmental samples using a solid phase spectrophotometric method." *Spectrochimica Acta Part A: Molecular and Biomolecular Spectroscopy* 96 (2012): 541–547.

Amin, Hira, Basir Ahmed Arain, Muhammad Sadiq Abbasi, Farah Amin, Taj Muhammad Jahangir, and Noor-ul-Ain Soomro. "Evaluation of chromium phytotoxicity, phyto-tolerance, and phyto-accumulation using biofuel plants for effective phytoremediation." *International Journal of Phytoremediation* 21, no. 4 (2019): 352–363.

Amoozegar, Mohammad Ali, Ali Ghasemi, Mohammad Reza Razavi, and Saied Naddaf. "Evaluation of hexavalent chromium reduction by chromate-resistant moderately halophile, Nesterenkonia sp. strain MF2." *Process Biochemistry* 42, no. 10 (2007): 1475–1479.

Antil, R. S. "Integrated plant nutrient supply for sustainable soil health and crop productivity." *Focus Global Reporter* 3 (2012).

Ao, Ming, Xiaoting Chen, Tenghaobo Deng, Shengsheng Sun, Yetao Tang, Jean Louis Morel, Rongliang Qiu, and Shizhong Wang. "Chromium biogeochemical behaviour in soil-plant systems and remediation strategies: A critical review." *Journal of Hazardous Materials* 424 (2022): 127233.

Augustynowicz, Joanna, Ewa Sitek, Tomasz Bryniarski, Agnieszka Baran, Beata Ostachowicz, Małgorzata Urbańska-Stopa, and Marek Szklarczyk. "The use of *Callitriche cophocarpa Sendtn.* for the reclamation of Cr-contaminated freshwater habitat: Benefits and limitations." *Environmental Science and Pollution Research* 27 (2020): 25510–25522.

Aziz, H. A., N. Othman, M. S. Yusuff, D. R. H. Basri, F. A. H. Ashaari, M. N. Adlan, F. Othman, M. Johari, and M. Perwira. "Removal of copper from water using limestone filtration technique: Determination of mechanism of removal." *Environment International* 26, no. 5–6 (2001): 395–399.

Balamurugan, D., C. Udayasooriyan, K. Vinoth-Kumar, RM Jayabalakrishman, and Natesan. R. "Removal of hexavalent chromium [Cr (VI)] from spiked soil using Na Y (Nano sodium) zeolite supported zero valent iron nanoparticles." *Environment and Ecology Research* 2, no. 8 (2014): 291–300.

Banerjee, Soumya, Biswajit Kamila, Sanghamitra Barman, S. R. Joshi, Tamal Mandal, and Gopinath Halder. "Interlining Cr (VI) remediation mechanism by a novel bacterium *Pseudomonas brenneri* isolated from coalmine wastewater." *Journal of Environmental Management* 233 (2019): 271–282.

Bhattacharya, Amrik, and Anshu Gupta. "Evaluation of *Acinetobacter* sp. B9 for Cr (VI) resistance and detoxification with potential application in bioremediation of heavy-metals-rich industrial wastewater." *Environmental Science and Pollution Research* 20 (2013): 6628–6637.

Bokhari, Syeda Huma, Ghazala Nawaz, Azizullah Azizullah, Muhammad Mahmood-Ul-Hassan, and Zeshan Ali. "Heavy metals phytofiltration potential of *Hydrocotyle umbellata* from Nullah Lai wastewater and its environmental risk." *International Journal of Phytoremediation* 24, no. 14 (2022): 1465–1474.

Bolan, Nanthi S., Jin Hee Park, Brett Robinson, Ravi Naidu, and Keun Young Huh. "Phytostabilization: A green approach to contaminant containment." *Advances in Agronomy* 112 (2011): 145–204.

Branzini, Agustina, and Marta Susana Zubillaga. "Comparative use of soil organic and inorganic amendments in heavy metals stabilization." *Applied and Environmental Soil Science* 2012 (2012): 721032.

Caldelas, Cristina, Jordi Bort, and Anna Febrero. "Ultrastructure and subcellular distribution of Cr in *Iris pseudacorus* L. using TEM and X-ray microanalysis." *Cell Biology and Toxicology* 28 (2012): 57–68.

Center for Public Environmental Oversight (CPEO) (2021). Phytoremediation.

Cervantes, Carlos, Jesús Campos-García, Silvia Devars, Félix Gutiérrez-Corona, Herminia Loza-Tavera, Juan Carlos Torres-Guzmán, and Rafael Moreno-Sánchez. "Interactions of chromium with microorganisms and plants." *FEMS Microbiology Reviews* 25, no. 3 (2001): 335–347.

Chai, Liyuan, Shunhong Huang, Zhihui Yang, Bing Peng, Yan Huang, and Yuehui Chen. "Cr (VI) remediation by indigenous bacteria in soils contaminated by chromium-containing slag." *Journal of Hazardous Materials* 167, no. 1–3 (2009): 516–522.

Chandra, P. R., A. K. Abdussalam, and S. Nabeesa. "Distribution of bioaccumulated Cd and Cr in two Vigna species and the associated histological variations." *Journal of Stress Physiology & Biochemistry* 6 (2010): 4–12.

Chaudhary, Khushboo, Swati Agarwal, and Suphiya Khan. "Role of phytochelatins (PCs), metallothioneins (MTs), and heavy metal ATPase (HMA) genes in heavy metal tolerance." *Mycoremediation and Environmental Sustainability* 2 (2018): 39–60.

Chen, Z., S. Song, and Y Wen. "Reduction of Cr(VI) into Cr(III) by organelles of *Chlorella vulgaris* in aqueous solution: An organelle-level attempt." *Sci. Total Environ* 572 (2016): 361–368.

Chen, Huixia, Junfeng Dou, and Hongbin Xu. "Remediation of Cr (VI)-contaminated soil with co-composting of three different biomass solid wastes." *Journal of Soils and Sediments* 18 (2018): 897–905.

Chen, Huixia, Junfeng Dou, and Hongbin Xu. "The effect of low-molecular-weight organic-acids (LMWOAs) on treatment of chromium-contaminated soils by compost-phytoremediation: Kinetics of the chromium release and fractionation." *Journal of Environmental Sciences* 70 (2018): 45–53.

Chen, Yimeng, Hualin Chen, Ronald W. Thring, Huan Liu, Jiangmin Zhou, Yueliang Tao, and Jianbing Li. "Immobilization of chromium contaminated soil by co-pyrolysis with rice straw." *Water, Air, & Soil Pollution* 231 (2020): 1–13.

Chibuike, Grace U., and Smart C. Obiora. "Heavy metal polluted soils: Effect on plants and bioremediation methods." *Applied and Environmental Soil Science* 1 (2014): 1–12.

Chien, C. C., Z. H. Yang, W. Z. Cao, Y. T. Tu, and C. M. Kao. "Application of an aquatic plant ecosystem for swine wastewater polishment: A full-scale study." *Desalination and Water Treatment* 57, no. 45 (2016): 21243–21252.

Chintakovid, Watchara, Pornsawan Visoottiviseth, Somkiat Khokiattiwong, and Siriporn Lauengsuchonkul. "Potential of the hybrid marigolds for arsenic phytoremediation and income generation of remediators in Ron Phibun District, Thailand." *Chemosphere* 70, no. 8 (2008): 1532–1537.

Chintani, Yuanita Sekar, Erni Saurmalinda Butarbutar, Andhika Puspito Nugroho, and Tarzan Sembiring. "Uptake and release of chromium and nickel by Vetiver grass (*Chrysopogon zizanioides* (L.) Roberty)." *SN Applied Sciences* 3 (2021): 1–13.

Chug, Ravneet, Vinod Singh Gour, Shruti Mathur, and S. L. Kothari. "Optimization of extracellular polymeric substances production using Azotobacter beijreinckii and *Bacillus subtilis* and its application in chromium (VI) removal." *Bioresource Technology* 214 (2016): 604–608.

Cornelis, Rita, Joe Caruso, Helen Crews, and Klaus Heumann. "Handbook of Elemental Speciation II–Species in the Environment, Food, Medicine and Occupational Health." (2005).

DalCorso, Giovanni, Elisa Fasani, Anna Manara, Giovanna Visioli, and Antonella Furini. "Heavy metal pollutions: State of the art and innovation in phytoremediation." *International Journal of Molecular Sciences* 20, no. 14 (2019): 3412.

de los Angeles Beltrán-Nambo, Maria, Nancy Rojas-Jacuinde, Miguel Martinez-Trujillo, Pablo Fabián Jaramillo-López, Mariela Gomez Romero, and Yazmín Carreón-Abud. "Differential strategies of two species of arbuscular mycorrhizal fungi in the protection of maize plants grown in chromium-contaminated soils." *BioMetals* 34, no. 6 (2021): 1247–1261.

De Rossi, Andrea, Magali Rejane Rigon, Munise Zaparoli, Rafael Dalmas Braido, Luciane Maria Colla, Guilherme Luiz Dotto, and Jeferson Steffanello Piccin. "Chromium (VI) biosorption by *Saccharomyces cerevisiae* subjected to chemical and thermal treatments." *Environmental Science and Pollution Research* 25 (2018): 19179–19186.

de Souza, Tamara Daiane, Alisson Carraro Borges, Amanda Fernandes Braga, Renato Welmer Veloso, and Antonio Teixeira de Matos. "Phytoremediation of arsenic-contaminated water by Lemna Valdiviana: An optimization study." *Chemosphere* 234 (2019): 402–408.

Desai, Chirayu, Kunal Jain, and Datta Madamwar. "Hexavalent chromate reductase activity in cytosolic fractions of *Pseudomonas* sp. G1DM21 isolated from Cr (VI) contaminated industrial landfill." *Process Biochemistry* 43, no. 7 (2008): 713–721.

Dheeba, B., P. Sampathkumar, and K. Kannan. "Fertilizers and mixed crop cultivation of chromium tolerant and sensitive plants under chromium toxicity." *Journal of Toxicology* 2015 (2015): 367217.

Din, Bashir Ud, Amna, Mazhar Rafique, Muhammad Tariq Javed, Muhammad Aqeel Kamran, Shehzad Mehmood, Mursalin Khan, Tariq Sultan, Muhammad Farooq Hussain Munis, and Hassan Javed Chaudhary. "Assisted phytoremediation of chromium spiked soils by *Sesbania sesban* in association with *Bacillus xiamenensis* PM14: A biochemical analysis." *Plant Physiology and Biochemistry* 146 (2020): 249–258.

Ding, Changfeng, Xiaogang Li, Taolin Zhang, Yibing Ma, and Xingxiang Wang. "Phytotoxicity and accumulation of chromium in carrot plants and the derivation of soil thresholds for Chinese soils." *Ecotoxicology and Environmental Safety* 108 (2014): 179–186.

Escudero Oñate, Carlos, Núria Fiol Santaló, Isabel Villaescusa Gil, and Jean Claude Bollinger. "Effect of chromium speciation on its sorption mechanism onto grape stalks entrapped into alginate beads." *Arabian Journal of Chemistry* 10, no. 1 (2017): S1293–S1302.

Essahale, A., M. Malki, I. Marin, and M. Moumni. "Hexavalent chromium reduction and accumulation by Acinetobacter AB1 isolated from Fez Tanneries in Morocco." *Indian Journal of Microbiology* 52 (2012): 48–53.

Fahad, Shah, Saddam Hussain, Asghari Bano, Shah Saud, Shah Hassan, Darakh Shan, and Faheem Ahmed Khan et al. "Potential role of phytohormones and plant growth-promoting rhizobacteria in abiotic stresses: Consequences for changing environment." *Environmental Science and Pollution Research* 22 (2015): 4907–4921.

Farid, Mujahid, Shafaqat Ali, Rashid Saeed, Muhammad Rizwan, Syed Asad Hussain Bukhari, Ghulam Hassan Abbasi, Afzal Hussain, Basharat Ali, Muhammad Shahid Ibni Zamir, and Irfan Ahmad. "Combined application of citric acid and 5-aminolevulinic acid improved biomass, photosynthesis and gas exchange attributes of sunflower (*Helianthus annuus* L.) grown on chromium contaminated soil." *International Journal of Phytoremediation* 21, no. 8 (2019): 760–767.

Feki, Kaouthar, Sana Tounsi, Moncef Mrabet, Haythem Mhadhbi, and Faiçal Brini. "Recent advances in physiological and molecular mechanisms of heavy metal accumulation in plants." *Environmental Science and Pollution Research* (2021): 1–20.

Gao, Yang, and Jun Xia. "Chromium contamination accident in China: Viewing environment policy of China." (2011): 8605–8606.

Gao, Z., Y. Geng, X. Zeng, X. Tian, T. Yao, X. Song, and C. Su. "Evolution of the anthropogenic chromium cycle in China." *Journal of Industrial Ecology* 26, no. 2 (2022): 592–608.

Ghosh, Mridul, and S. P. Singh. "A review on phytoremediation of heavy metals and utilization of it's by products." *Asian J Energy Environ* 6, no. 4 (2005): 18.

Grace Pavithra, Kirubanandam, V. Jaikumar, P. Senthil Kumar, and PanneerSelvam SundarRajan. "A review on cleaner strategies for chromium industrial wastewater: Present research and future perspective." *Journal of Cleaner Production* 228 (2019): 580–593.

Gu, Yanling, Weihua Xu, Yunguo Liu, Guangming Zeng, Jinhui Huang, Xiaofei Tan, Hao Jian, Xi Hu, Fei Li, and Dafei Wang. "Mechanism of Cr (VI) reduction by *Aspergillus niger*: Enzymatic characteristic, oxidative stress response, and reduction product." *Environmental Science and Pollution Research* 22 (2015): 6271–6279.

Gupta, Neha, Hari Ram, and Balwinder Kumar. "Mechanism of zinc absorption in plants: Uptake, transport, translocation and accumulation." *Reviews in Environmental Science and Bio/Technology* 15 (2016): 89–109.

Gupta, Pratishtha, Rupa Rani, Avantika Chandra, and Vipin Kumar. "Potential applications of *Pseudomonas* sp. (strain CPSB21) to ameliorate Cr6+ stress and phytoremediation of tannery effluent contaminated agricultural soils." *Scientific Reports* 8, no. 1 (2018): 4860.

Haokip, Nengpilam, and Abhik Gupta. "Phytoremediation of chromium and manganese by *Ipomoea aquatica* Forssk. from aqueous medium containing chromium-manganese mixtures in microcosms and mesocosms." *Water and Environment Journal* 35, no. 3 (2021): 884–891.

Hasan, Shaikh Md Mahady, Md Ali Akber, Md Mezbaul Bahar, Md Azharul Islam, Md Ahedul Akbor, Md Abu Bakar Siddique, and Md Atikul Islam. "Chromium contamination from tanning industries and phytoremediation potential of native plants: A study of Savar Tannery Industrial Estate in Dhaka, Bangladesh." *Bulletin of Environmental Contamination and Toxicology* 106, no. 6 (2021): 1024–1032.

Hoffman, Emma, Meagan Bernier, Brenden Blotnicky, Peter G. Golden, Jeffrey Janes, Allison Kader, Rachel Kovacs-Da Costa, Shauna Pettipas, Sarah Vermeulen, and Tony R. Walker. "Assessment of public perception and environmental compliance at a pulp and paper facility: A Canadian case study." *Environmental Monitoring and Assessment* 187 (2015): 1–13.

Hsu, Liang-Ching, Yu-Ting Liu, and Yu-Min Tzou. "Comparison of the spectroscopic speciation and chemical fractionation of chromium in contaminated paddy soils." *Journal of Hazardous Materials* 296 (2015): 230–238.

Irshad, Sana, Zuoming Xie, Sajid Mehmood, Asad Nawaz, Allah Ditta, and Qaisar Mahmood. "Insights into conventional and recent technologies for arsenic bioremediation: A systematic review." *Environmental Science and Pollution Research* 28 (2021): 18870–18892.

Kabata-Pendias, Alina. *Trace elements in soils and plants.* CRC press, 2000.

Kafle, Arjun, Anil Timilsina, Asmita Gautam, Kaushik Adhikari, Anukul Bhattarai, and Niroj Aryal. "Phytoremediation: Mechanisms, plant selection and enhancement by natural and synthetic agents." *Environmental Advances* 8 (2022): 100203.

Kalam, Saqib Ul, Fauzia Naushin, Fareed Ahmad Khan, and Nishanta Rajakaruna. "Long-term phytoremediating abilities of *Dalbergia sissoo* Roxb. (Fabaceae)." *SN Applied Sciences* 1 (2019): 1–8.

Kalola, Vidhi, and Chirayu Desai. "Biosorption of Cr (VI) by *Halomonas* sp. DK4, a halo-tolerant bacterium isolated from chrome electroplating sludge." *Environmental Science and Pollution Research* 27, no. 22 (2020): 27330–27344.

Kanwar, Varinder Singh, Ajay Sharma, Arun Lal Srivastav, and Lata Rani. "Phytoremediation of toxic metals present in soil and water environment: A critical review." *Environmental Science and Pollution Research* 27 (2020): 44835–44860.

Karthik, Chinnannan, Mohammad Oves, R. Thangabalu, Ranandkumar Sharma, S. B. Santhosh, and P. Indra Arulselvi. "*Cellulosimicrobium funkei*-like enhances the growth of *Phaseolus vulgaris* by modulating oxidative damage under Chromium(VI) toxicity." *Journal of Advanced Research* 7, no. 6 (2016): 839–850.

Kim, Young Jin, Jong Hoon Kim, Chang Eun Lee, Young Geun Mok, Jong Soon Choi, Hyoung Sun Shin, and Seongbin Hwang. "Expression of yeast transcriptional activator MSN1 promotes accumulation of chromium and sulfur by enhancing sulfate transporter level in plants." *FEBS Letters* 580, no. 1 (2006): 206–210.

Kullu, Bandana, Deepak Kumar Patra, Srinivas Acharya, Chinmay Pradhan, and Hemanta Kumar Patra. "AM fungi mediated bioaccumulation of hexavalent chromium in *Brachiaria mutica* - a mycorrhizal phytoremediation approach." *Chemosphere* 258 (2020): 127337.

Kumar, PBA Nanda, Viatcheslav Dushenkov, Harry Motto, and Ilya Raskin. "Phytoextraction: The use of plants to remove heavy metals from soils." *Environmental Science & Technology* 29, no. 5 (1995): 1232–1238.

Lakshmi, S., and P. Sundaramoorthy. "Effect of chromium on germination and seedling growth of vegetable crops." *Asian Journal of Science and Technology* 1 (2010): 28–31.

Latha, S., G. Vinothini, and D. Dhanasekaran. "Chromium [Cr (VI)] biosorption property of the newly isolated actinobacterial probiont *Streptomyces werraensis* LD22." *3 Biotech* 5 (2015): 423–432.

Levizou, Efi, Anna A. Zanni, and Vasileios Antoniadis. "Varying concentrations of soil chromium (VI) for the exploration of tolerance thresholds and phytoremediation potential of the oregano (*Origanum vulgare*)." *Environmental Science and Pollution Research* 26 (2019): 14–23.

Lilli, Maria A., Daniel Moraetis, Nikolaos P. Nikolaidis, George P. Karatzas, and Nicolas Kalogerakis. "Characterization and mobility of geogenic chromium in soils and river bed sediments of Asopos basin." *Journal of Hazardous Materials* 281 (2015): 12–19.

Liu, Donghua, Jinhua Zou, Min Wang, and Wusheng Jiang. "Hexavalent chromium uptake and its effects on mineral uptake, antioxidant defence system and photosynthesis in *Amaranthus viridis* L." *Bioresource Technology* 99, no. 7 (2008): 2628–2636.

Long, Bibo, Binhui Ye, Qinglin Liu, Shu Zhang, Jien Ye, Lina Zou, and Jiyan Shi. "Characterization of *Penicillium oxalicum* SL2 isolated from indoor air and its application to the removal of hexavalent chromium." *PLoS One* 13, no. 1 (2018): e0191484.

Lotlikar, Nikita P., Samir R. Damare, Ram Murti Meena, P. Linsy, and Brenda Mascarenhas. "Potential of marine-derived fungi to remove hexavalent chromium pollutant from culture broth." *Indian Journal of Microbiology* 58 (2018): 182–192.

Lukina, A. O., C. Boutin, O. Rowland, and D. J. Carpenter. "Evaluating trivalent chromium toxicity on wild terrestrial and wetland plants." *Chemosphere* 162 (2016): 355–364.

Ma, Lena Q., Kenneth M. Komar, Cong Tu, Weihua Zhang, Yong Cai, and Elizabeth D. Kennelley. "A fern that hyperaccumulates arsenic." *Nature* 409, no. 6820 (2001): 579–579.

Ma, Yibing, and Peter S. Hooda. "Chromium, nickel and cobalt." *Trace Elements in Soils* (2010): 461–479. https://doi.org/10.1002/9781444319477.ch19

Mahmoud, M. S., and Samah A. Mohamed. "Calcium alginate as an eco-friendly supporting material for Baker's yeast strain in chromium bioremediation." *HBRC Journal* 13, no. 3 (2017): 245–254.

Majumder, Rajib, Lubna Sheikh, Animesh Naskar, Vineeta, Manabendra Mukherjee, and Sucheta Tripathy. "Depletion of Cr(VI) from aqueous solution by heat dried biomass of a newly isolated fungus *Arthrinium malaysianum*: A mechanistic approach." *Scientific Reports* 7, no. 1 (2017): 11254.

Mamais, Daniel, Constantinos Noutsopoulos, Ioanna Kavallari, Eleni Nyktari, Apostolos Kaldis, Eleni Panousi, George Nikitopoulos, Kornilia Antoniou, and Maria Nasioka. "Biological groundwater treatment for chromium removal at low hexavalent chromium concentrations." *Chemosphere* 152 (2016): 238–244.

Marieschi, Matteo, Gessica Gorbi, Corrado Zanni, A. Sardella, and Anna Torelli. "Increase of chromium tolerance in *Scenedesmus acutus* after sulfur starvation: Chromium uptake and compartmentalization in two strains with different sensitivities to Cr(VI)." *Aquatic Toxicology* 167 (2015): 124–133.

Marques, Ana PGC, Antonio OSS Rangel, and Paula ML Castro. "Remediation of heavy metal contaminated soils: An overview of site remediation techniques." *Critical Reviews in Environmental Science and Technology* 41, no. 10 (2011): 879–914.

Martí, E., J. Sierra, J. Cáliz, G. Montserrat, X. Vila, and M.A Garau. "Ecotoxicity of Cr, Cd, and Pb on two Mediterranean soils." *Archives of Environmental Contamination and Toxicology* 64 (2013): 377–387.

McLean, Jeff, and Terry J. Beveridge. "Chromate reduction by a pseudomonad isolated from a site contaminated with chromated copper arsenate." *Applied and Environmental Microbiology* 67, no. 3 (2001): 1076–1084.

Miller, Ralinda Rae. *Artificially-induced or blast-enhanced fracturing.* Ground-Water Remediation Technologies Analysis Center, 1996.

Mishra, K., K. Gupta, and U. N Rai. "Bioconcentration and phytotoxicity of chromium in *Eichhornia crassipes*." *Journal of Environmental Biology* 30 (2009): 521–526.

Mohammadi, Hamid, Ali Reza Amani-Ghadim, Amir Abbas Matin, and Mansour Ghorbanpour. "Fe 0 nanoparticles improve physiological and antioxidative attributes of sunflower (*Helianthus annuus*) plants grown in soil spiked with hexavalent chromium." *3 Biotech* 10 (2020): 1–11.

Mohammadi, Hamid, Mehrnaz Hatami, Khatoon Feghezadeh, and Mansour Ghorbanpour. "Mitigating effect of nano-zerovalent iron, iron sulfate and EDTA against oxidative stress induced by chromium in *Helianthus annuus* L." *Acta Physiologiae Plantarum* 40 (2018): 1–15.

Molokwane, Pulane E., Kakonge C. Meli, and Evans M. Nkhalambayausi-Chirwa. "Chromium (VI) reduction in activated sludge bacteria exposed to high chromium loading: Brits culture (South Africa)." *Water Research* 42, no. 17 (2008): 4538–4548.

Mondal, N. K., and P. Nayek. "Hexavalent chromium accumulation kinetics and physiological responses exhibited by *Eichhornia* sp. and *Pistia* sp." *International Journal of Environmental Science and Technology* 17 (2020): 1397–1410.

Morkunas, Iwona, Agnieszka Woźniak, Van Chung Mai, Renata Rucińska-Sobkowiak, and Philippe Jeandet. "The role of heavy metals in plant response to biotic stress." *Molecules* 23, no. 9 (2018): 2320.

Murphy, Timothy F., Charmaine Kirkham, Antoinette Johnson, Aimee L. Brauer, Mary Koszelak-Rosenblum, and Michael G. Malkowski. "Sulfate-binding protein, CysP, is a candidate vaccine antigen of *Moraxella catarrhalis*." *Vaccine* 34, no. 33 (2016): 3855–3861.

Murthy, M.K., P. Khandayataray, and D. Samal. "Chromium toxicity and its remediation by using endophytic bacteria and nanomaterials: A review." *Journal of Environmental Management* 318 (2022): 115620.

Naveed, Muhammad, Bisma Tanvir, Wang Xiukang, Martin Brtnicky, Allah Ditta, Jiri Kucerik, Zinayyera Subhani et al. "Co-composted biochar enhances growth, physiological, and phytostabilization efficiency of *Brassica napus* and reduces associated health risks under chromium stress." *Frontiers in Plant Science* 12 (2021): 775785.

Nayak, A. K., S. S. Panda, A. Basu, and N. K. Dhal. "Enhancement of toxic Cr (VI), Fe, and other heavy metals phytoremediation by the synergistic combination of native *Bacillus cereus* strain and *Vetiveria zizanioides* L." *International Journal of Phytoremediation* 20, no. 7 (2018): 682–691.

Nriagu, Jerome O., and Evert Nieboer, eds. *Chromium in the natural and human environments.* Vol. 20. John Wiley & Sons, 1988.

Ontañon, Ornella M., Paola S. González, and Elizabeth Agostini. "Biochemical and molecular mechanisms involved in simultaneous phenol and Cr (VI) removal by *Acinetobacter guillouiae* SFC 500-1A." *Environmental Science and Pollution Research* 22 (2015): 13014–13023.

Owlad, Mojdeh, Mohamed Kheireddine Aroua, Wan Ashri Wan Daud, and Saeid Baroutian. "Removal of hexavalent chromium-contaminated water and wastewater: A review." *Water, Air, and Soil Pollution* 200 (2009): 59–77.

Patra, D. K., C. Pradhan, and H. K Patra. "Chelate based phytoremediation study for attenuation of chromium toxicity stress using lemongrass *Cymbopogon flexuosus*." *International Journal of Phytoremediation* 20, no. 13 (2018): 1324–1329.

Patra, Deepak Kumar, Chinmay Pradhan, and Hemanta Kumar Patra. "An in situ study of growth of Lemongrass *Cymbopogon flexuosus* (Nees ex Steud.) W. Watson on varying concentration of Chromium (Cr+ 6) on soil and its bioaccumulation: Perspectives on phytoremediation potential and phytostabilisation of chromium toxicity." *Chemosphere* 193 (2017): 793–799.

Patra, Deepak Kumar, Chinmay Pradhan, Jagdish Kumar, and Hemanta Kumar Patra. "Assessment of chromum phytotoxicity, phytoremediation and tolerance potential of *Sesbania sesban* and *Brachiaria mutica* grown on chromite mine overburden dumps and garden soil." *Chemosphere* 252 (2020): 126553.

Pattnaik, Swati, Debasis Dash, Swati Mohapatra, Matrujyoti Pattnaik, Amit K. Marandi, Surajit Das, and Devi P. Samantaray. "Improvement of rice plant productivity by native Cr (VI) reducing and plant growth promoting soil bacteria *Enterobacter cloacae*." *Chemosphere* 240 (2020): 124895.

Peng, He, Ke Liang, Huanyan Luo, Huayan Huang, Shihua Luo, AKang Zhang, Heng Xu, and Fei Xu. "A *Bacillus* and *Lysinibacillus* sp. bio-augmented *Festuca arundinacea* phytoremediation system for the rapid decontamination of chromium influenced soil." *Chemosphere* 283 (2021): 131186.

Perotti, Romina, Cintia Elizabeth Paisio, Elizabeth Agostini, María Inés Fernandez, and Paola Solange González. "CR (VI) phytoremediation by hairy roots of *Brassica napus*: Assessing efficiency, mechanisms involved, and post-removal toxicity." *Environmental Science and Pollution Research* 27 (2020): 9465–9474.

Pilipović, Andrej, Saša Orlović, Vladislava Galović, P. Poljaković, G. Leopold, V. Zoran, and V. Verica. "Environmental application of forest tree species in phytoremediation and reclamation." Needs and priorities for research and education in biotechnology applied to emerging environmental challenges in SEE countries 39 (2008).

Polti, Marta A., María Julia Amoroso, and Carlos M. Abate. "Intracellular chromium accumulation by *Streptomyces* sp. MC1." *Water, Air, & Soil Pollution* 214 (2011): 49–57.

Polti, Marta A., Roberto O. García, María J. Amoroso, and Carlos M. Abate. "Bioremediation of chromium (VI) contaminated soil by *Streptomyces* sp. MC1." *Journal of Basic Microbiology* 49, no. 3 (2009): 285–292.

Pootakham, Wirulda, David Gonzalez-Ballester, and Arthur R. Grossman. "Identification and regulation of plasma membrane sulfate transporters in Chlamydomonas." *Plant Physiology* 153, no. 4 (2010): 1653–1668.

Prado, Carolina, Eduardo Pagano, Fernando Prado, and Mariana Rosa. "Detoxification of Cr (VI) in *Salvinia minima* is related to seasonal-induced changes of thiols, phenolics and antioxidative enzymes." *Journal of Hazardous Materials* 239 (2012): 355–361.

Pushkar, Bhupendra, Pooja Sevak, Sejal Parab, and Nikita Nilkanth. "Chromium pollution and its bioremediation mechanisms in bacteria: A review." *Journal of Environmental Management* 287 (2021): 112279.

Qadir, Muhammad, Anwar Hussain, Muhammad Hamayun, Mohib Shah, Amjad Iqbal, and Waheed Murad. "Phytohormones producing rhizobacterium alleviates chromium toxicity in *Helianthus annuus* L. by reducing chromate uptake and strengthening antioxidant system." *Chemosphere* 258 (2020): 127386.

Qianqian, Ma, Fasih Ullah Haider, Muhammad Farooq, Muhammad Adeel, Noman Shakoor, Wu Jun, Xu Jiaying, Xu Wang Wang, Luo Panjun, and Liqun Cai. "Selenium treated foliage and biochar treated soil for improved lettuce (*Lactuca sativa* L.) growth in Cd-polluted soil." *Journal of Cleaner Production* 335 (2022): 130267.

Quantin, Cécile, Vojtěch Ettler, Jérémie Garnier, and Ondřej Šebek. "Sources and extractibility of chromium and nickel in soil profiles developed on Czech serpentinites." *Comptes Rendus Geoscience* 340, no. 12 (2008): 872–882.

Radziemska, Maja, Agnieszka Bęś, Zygmunt M. Gusiatin, Łukasz Sikorski, Martin Brtnicky, Grzegorz Majewski, Ernesta Liniauskienė et al. "Successful outcome of phytostabilization in Cr (VI) contaminated soils amended with alkalizing additives." *International Journal of Environmental Research and Public Health* 17, no. 17 (2020): 6073.

Radziemska, Maja, Eugeniusz Koda, Ayla Bilgin, and Mgdalena D. Vaverková. "Concept of aided phytostabilization of contaminated soils in postindustrial areas." *International Journal of Environmental Research and Public Health* 15, no. 1 (2018): 24.

Rai, Upendra Nath, Rudra Deo Tripathi, and Nikhil Kumar. "Bioaccumulation of chromium and toxicity on growth, photosynthetic pigments, photosynthesis, invivo nitrate reductase activity and protein content in a chlorococcalean green alga Glaucocystis nostochinearum Itzigsohn." *Chemosphere* 25, no. 11 (1992): 1721–1732.

Ranieri, Ezio, Angelo Tursi, Silvia Giuliano, Vincenzo Spagnolo, Ada Cristina Ranieri, and Andrea Petrella. "Phytoextraction from chromium-contaminated soil using Moso Bamboo in mediterranean conditions." *Water, Air, & Soil Pollution* 231 (2020): 1–12.

Ranieri, Ezio, Gianfranco D'Onghia, Francesca Ranieri, Andrea Petrella, Vincenzo Spagnolo, and Ada Cristina Ranieri. "Phytoextraction of Cr (VI)-contaminated soil by *Phyllostachys pubescens*: A case study." *Toxics* 9, no. 11 (2021): 312.

Rath, Ayushee, and Anath Bandhu Das. "Chromium stress induced oxidative burst in *Vigna mungo* (L.) Hepper: Physio-molecular and antioxidative enzymes regulation in cellular homeostasis." *Physiology and Molecular Biology of Plants* 27, no. 2 (2021): 265–279.

Rizwan, Muhammad Shahid, Muhammad Imtiaz, Jun Zhu, Balal Yousaf, Mubshar Hussain, Liaqat Ali, Allah Ditta et al. "Immobilization of Pb and Cu by organic and inorganic amendments in contaminated soil." *Geoderma* 385 (2021): 114803.

Sahoo, Ranjan Kumar, Varsha Rani, and Narendra Tuteja. "*Azotobacter vinelandii* helps to combat chromium stress in rice by maintaining antioxidant machinery." *3 Biotech* 11, no. 6 (2021): 275.

Said, W. A., and D. L. Lewis. "Quantitative assessment of the effects of metals on microbial degradation of organic chemicals." *Applied and Environmental Microbiology* 57, no. 5 (1991): 1498–1503.

Salt, David E., Michael Blaylock, Nanda PBA Kumar, Viatcheslav Dushenkov, Burt D. Ensley, Ilan Chet, and Ilya Raskin. "Phytoremediation: A novel strategy for the removal of toxic metals from the environment using plants." *Nature Biotechnology* 13, no. 5 (1995): 468–474.

Samreen, Sayma, Athar Ali Khan, Manzoor R. Khan, Shamim Akhtar Ansari, and Adnan Khan. "Assessment of phytoremediation potential of seven weed plants growing in chromium-and nickel-contaminated soil." *Water, Air, & Soil Pollution* 232 (2021): 1–18.

Samuel, Jastin, Madona L. Paul, Mrudula Pulimi, M. Joyce Nirmala, Natarajan Chandrasekaran, and Amitava Mukherjee. "Hexavalent chromium bioremoval through adaptation and consortia development from Sukinda chromite mine isolates." *Industrial & Engineering Chemistry Research* 51, no. 9 (2012): 3740–3749.

Sánchez-Galván, Gloria, Eugenia J. Olguín, Francisco J. Melo, David Jiménez-Moreno, and Víctor J. Hernández. "*Pontederia sagittata* and *Cyperus papyrus* contribution to carbon storage in floating treatment wetlands established in subtropical urban ponds." *Science of The Total Environment* 832 (2022): 154990.

Sandrin, Todd R., and Raina M. Maier. "Impact of metals on the biodegradation of organic pollutants." *Environmental Health Perspectives* 111, no. 8 (2003): 1093–1101.

Santos, Conceição, and Eleazar Rodriguez. "Review on some emerging endpoints of chromium (VI) and lead phytotoxicity." *Botany* (2012): 61–82.

Sarwar, Nadeem, Muhammad Imran, Muhammad Rashid Shaheen, Wajid Ishaque, Muhammad Asif Kamran, Amar Matloob, Abdur Rehim, and Saddam Hussain. "Phytoremediation strategies for soils contaminated with heavy metals: Modifications and future perspectives." *Chemosphere* 171 (2017): 710–721.

Schnoor, Jerald L. "Phytoremediation: Technology evaluation report." Ground-water Remediation Technologies Analysis Center (1997): 8–18.

Schnoor, Jerald L., Louis A. Light, Steven C. McCutcheon, N. Lee Wolfe, and Laura H. Carreia. "Phytoremediation of organic and nutrient contaminants." *Environmental Science & Technology* 29, no. 7 (1995): 318A–323A.

Shadreck, Mandina, and Tawanda Mugadza. "Chromium, an essential nutrient and pollutant: A review." *African Journal of Pure and Applied Chemistry* 7, no. 9 (2013): 310–317.

Shahid, Muhammad, Camille Dumat, Muhammad Aslam, and Eric Pinelli. "Assessment of lead speciation by organic ligands using speciation models." *Chemical Speciation & Bioavailability* 24, no. 4 (2012): 248–252.

Shahid, Muhammad, Camille Dumat, Sana Khalid, Eva Schreck, Tiantian Xiong, and Nabeel Khan Niazi. "Foliar heavy metal uptake, toxicity and detoxification in plants: A comparison of foliar and root metal uptake." *Journal of Hazardous Materials* 325 (2017): 36–58.

Shanker, Arun K., Carlos Cervantes, Herminia Loza-Tavera, and S. Avudainayagam. "Chromium toxicity in plants." *Environment International* 31, no. 5 (2005): 739–753.

Sharma, Jitendra Kumar, Nitish Kumar, Nater Pal Singh, and Anita Rani Santal. "Phytoremediation technologies and their mechanism for removal of heavy metal from contaminated soil: An approach for a sustainable environment." *Frontiers in Plant Science* 14 (2023): 1076876.

Sharma, Manik, Vinod Kumar, Renu Bhardwaj, and Ashwani Kumar Thukral. "Tartaric acid mediated Cr hyperaccumulation and biochemical alterations in seedlings of *Hordeum vulgare* L." *Journal of Plant Growth Regulation* 39 (2020): 1–14.

Sharma, Manik, Vinod Kumar, Sonia Mahey, Renu Bhardwaj, and Ashwani Kumar Thukral. "Antagonistic effects of EDTA against biochemical toxicity induced by Cr (VI) in *Hordeum vulgare* L. seedlings." *Physiology and Molecular Biology of Plants* 26 (2020): 2487–2502.

Shiyab, Safwan. "Morphophysiological effects of chromium in sour orange (*Citrus aurantium* L.)." *HortScience* 54, no. 5 (2019): 829–834.

Shreya, Desai, Hardik Naik Jinal, Vinodbhai Patel Kartik, and Natarajan Amaresan. "Amelioration effect of chromium-tolerant bacteria on growth, physiological properties and chromium mobilization in chickpea (*Cicer arietinum*) under chromium stress." *Archives of Microbiology* 202 (2020): 887–894.

Shukla, O. P., S. Dubey, and U. N. Rai. "Preferential accumulation of cadmium and chromium: Toxicity in *Bacopa monnieri* L. under mixed metal treatments." *Bulletin of Environmental Contamination and Toxicology* 78 (2007): 252–257.

Singh, N. P., and Anita Rani Santal. "Phytoremediation of heavy metals: The use of green approaches to clean the environment." *Phytoremediation: Management of Environmental Contaminants* 2 (2015): 115–129.

Srivastava, Dipali, Madhu Tiwari, Prasanna Dutta, Puja Singh, Khushboo Chawda, Monica Kumari, and Debasis Chakrabarty. "Chromium stress in plants: Toxicity, tolerance and phytoremediation." *Sustainability* 13, no. 9 (2021): 4629.

Srivastava, Shaili, and Indu Shekhar Thakur. "Biosorption potency of *Aspergillus niger* for removal of chromium (VI)." *Current Microbiology* 53 (2006): 232–237.

Suman, J., O. Uhlik, J. Viktorova, and T. Macek. "Phytoextraction of heavy metals: A promising tool for clean-up of polluted environment?." *Frontiers in Plant Science* 9 (2018): 1476.

Suresh, Gopal, Balamuralikrishnan Balasubramanian, Nagaiya Ravichandran, Balasubramanian Ramesh, Hesam Kamyab, Palanivel Velmurugan, Ganesan Vijaiyan Siva, and Arumugam Veera Ravi. "Bioremediation of hexavalent chromium-contaminated wastewater by *Bacillus thuringiensis* and *Staphylococcus capitis* isolated from tannery sediment." *Biomass Conversion and Biorefinery* 11 (2021): 383–391.

Taghipour, Marzieh, and Mohsen Jalali. "Influence of organic acids on kinetic release of chromium in soil contaminated with leather factory waste in the presence of some adsorbents." *Chemosphere* 155 (2016): 395–404.

Tariq, Muhammad, Muhammad Waseem, Muhammad Hidayat Rasool, Muhammad Asif Zahoor, and Irshad Hussain. "Isolation and molecular characterization of the indigenous *Staphylococcus aureus* strain K1 with the ability to reduce hexavalent chromium for its application in bioremediation of metal-contaminated sites." *PeerJ* 7 (2019): e7726.

Tauqeer, Hafiz Muhammad, Sabir Hussain, Farhat Abbas, and Muhammad Iqbal. "The potential of an energy crop "*Conocarpus erectus*" for lead phytoextraction and phytostabilization of chromium, nickel, and cadmium: An excellent option for the management of multi-metal contaminated soils." *Ecotoxicology and Environmental Safety* 173 (2019): 273–284.

Ullah, Sana, Sajid Mahmood, Rehmat Ali, Muhammad Rizwan Khan, Kalsoom Akhtar, and Nizamuddin Depar. "Comparing chromium phyto-assessment in *Brachiaria mutica* and Leptochloa fusca growing on chromium polluted soil." *Chemosphere* 269 (2021): 128728.

Upadhyay, Neha, Kanchan Vishwakarma, Jaspreet Singh, Mitali Mishra, Vivek Kumar, Radha Rani, Rohit K. Mishra, Devendra K. Chauhan, Durgesh K. Tripathi, and Shivesh Sharma. "Tolerance and reduction of chromium (VI) by *Bacillus* sp. MNU16 isolated from contaminated coal mining soil." *Frontiers in Plant Science* 8 (2017): 778. USGS, 2023. United States Geological Survey.

USEPA. Guidelines for Ecological Risk Assessment (1998). Report No. EPA/630/R-95/002F, USEPA, Washington DC..

Usman, Kamal, Mohammad A. Al-Ghouti, and Mohammed H. Abu-Dieyeh. "The assessment of cadmium, chromium, copper, and nickel tolerance and bioaccumulation by shrub plant *Tetraena qataranse*." *Scientific Reports* 9, no. 1 (2019): 5658.

Van Den Bos, Amelie. "Phytoremediation of volatile organic compounds in groundwater: Case studies in plume control." *Draft Report Prepared for the US EPA Technology Innovation Office under a National Network for Environmental Management Studies Fellowship* (2002). Viewed 2, June, 2021. https://clu-in.org/download/studentpapers/vandenbos.pdf

Vishnupradeep, R., L. Benedict Bruno, Zarin Taj, Chinnannan Karthik, Dinakar Challabathula, Adarsh Kumar, Helena Freitas, and Mani Rajkumar. "Plant growth promoting bacteria improve growth and phytostabilization potential of *Zea mays* under chromium and drought stress by altering photosynthetic and antioxidant responses." *Environmental Technology & Innovation* 25 (2022): 102154.

Wakeel, Abdul, Ming Xu, and Yinbo Gan. "Chromium-induced reactive oxygen species accumulation by altering the enzymatic antioxidant system and associated cytotoxic, genotoxic, ultrastructural, and photosynthetic changes in plants." *International Journal of Molecular Sciences* 21, no. 3 (2020): 728.

Wang, Pi-Chao, Tsukasa Mori, Kohya Komori, Masanori Sasatsu, Kiyoshi Toda, and Hisao Ohtake. "Isolation and characterization of an *Enterobacter cloacae* strain that reduces hexavalent chromium under anaerobic conditions." *Applied and Environmental Microbiology* 55, no. 7 (1989): 1665–1669.

Wang, Shan-Shan, Shu-Lin Ye, Yong-He Han, Xiao-Xia Shi, Deng-Long Chen, and Min Li. "Biosorption and bioaccumulation of chromate from aqueous solution by a newly isolated *Bacillus mycoides* strain 200AsB1." *RSC Advances* 6, no. 103 (2016): 101153–101161.

Wani, Parvaze Ahmad, Javid Ahmad Wani, and Shazia Wahid. "Recent advances in the mechanism of detoxification of genotoxic and cytotoxic Cr (VI) by microbes." *Journal of Environmental Chemical Engineering* 6, no. 4 (2018): 3798–3807.

Wani, Parvaze Ahmad, Said Hussaini Garba, Shazia Wahid, Nuhu Abubakar Hussaini, and Kareem Abiola Mashood. "Prevention of oxidative damage and phytoremediation of Cr (VI) by chromium (VI) reducing *Bacillus subtilus* PAW3 in cowpea plants." *Bulletin of Environmental Contamination and Toxicology* 103 (2019): 476–483.

Wei, Yuzhen, Muhammad Usman, Muhammad Farooq, Muhammad Adeel, Fasih Ullah Haider, Zhandong Pan, Weiwei Chen, Hongyan Liu, and Liqun Cai. "Removing hexavalent chromium by nano zero-valent iron loaded on attapulgite." *Water, Air, & Soil Pollution* 233, no. 2 (2022): 48.

World Health Organization (WHO). "Permissible limits of heavy metals in soil and plants." *Geneva, Switzerland* (1996).

Wu, Chan-Cui, Jie Liu, Xue-Hong Zhang, and Shi-Guang Wei. "Phosphorus enhances Cr (VI) uptake and accumulation in *Leersia hexandra* Swartz." *Bulletin of Environmental Contamination and Toxicology* 101 (2018): 738–743.

XI, M-z, BAI Z-k, and ZHAO Z-q. "Advances on the study of chelate-enhanced phytoremediation for heavy metal contaminated soils." *Soil and Fertilizer Sciences in China* 5 (2008): 003.

Xiao, Wendan, Xuezhu Ye, Xiaoe Yang, Zhiqiang Zhu, Caixia Sun, Qi Zhang, and Ping Xu. "Isolation and characterization of chromium (VI)-reducing *Bacillus* sp. FY1 and *Arthrobacter* sp. WZ2 and their bioremediation potential." *Bioremediation Journal* 21, no. 2 (2017): 100–108.

Xiao, Zhengli, Haidong Zhang, Yan Xu, Min Yuan, Xiaolian Jing, Jiale Huang, Qingbiao Li, and Daohua Sun. "Ultra-efficient removal of chromium from aqueous medium by biogenic iron based nanoparticles." *Separation and Purification Technology* 174 (2017): 466–473.

Xu, Xin, Sheng Nie, Hanying Ding, and Fan Fan Hou. "Environmental pollution and kidney diseases." *Nature Reviews Nephrology* 14, no. 5 (2018): 313–324.

Xu, Y., and Z. Fang. "The research progress of remediating the heavy metal-contaminated soil with biochar." *Chin. J. Environ Eng* 35 (2015): 156–159.

Yaashikaa, P. R., P. Senthil Kumar, VP Mohan Babu, R. Kanaka Durga, V. Manivasagan, K. Saranya, and A. Saravanan. "Modelling on the removal of Cr (VI) ions from aquatic system using mixed biosorbent (*Pseudomonas stutzeri* and acid treated Banyan tree bark)." *Journal of Molecular Liquids* 276 (2019): 362–370.

Yan, An, Yamin Wang, Swee Ngin Tan, Mohamed Lokman Mohd Yusof, Subhadip Ghosh, and Zhong Chen. "Phytoremediation: A promising approach for revegetation of heavy metal-polluted land." *Frontiers in Plant Science* 11 (2020): 359.

Yan, X., J. Wang, H. Song, Y. Peng, S. Zuo, T. Gao, X. Duan, D. Qin, and J. Dong. "Evaluation of the phytoremediation potential of dominant plant species growing in a chromium salt–producing factory wasteland, China." *Environmental Science and Pollution Research* 27 (2020): 7657–7671.

Yanitch, Aymeric, Hafssa Kadri, Cédric Frenette-Dussault, Simon Joly, Frederic E. Pitre, and Michel Labrecque. "A four-year phytoremediation trial to decontaminate soil polluted by wood preservatives: Phytoextraction of arsenic, chromium, copper, dioxins and furans." *International Journal of Phytoremediation* 22, no. 14 (2020): 1505–1514.

Ye, Li-Li, Yong-Shan Chen, Yu-Dao Chen, Lian-Wen Qian, Wen-Li Xiong, Jing-Hua Xu, and Jin-Ping Jiang. "Phytomanagement of a chromium-contaminated soil by a high-value plant: Phytostabilization of heavy metal contaminated sites." *BioResources* 15, no. 2 (2020): 3545–3565.

Zanganeh, Fahimeh, Ava Heidari, Adel Sepehr, and Abbas Rohani. "Bioaugmentation and bioaugmentation–assisted phytoremediation of heavy metal contaminated soil by a synergistic effect of cyanobacteria inoculation, biochar, and purslane (*Portulaca oleracea* L.)." *Environmental Science and Pollution Research* 29, no. 4 (2022): 6040–6059.

Zayed, Adel M., and Norman Terry. "Chromium in the environment: Factors affecting biological remediation." *Plant and Soil* 249 (2003): 139–156.

Zhou, Ben-Jun, and Tian-Hu Chen. "Biodegradation of phenol with chromium (VI) reduction by the *Pseudomonas* sp. strain JF122." *Desalination and Water Treatment* 57, no. 8 (2016): 3544–3551.

Ziagova, M. G., A. I. Koukkou, and M. Liakopoulou-Kyriakides. "Optimization of cultural conditions of *Arthrobacter* sp. Sphe3 for growth-associated chromate (VI) reduction in free and immobilized cell systems." *Chemosphere* 95 (2014): 535–540.

Zulfiqar, U., W. Jiang, X. Wang, S. Hussain, M. Ahmad, and M. F. Maqsood. "Insights into the plant-microbe interaction, soil amendments and advanced genetic approaches for cadmium remediation: A review." *Frontiers in Plant Science* 13 (2022)

Zulfiqar, Usman, Fasih Ullah Haider, Muhammad Ahmad, Saddam Hussain, Muhammad Faisal Maqsood, Muhammad Ishfaq, Babar Shahzad et al. "Chromium toxicity, speciation, and remediation strategies in soil-plant interface: A critical review." *Frontiers in Plant Science* 13 (2023): 1081624.

Zulfiqar, Usman, Muhammad Farooq, Saddam Hussain, Muhammad Maqsood, Mubshar Hussain, Muhammad Ishfaq, Muhammad Ahmad, and Muhammad Zohaib Anjum. "Lead toxicity in plants: Impacts and remediation." *Journal of Environmental Management* 250 (2019): 109557.

3 Recent Advances in Phytoremediation of Heavy Metal-Contaminated Soil
Application and Its Limitations

S. Praveen Kumar, S. Sathiyamurthi, S. Nalini,
M. Sivasakthi, M. Santhosh Kumar,
R. Ragavaraj, and J. Prabakaran

INTRODUCTION

Over 78% of the elements that are now known in the periodic table have properties that are typical of metals. There are a total of 118 elements. Micronutrients like, Co, Mn, Fe, Mo, Cu, and Zn are essential for plant growth. Heavy metals such as As, Se, V, Cd, Pb, Cr, Hg, Ni, U, and W are nonessential for plant growth, although some are beneficial to plants and microbes. Heavy metals can be defined as elements that either have a higher density or high atomic mass (more than 20). These metals are naturally occurring elements that can also be discharged into the environment via industrial processes, mining, agriculture, and improper waste disposal. HMs cause significant environmental concern owing to their toxicity and their potential adverse effects on anthropoid health and ecosystems. When released into soil, water, or air, they can persist for extended periods and accumulate, posing hazards to living organisms. Many studies have been done to figure out how to get these heavy metals out of the soils. Many different approaches have been proposed by various researchers to resolve the problem of soil contamination (SC) by HM. Among the several methodologies considered, phytoremediation (PR) has emerged as a highly efficacious and economically feasible option for the extraction of HM from soil. Phytoremediation is an environmentally sustainable approach that utilizes the innate capacities of plants to address, alleviate, or stabilize pollutants found in water, soil, or air.

Currently, a comprehensive collection of about 400 species of hyperaccumulator (HA) plants has been identified, capable of accumulating elements such as arsenic, cadmium, manganese, nickel, and zinc, among others. As an illustration,

DOI: 10.1201/9781003442295-3

57

certain hyperaccumulator plants have been found to exhibit notable accumulation concentrations (ppm of Dry Weight) for various metals. For instance, *Thlaspi caerulescens* demonstrates accumulation concentrations of 51,600 mg/kg for zinc and 18,000 mg/kg for cadmium. *Ipomea alpine* exhibits a concentration of 12,300 mg/kg for Cu, while *Psychotria douarrei* demonstrates a concentration of 47,500 mg/kg for Ni. *Thlaspi rotundifolium* exhibits a concentration of 8200 mg/kg for Pb, *Astragalus racemosus* demonstrates a concentration of 14,900 mg/kg for Se, and *Pteris vitatta* exhibits a concentration of 20,000 mg/kg for As (Ingole and Dhawale, 2021). Plants such as *Brassica juncea*, willow, hybrid poplars, duckweed, *Zea mays*, *Medicago sativa*, and *Lolium* spp. are among the most regularly used plants for phytoremediation of a wide variety of metals, metalloids, nonmetals, nutrients, and organic pollutants. Other commonly used plants are *Brassica oleracea* and *Lolium* spp. Chelating agents and surfactants, such as citric acid, EDTA, CDTA, DTPA, EGTA, EDDHA, and NTA, are incorporated into soil to enhance the mobility of metals in the soil, allowing plants to absorb these elements more efficiently. The use of genetically modified plants entails the manipulation of high biomass plant species in order to augment their capacity for extracting greater quantities of metals from soil. One possible approach to accomplish this objective is enhancing the biomass output of hyperaccumulator plants or employing molecular cloning and gene expression techniques to activate genes responsible for heavy metal accumulation and xenobiotic degradation enzymes. These genetic modifications seek to improve the efficacy of PR, which uses flora to eliminate the environmental pollutants.

Phytoremediation is particularly well-suited for implementation in countries like India due to its cost-effective nature and dependence on the inherent capacity of plants to take up metal ions from soil. Plants with a higher propensity for accumulating HMs are employed in the remediation of areas polluted with these elements (Dheri et al., 2007). Depending on the types, forms, and media of the contaminants, each plant uses a different way or a mix of methods to clean up the soil and water. Rhizofiltration phytodegradation, phytodegradation, rhizodegradation, and phytovolatilization are all ways to clean up contaminated groundwater. Rhizofiltration, phytodegradation, or rhizodegradation can be used to clean up surface and wastewater pollution. Soil, sediments, or sludges that are polluted can be cleaned up by phytoextraction, phytodegradation, phytostabilization, rhizodegradation, or phytovolatilization. In this book, chapters discuss the recent advances in PR of heavy metal-contaminated soil and their limitations and applications (Figure 3.1).

MECHANISMS OF PHYTOREMEDIATION

Phytoremediation involves a number of distinct mechanisms.

1. **Phytostabilization:** The purpose of phytostabilization is to immobilize or reduce the mobility of environmental contaminants. Certain plant species exhibit the capacity to absorb and isolate contaminants from the soil, therefore hindering their dispersion into adjacent ecosystems, such as groundwater. This procedure entails the accumulation of HMs within the root system

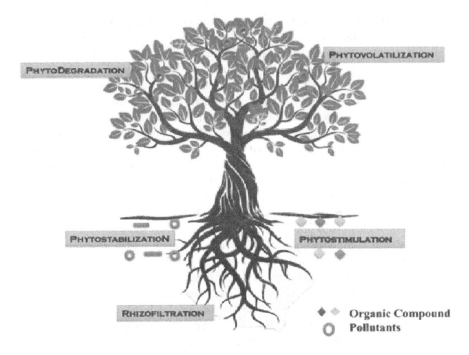

FIGURE 3.1   Mechanisms of phytoremediation (Tang 2016).

of the plant. This method is frequently applied to heavy metals and metalloids. Phytostabilization plants typically have extensive root systems that
can reach contaminated soil layers.

2. **Phytoextraction:** Through this process, plants absorb contaminants from
   the water or sediment through their roots, which are then transferred to
   their above-ground tissues, such as their leaves and stalks. Heavy metals,
   chemical molecules, or radioactive substances could all play a role in contamination. Once the toxins have accumulated in the plant's aerial portions,
   they can be harvested and removed from their location. Phytoextraction
   can be effectively carried out using certain plant species that possess the
   intrinsic capacity to hyperaccumulate specific metals.

3. **Phytodegradation:** Utilizing plants and the microorganisms that are naturally associated with them to degrade or assimilate pollutants is an illustration of phytodegradation. Several plant species can generate enzymes
   or stimulate microbial activity in the rhizosphere and break down organic
   pollutants into less hazardous forms. This technique is particularly useful
   for organic compounds such as petroleum hydrocarbons and insecticides
   due to its high efficacy.

4. **Rhizofiltration:** In rhizofiltration, the roots of plants are used as a medium
   to filter and remove contaminants from water. This method is also referred
   to as "vegetative filtration." When serving this function, it is common practice to make use of plants like water hyacinth and reed beds, both of which

have extensive root systems. When contaminated water passes through the root zone, plants assimilate and store the toxins in their roots. This process occurs when the water moves through the root zone. This method is effective at removing a wide variety of pollutants, such as organic molecules, HMs, and nutrients. Rhizofiltration is able to effectively remove numerous elements (including silver chromium, cobalt, copper, cadmium, mercury, nickel, manganese, lead, zinc, and molybdenum), metalloids (including arsenic and selenium), and radioactive substances (including cesium-137, strontium-90, plutonium-239, and uranium-239). Significant botanical species employed in this procedure encompass sunflower, Indian mustard, tobacco, rye, and spinach. Terrestrial plants are commonly employed in this approach due to their rapid growth and possession of fibrous roots.

5. **Phytovolatilization:** Phytovolatilization refers to process by which plants absorb toxins from polluted areas by their roots, metabolize the most hazardous components into less damaging forms inside their tissues, and subsequently release these transformed compounds into the atmosphere through their leaves. This process takes place over the course of several stages. The cleanup of mercury, selenium, and organic solvent contaminants are the most prominent uses for this approach, which is predicated on the evapotranspiration process. According to Sinha et al. (2007), increasing the pH (liming) or adding organic matter (sewage sludge, compost), anions (phosphate and sulfate), and hydroxides and metallic oxides (iron oxides) might improve the transformation of metals into innocuous forms. Increasing the pH can be accomplished via liming. Alterations in the redox conditions of the soil, as well as the release of anions and/ or lignin by the plants themselves, may play a part in the process. Sinha and coworkers (2007) reported that plants exhibit characteristics of both "accumulators" and "excluders." Contaminants are broken down or changed into harmless forms in their cells. The excluders stop contaminants from getting into the plants' material. Barley (*Hordeum vulgare*) is a good example.

IDENTIFICATION OF HEAVY METAL-CONTAMINATED SOIL

Several advanced methods, such as sequential fractionation, XAFS spectroscopy, and XRD powder diffraction, can find heavy metal pollution in soil, but using plants is one of the easiest and least expensive ways to do this. Some plant species are especially sensitive to the effects of HM, and one of their distinguishing characteristics is that they can serve as indicators of the presence of heavy metals (Table 3.1).

Sources of HM in Soil

Soil contamination with HM has two sources.

1. Natural
2. Anthropogenic

TABLE 3.1
Some Plants Species that Are Sensitive to Heavy Metals (Tangahu et al., 2011)

S. No	Plant Name	Scientific Name	Metals/ Element Name
1	Barley	*Hordeum vulgare*	Cadmium, lead, and copper
2	Lettuce	*Lactuca sativa*	Cadmium, lead, and copper
3	Radish	*Raphanus sativus*	Cadmium and lead
4	Tomatoes	*Solanum lycopersicum*	Cadmium, lead, and mercury
5	Ryegrass	*Lolium perenne*	Cadmium, lead, and copper
6	Cress	*Arabidopsis thaliana*	Cadmium, lead, and zinc

NATURAL SOURCES

1. **Weathering of rocks and minerals:** As rocks and minerals break down over time due to biological, chemical, and physical weathering, small amounts of HM are released into the soil and water.
2. **Sulfide minerals:** Weathering of sulfide minerals like pyrite (iron sulfide) can release metals such as lead, iron, and zinc.
3. **Clay minerals:** Some clay minerals can naturally contain heavy metals like chromium and nickel, which can be released into the environment as these minerals weather.
4. **Volcanic activity:** Volcanic eruptions release a significant amount of heavy metal from the earth's mantle and crust. Volcanic ash, lava, and gases can carry metals like mercury, lead, and cadmium into the atmosphere and onto the soil.
5. **Erosion and sedimentation:** Natural erosion of rocks and soils by wind, water, and ice can transport heavy metals over long distances and deposit them in new locations. These metals are eventually incorporated into the soil through sedimentation.
6. **Hydrothermal activity:** Hydrothermal fluids from deep within the earth's crust can carry heavy metals to the surface. When these fluids come into contact with cooler environments, such as oceans or lakes, they can deposit minerals containing heavy metals on the seafloor or lake beds.
7. **Mineral deposits:** Certain mineral deposits, such as those containing ores of HMs like zinc, lead, nickel, and copper, can naturally release these metals into the environment as they undergo weathering and erosion.
8. **Natural earth processes:** Heavy metals can get into the atmosphere through geological processes such as tectonic activity, metamorphism, and magmatism.

ANTHROPOGENIC SOURCES

1. **Industrial activities:** HMs enter the environment via industrial processes like mining, refining, and metal production.
2. **Agricultural practices:** Fertilizers, pesticides, and herbicides containing metals can introduce heavy metals into soil. Animal manure from livestock exposed to heavy metals can also contribute to pollution.

3. **Waste disposal and landfills:** Heavy metals can leach into the soil if electronic waste (e-waste), batteries, and other metal-containing refuse are disposed of improperly.
4. **Urbanization and construction:** Urban development and construction activities can disturb soil, releasing heavy metals from building materials, road surfaces, and other sources.
5. **Atmospheric deposition:** Heavy metals can be deposited onto soil from the atmosphere through air pollution and industrial emissions.
6. **Mining and smelting:** As a consequence of these actions, a substantial amount of HM could enter the soil and surrounding environment.
7. **Accidental spills and leaks:** Accidental releases of heavy metals from industries or transportation (e.g., oil spills) can contaminate soil.
8. **Wastewater and sewage sludge:** HM can enter soil through the disposal of untreated or improperly treated effluent and the addition of sewage sediment to farmland (Table 3.2).

HEAVY METAL CONTAMINATION AND ITS PHYTOREMEDIATION TECHNIQUES

ARSENIC

Arsenic (As) is a naturally occurring element, atomic number 33, that is extensively distributed in earth's crust. There are two forms of arsenic in soil—organic and inorganic. To remove the arsenic from soil, phytoextraction and phytoimmobilization are cost-effective green technologies that have minimal impact on soil fertility and biodiversity. Multiple studies have identified *Pteris vittata* L. as a highly promising option for phytoremediation of As contamination (Ronzan et al., 2017). This fern species exhibits a notable capacity to endure soil arsenic concentrations of up to 1500 parts per million (ppm), while concurrently demonstrating a rapid uptake of this metalloid inside its fronds. Furthermore, this phenomenon is distinguished by its robust expansion accompanied by a substantial accumulation of biomass, an extensive proliferation of root systems, and a notable pace of growth. The aforementioned characteristics render *P. vittata* a highly suitable plant for As remediation. Nevertheless, as evidenced by proteome investigations, the tolerance of the organism to arsenic is not entirely comprehensive. Bona et al. (2010) reported that the occurrence of As has been found to impede the activity of enzymes associated with sucrose metabolism, bioenergetics, photosynthesis, and carbon fixation in the fronds. Arbuscular mycorrhizal fungi (AMF) play a significant role in phytoremediation, which is the process of using plants to clean up soil, air, and water contaminated with hazardous contaminants. AMF form symbiotic relationships with the roots of most terrestrial plants, including many hyperaccumulators used in phytoremediation efforts. AMF are recognized for their significant involvement in the mobilization and immobilization mechanisms of metal ions, potentially influencing the availability of these ions to plants. The potential advantages of AMF can be ascribed to two mechanisms: alterations in the solubility of metals resulting from pH fluctuations in the soil solution, and the confinement of HMs inside and

TABLE 3.2

List of Some Prevalent HM and Plant Species Known for Their Phytoremediation Abilities (Bhat et al., 2022)

S. No	Heavy Metals	Plant Species
1	Lead (Pb)	• Oriental mustard (*Brassica juncea*) • Hemp (*Cannabis sativa*) • Vetiver grass (*Chrysopogon zizanioides*) • Sunflower (*Helianthus annuus*)
2	Cadmium (Cd)	• Oriental mustard (*Brassica juncea*) • Alpine pennycress (*Thlaspi caerulescens*) • Water hyacinth (*Eichhornia crassipes*) • Sunflower (*Helianthus annuus*)
3	Arsenic (As)	• Oriental mustard (*Brassica juncea*) • Ferns (*Pteris vittata*) • Cattails (*Typha* spp.) • Sunflower (*Helianthus annuus*)
4	Chromium (Cr)	• Oriental mustard (*Brassica juncea*) • Sunflower (*Helianthus annuus*) • Alpine pennycress (*Thlaspi caerulescens*) • Hemp (*Cannabis sativa*)
5	Copper (Cu)	• Oriental mustard (*Brassica juncea*) • Sunflower (*Helianthus annuus*) • Willow tree (*Salix* spp.) • Poplar tree (*Populus* spp.)
6	Nickel (Ni)	• Indian mustard (*Brassica juncea*) • Alyssum (*Alyssum spp.*) • Pennycress (*Thlaspi spp.*) • Sunflower (*Helianthus annuus*)
7	Mercury (Hg)	• Oriental mustard (*Brassica juncea*) • Water hyacinth (*Eichhornia crassipes*) • Cattails (*Typha* spp.) • Sunflower (*Helianthus annuus*)
8	Selenium (Se)	• Oriental mustard (*Brassica juncea*) • Stanleya (*Stanleya* spp.) • Astragalus (*Astragalus* spp.) • Sunflower (*Helianthus annuus*)
9	Aluminum (Al)	• Tea plant (*Camellia sinensis*) • Rice (*Oryza sativa*) • Barley (*Hordeum vulgare*) • Blueberry (*Vaccinium* spp.)

on fungi biomass. According to the findings of Cantamessa et al. (2020), the use of PR techniques at the metallurgical site would necessitate a minimum duration of four years in order to achieve a reduction in the concentration of As to 50 mg/kg. The metallurgical site discussed here is an industrial site located in northwestern Italy. This site is highly contaminated with arsenic and several heavy metals due to past metallurgical activities, including the processing of metals such as copper, gold, lead, and zinc.

Phytoremediation research has also shown the potential utilization of emergent aquatic flora, specifically *Vetiveria zizanioides* and *Cyperus vaginatus*. The use of a moderate arsenic accumulator possessing a substantial biomass has the potential to yield a higher annual arsenic removal rate compared to employing a HA with a limited biomass. As an example, the yearly removal of As by Sesuvium at a concentration of 500 M As was found to be 1955 g As/ha/y, which exceeded the calculated annual As removal by Pteris (ranging from 525 to 1470 g As/ha/y) (Figure 3.2).

Uddin and colleagues (2016) conducted an experiment to determine the efficacy of three crop species in the remediation of arsenic-polluted soil: *Corchorus capsularis* (jute), *Hibiscus cannabinus* (kenaf), and *Hibiscus sabdariffa* (mesta). The soils that were found to be tainted originated from a specific location within the district of Mymensingh in Bangladesh. A quantity of 10 kg of contaminated soil was divided and distributed evenly among pots of equal size. Subsequently, a total of 10 uncontaminated seeds from each of the three plant species were sown within the designated pots. After a duration of one week, 40% of the seedlings were removed from their respective pots. Nitrogen, phosphorus, and potassium, which are essential plant

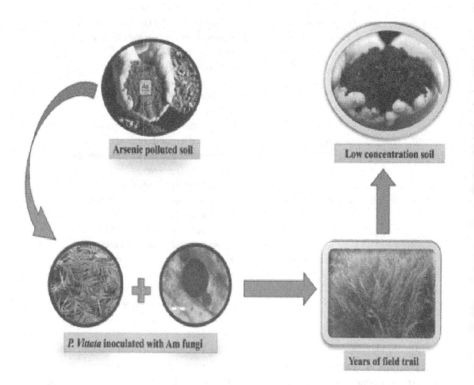

FIGURE 3.2   Use of *P. vittata* and Am fungi to reduce the As content in soil. (*Source*: Cantamessa et al., 2020.)

nutrients, were added to the containers. The contaminated soil was extracted using hydrogen peroxide (H_2O_2) and nitric acid. Atomic absorption spectroscopy (AAS) is a common analytical methodology in the field of chemistry. The study utilized a spectrophotometer paired with the hydride generation method, known as HGAAS, to determine the presence of arsenic in the samples. The roots were better at taking in and holding onto As than the shoots. Overall, phytoremediation of arsenic in the polluted soil was found to work best with kenaf. In fact, it was better because it made more energy.

CHROMIUM

Chromium (Cr) is the metal extensively utilized in several industries such as electroplating, metallurgy, paint and pigment manufacture, tanning, timber preservation, chromium compound manufacturing, and pulp and paper production. The tanning industry significantly contributes to the pollution of water resources, particularly in terms of Cr contamination (Ramana et al., 2012). The trivalent and hexavalent forms of the heavy metal chromium can be found in large quantity in soils, rocks, and volcanic debris as well as in groundwaters (Coelho et al., 2017). The Cruciferae family, which includes species such as *B. juncea* and *R. sativus*, has been extensively documented for its notable capacity to accumulate metals or metalloids (Kanwar et al., 2012). *Brassica juncea* and *B. nigra* are better at taking HMs from the soil and putting them in their stems where they are more concentrated. For phytoextraction to work, the plants must be able to handle heavy metals and store a lot of them in their parts that can be harvested. They must have a fast growth rate and be able to make a lot of energy. In addition to hyperaccumulator plant species like *Zea mays* (corn), *Salix viminalis* (willow), *Brassica juncea* (Indian mustard), *Lolium perenne* (perennial ryegrass), and *Helianthus annuus* (sunflower), which have a high capacity for heavy metal uptake and remediation, other species may also be considered for phytoremediation efforts. These plant species possess the capacity to endure the adverse effects of industrial effluents. For plants to keep getting chromium from their supports, the chromium must be able to move from their roots to their shoots. It can lower the amount of chromium in the soil, making the rootls poisonous (Ghosh and Singh, 2005). Depending on the plant species, heavy metal tolerance can be induced by preventing their entry, allowing them in, or retaining them. Sundaramoorthy et al. (2010) discovered that it was advantageous for businesses to cultivate naturally occurring weeds and grasses to enhance soil on a larger scale and in less time.

LEAD (PB)

Lead, following arsenic, is known to be the most perilous HM due to its lethal effects on various forms of life (Zhou et al., 2014). Lead is a heavy chemical that is toxic and not needed for life. However, even small amounts of lead can be harmful if they are eaten. Due to its widespread dispersal, lead (Pb) is the most prevalent environmental pollutant (Mangkoedihardjo and Surahmaida, 2008). According to Khan et al. (2010), *Eucalyptus rostrata* is known as the red river gum. This is an evergreen tree with creamy bark with brown or reddish tones. This bark is shed annually, and it peels off to reveal fresh bark underneath. The foliage of the plant exhibits a transition in coloration, with the leaves first appearing as grey-green during their early stages of development, and subsequently maturing into a shade of blue-green. The foliage can deter pests. Umbels of white to cream flowers in summer. Maximum height is 40 m with a 6-m span. Prefers solar exposure. Salt and sunlight tolerant. Bafeel (2008) conducted an experimental study to examine the impact of AMF on the absorption of lead (Pb) by *Eucalyptus rostrata* L. plants cultivated in lead-contaminated soil, both with and without the presence of *Phaseolus vulgaris*. The authors found that the presence of mycorrhizae can affect the concentration of lead in the branches of *Eucalyptus* grown in contaminated soil. Arbuscular mycorrhizal (AM) infection makes it simpler for plants to absorb metals by, among other things, increasing the available soil and facilitating more efficient hyphal transfer (Bafeel, 2008), which explains why mycorrhizal plants tend to have higher concentrations of these elements.

Coronopus didymus is classified as an annual herbaceous plant within the Brassicaceae family, notable for its uncontrolled growth habits and lack of culinary characteristics. This plant species demonstrates a worldwide distribution and has its origins in the South American continent (Yannitsaros, 1986). During the winter season, spanning from October to February, there is a documented proliferation of a certain plant species (*Lantana camara*) in the northern regions of India, notably in areas adjacent to roads and inside garden environments. Yannitsaros (1986) reported that plant has rapid growth, featuring a highly branched root and shoot system, a shorter duration for harvest, and substantial biomass generation. In their study, Sidhu and coworkers (2018) conducted a pot experiment for a period of 4 and 6 weeks. The goal of the experiment was to measure how much *C. didymus* plants grew, how much biomass they had, how well they used light to make food, how much lead (Pb) they took in, and how much they stored. The plants developed well and showed no signs of toxicity when they were subjected to various soil conditions, including large amounts of lead (Pb) in the form of lead nitrate at concentrations of 100, 350, 1500, and 2500 mg/kg. Following a period of four weeks, the quantity of lead found in the root was 1652 mg Pb/kg DW, whereas the amount found in the shoot was 502–527 ppm of Pb on dry weight basis. After six weeks, the maximum lead quantities were 3091 and 527 ppm of Pb on dry weight basis, respectively. These values were obtained from the DW. When plants were exposed to Pb for six weeks, they grew and made more biomass than when they were exposed to Pb for only four weeks.

The plant species known as *Helianthus annuus* L., commonly referred to as sunflower, possesses significant agronomic characteristics, including its ability to withstand exceedingly high temperatures and its ability to acclimatize to diverse soil and climatic conditions. Consequently, due to its rapid growth, significant biomass, and capacity to hyperaccumulate HM, it has emerged as a preferred crop for environmental remediation purposes. This is due to the fact that it is capable of doing all of these things. According to Al-Jobori and Khadim (2019), a large quantity of metals was found in the plant's roots, while the shoots exhibited a lesser degree of accumulation. Their findings also suggested that roots were able to absorb more of the metals. According to the research done by Adejube and colleagues (2017), the length of time that a sunflower plant was allowed to develop before being harvested had an effect on how effectively heavy metals were removed from polluted soil. It was discovered that there was a general downward tendency as the period increased from 30 days to 60 days to 90 days.

CADMIUM

The pretreated biomass of *Durvillaea potatorum* (DP95Ca) can be used as an efficient bioabsorbent for Cd pollution in water bodies. The cadmium absorption capacity of DP95Ca was higher than for other absorbent materials (Matheickal et al., 1999). It was said to be that the maximum amount of cadmium could be removed by the biomass of *Padina* sp. when the pH was set to 5. Within 35 minutes, the kinetics of cadmium absorption become 90% more rapid. It has been demonstrated that the usage of biomass for the removal of cadmium by absorption is an effective method for the reclamation of wastewaters. As a sorbent for the removal of cadmium and Hg from aqueous solutions, Vazquez and coworkers (2002) utilized *Pinus pinaster* bark that had been pretreated with formaldehyde. This preparation of the bark was done prior to its use. The cadmium HA plants are alpine pennygrass (*Thlaspi caerulescens*) (Sheoran et al., 2009) and *Solanum photeinocarpum* (Zhang et al., 2011). According to Rascio and Navari-Izzo (2011), the alpine pennygrass (*Thlaspi caerulescens*) plant can collect cadmium at levels as high as 1800 mg/kg of plant dry weight.

These types of HAs of cadmium are effectively utilized in cadmium-polluted sites to control cadmium toxicity in soil, air, and soil water (Suman et al., 2018). The research that was carried out by Ravanbakhsh and coworkers (2016) examined the capacity of *Noccaea caerulescens, N. praecox*, and *Arabidopsis haller* to accumulate cadmium. The phytoremediation processes of phytoextraction and phytostabilization often utilize several key plant species. These species include *Rorippa globosa*, with a concentration of over 100 mg/kg, *Solanum photeinocarpum* at 158 mg/kg, *Deschampsia cespitosa* at 236 mg/kg, *Eleocharis acicularis* at 239 mg/kg, *Thlaspic aerulescens* with a range of 263–5000 mg/kg, and *Azolla pinnata* at 740 mg/kg. One of the plant species that has received significant attention in scientific research as a HA is *Thlaspic aerulescens*, as noted by Koptsik (2014). Al-Khafaji and coworkers (2018) and Dogan and coworkers (2018) identified *L. minor, T. angustifolia, C. demersum*, and *E. canadensis* as noteworthy plant species that possess the capacity for HA of cadmium.

Yadav et al. (2018) stated that cadmium removal by higher plants is called phytoremediation, and varied research has demonstrated the bioremediation of cadmium through various plant types. Processes including phytostabilization, phytodegradation, phytovolatilization, phytostimulation, and rhizodegradation are involved in the different phytoremediation methods for Cd. Several studies have documented the intrinsic capacity of plants for heavy metal cleanup (Yadav et al., 2018). *Pennisetum sinese*, often known as king grass, possesses the capacity to collect cadmium in many parts of its anatomy, including leaves, stems, and roots, by uptake from the surrounding soil. The observed amounts of cadmium buildup in both the shoots and roots of this particular type of grass varied between 0.22 and 28.87 mg/kg and 34.01 mg/kg, respectively. The plant's cadmium extraction yielded a range of 0.22–1.86 mg/plant, as reported by Li et al. (2016).

MERCURY

The global concern regarding the remediation of mercury-polluted soils (Hg) originates from the bioaccumulative properties of methylmercury, a highly toxic chemical, in the food web and the environment. Biophysical and chemical remediation technologies are currently the most widely used techniques for the restoration of soil polluted with HMs (Khalid et al., 2017). Siahaan et al. (2014) asserts that chlorosis, characterized by the yellowing of plant foliage, represents the primary symptom exhibited by plants affected by mercury toxicity. The phenomenon of mercury poisoning has been seen to elicit the discoloration of plant roots and a decrease in the overall amount of leaves and roots, and to inflict harm upon the root caps. The presence of mercury toxicity in plants can be ascribed to multiple factors, encompassing alterations in the permeability of cellular membranes, interactions involving sulfhydryl (−SH) groups and cations, the propensity to interact with active phosphate and ADP or ATP groups, and the substitution of vital elements, particularly macro elements. Over the course of the last two decades, numerous academics hailing from various nations have endeavored to identify a hyperaccumulator of mercury (Hg). Through their efforts, over 200 plant species have been subjected to examination in order to assess their capacity for Hg accumulation and transfer. It is noteworthy to mention that the predominant plant species employed in these experimental investigations are categorized as herbs. Perennial plants are frequently used in the practice of PR in soils contaminated with mercury (Hg) due to their extended lifespan and cost effectiveness in terms of time and economics. Chamba et al. (2017) and Qian and coworkers (2018) have done research that identified *Eremochloa ciliaris* and *Erato polymnioides* as plant species with the ability to hyperaccumulate mercury.

During the process of phytoremediation, plants possess the capacity to either sequester or metabolize noxious substances present in the soil. In the specific context of *L. perenne*, it has been observed that under conditions of HM exposure, there is a propensity for the accumulation of these metals within different plant structures, such as the roots, leaves, and stems (Leudo et al., 2020). According to Cummins and coworkers (2011), certain plants have the ability to release metals that have built up in their aboveground tissues. These metals include selenium (Se), mercury (Hg), and arsenic (As). In recent years, *Miscanthus* has been gaining a significant amount of awareness across a wide variety of fields. The plant is able to flourish in naturally existing deteriorated and marginal lands, such as mining areas, wasteland, and hillsides (Sun et al., 2011). The plant has a significant ability to endure varied environmental circumstances and thrive in naturally existing degraded and marginal lands. The *Miscanthus* genus comprises a total of 20 species, all of which exhibit rhizomatous growth patterns and possess characteristics similar to bamboo. These plants are classified as perennial C4 grasses, as described by Lewandowski et al. (2003). Out of the mentioned species, only four possess notable commercial significance, namely *M. sinensis, M. sacchariflorus,*

M. floridulus, and *M.x giganteus*. The examined species exhibits a notable productivity, leading to a significant production of biomass. The biomass in question exhibits promise for many uses, such as the generation of biofuel, biobased chemicals, and fiber and pulp products. In addition to this, *Miscanthus* has shown a great deal of potential in the process of cleaning up contaminated environments. According to the findings of a number of studies, *Miscanthus* maintains sustained development in soils polluted with HM, organic molecules, and agrochemicals, despite the fact that the researchers observed biomass losses at exceptionally high pollutant concentrations (Antonkiewicz et al., 2016; Techer et al., 2012; Wanat et al., 2013). Several studies (Babu et al., 2015; Nsanganwimana et al., 2014; Wanat et al., 2013) have reported the presence of HM in *Miscanthus*. These studies primarily focused on tin, cadmium, arsenic, chromium, nickel, lead, and copper, with Hg being less frequently studied. According to the findings of the study, the higher concentrations of HMs were found in the roots and rhizome of *Miscanthus* plants, while the lowest concentrations were found in the stems and leaves. Furthermore, numerous studies have provided evidence that the root exudates of *Miscanthus* possess the capacity to augment the biodegradation mechanism of organic pollutants, encompassing polycyclic aromatic hydrocarbons (PAHs), petroleum compounds, and pesticides (Didier et al., 2012). *Miscanthus* has showed potential as both an energy crop and a phytoremediator for sites contaminated with various metals; however, there hasn't been a lot of research done on its ability to tolerate mercury and its phytoremediation capabilities specifically in polluted areas. (Zhao et al., 2019).

Copper (Cu)

Copper has been recognized as a substantial factor in the contamination of soil and water, mostly attributable to agricultural practices involving the application of fungicides and pesticides containing copper. The repeated utilization of fungicides containing copper on agricultural crops results in the buildup of residual fungicide compounds within the soil, hence causing soil pollution. It has been determined that the presence of copper pollution in soil is one of the factors that contribute to the degradation of soil. The presence of contamination causes copper phytotoxicity, which, in turn, has an adverse impact on the microbiological function and fertility of the soil. These negative consequences can be traced back to the existence of contamination. In addition to this, it causes an increase in the migration of copper into both the surface water and the groundwater (Apori et al., 2018). Ariyakanon and Winaipanich (2006), studied how well *Brassica juncea* (L.) Czern and *Bidens alba* (L.) DC. var. radiata take copper out of the soil. The findings of the study indicated that *Brassica juncea* (L.) Czern and *Bidens alba* (L.) DC. var. radiata exhibited the highest concentrations of copper, measuring 3771 ppm (dry weight) and 879 ppm (dry weight), respectively. These results were obtained using experimental pots containing soil with a copper concentration of 150 mg/kg. According to the findings, *Brassica juncea* (L.) Czern has the ability to act as a hyperaccumulator for the purpose of Cu remediation in soils that have been subjected to pollution.

INNOVATIVE METHODS FOR ENHANCING PLANT PERFORMANCE IN ORDER TO FACILITATE THE PHYTOREMEDIATION OF HEAVY METALS

PHYTOCHELATINS

Phytochelatins (PCs) are produced from glutathione, a tripeptide found in every living cell. The enzyme phytochelatin synthase catalyzes the addition of glutathione molecules, resulting in the formation of a chain of repeating units. The length of the chain is contingent upon the quantity of heavy metal present, whereby higher concentrations of HM result in the formation of longer chains. During their synthesis, phytochelatins can form coordination compounds with heavy metals. These complexes can be exported from the cell or stored in vacuoles due to their solubility in water. Heavy metals can damage proteins and other cellular components. Phytochelatins prevent this from occurring. The ability of phytochelatins to remove heavy metals is necessary for plants to survive in polluted environments. Additionally, scientists have discovered that phytochelatins can be used to treat humans. For instance, they have been investigated as a treatment for those who have been poisoned by heavy metals. However, a number of studies have shown that cultivating plants in contaminated soil can increase their phytochelatin levels. This suggests that phytochelatins may possess a function in reducing the bioavailability of heavy metals in the diet. PCs share the same structure as glutathione synthetase (GSH), and their general structural formula is (g-Glu-Cys)n-Aa, where n is between 2 and 11 and Aa is a C-terminal amino acid. Due to the extensive range of n, there are numerous structural types of PCs (Vershinina et al., 2022). In the majority of cases, the abbreviation Gly is used to represent the C-terminal amino acid. However, only a few plant species have been identified to possess C-terminal isophytochelatins consisting of the amino acids alanine, glutamic acid, and serine. The process of polymerization has a significant role in determining the phytoremediation capacity of polycyclic compounds. The evaluation of the phytoremediation capacity of the aquatic plant *Lemna minor* was conducted. According to Török et al. (2015), the levels of Cd accumulation were higher in polycarbonate species with a higher degree of polymerization (PC4, PC6, and PC7) in comparison to those with a lower degree of polymerization (PC2 and PC3).

GENETIC ENGINEERING

The application of genetic engineering (GE) has shown promise in enhancing plants' capacity to alleviate heavy metal pollution. In the process of genetic modification in plants, a specific gene is obtained from a separate organism, which could be another plant or microbes, or animal. This gene is then extracted and inserted into the genome of the desired plant species. After undergoing DNA recombination, plants undergo genetic change by the incorporation of a foreign gene, resulting in the acquisition of certain traits. In the realm of phytoremediation, genetic engineering presents a more efficient approach for imparting favorable traits to plants compared to traditional breeding methods. In addition, the discipline of GE enables the transfer of beneficial genes from hyperaccumulator plants to plant species that lack sexual compatibility. Traditional methods of reproduction, such as crossing, are inadequate in attaining this goal.

MICROBIAL-ENHANCED PHYTOREMEDIATION

The majority of terrestrial plants share their environments with a wide variety of microorganisms, including arbuscular mycorrhizal fungi (AMF) and plant growth-promoting rhizobacteria (PGPR). AMF have the capacity to increase both the growth of the host plant and its resistance to stress. Karimi et al. (2014) conducted a study to assess the availability of plant-essential trace elements (PTEs) in soil, and absorption and translocation of metals by plants. As a result of these circumstances, AMF have emerged as the preeminent mutualistic fungi renowned for their efficacy in phytoremediation. PGPR is a term used to describe a wide range of bacteria that reside in the rhizosphere, the region of soil around plant roots. These bacteria have the capacity to improve plant growth and development in soils that contain potentially toxic elements (PTXEs). These bacteria mitigate the adverse impacts of PTXEs on plants, hence facilitating their growth and development. Also, PGPR can improve the phytoremediation of PTXEs by moving nutrients in the soil, making many plant growth factors, making metals more bioavailable by releasing metabolites (like organic acids and siderophores), and causing oxidation/reduction reactions. Kamran et al. (2015) found that giving *Pseudomonas putida* to *Eruca sativa* increased the plant's shoot and root growth as well as its ability to take in Cd (Yang et al., 2016). The effectiveness of employing indigenous arbuscular mycorrhizal fungi and plant growth-promoting rhizobacteria in the repair of sterilized soils that have been intentionally polluted with PTXEs has been demonstrated. The effectiveness of microbial inoculation in augmenting the PR of potentially toxic elements is influenced by multiple factors, including the genetic makeup of the plant, the particular species of microbial inoculant used, the concentrations and types of PTXEs present, and the soil's physicochemical properties (Vimal et al., 2017).

CHELATE-MEDIATED PHYTOEXTRACTION

Either by the process of naturally extracting chelates or through the process of injecting them into the soil in the form of soil supplements, the availability of heavy metals to plants can be improved. For the purpose of metal cleanup using chelates, several examples have been recorded, such as *Pisum sativum*, *Brassica juncea*, and *Zea mays*, among others (Jiang et al., 2004). Chelates can exhibit either a natural or a synthetic origin, with their classification being contingent upon their biodegradability. Natural chelates can exist in either protein or nonprotein forms. Numerous subclasses can be identified under this category, including phytochelatins, metallothioneins, ferritins, and organic acids, among others. On the other hand, it should be noted that synthetic chelates include nonbiodegradable characteristics. However, it is important to acknowledge that they do exhibit a substantial improvement in the effectiveness of metal absorption. Various synthetic chelating agents, such as EDTA, EDDS, NTA, GEDTA, and DTPA, have been utilized as effective agents to increase the mobility and transport of heavy metals from polluted soil by employing *B. juncea* as a hyperaccumulator. EDTA, HEDTA, and EDDHA are well recognized as highly efficient labile chelators that possess the capacity to enhance the phytoextraction process for several elements, including Ni, Pb, Cd, Cr, Cu, and Zn.

LIMITATIONS

There are many benefits of using plant resources for reducing HM contamination in soil and water. Phytoremediation is a technological strategy that facilitates the implementation of remediation techniques in the original location, hence obviating the need for substantial soil displacement or excavation. This method preserves the integrity of the topsoil, allowing it to remain undisturbed and suitable for use. It utilizes solar energy and is generally simple to implement. Numerous metals and radionuclides are treatable. As for contaminated sites, phytoremediation is a useful method for sites that cannot be remedied with other techniques, such as those with a large surface area and low contaminant concentrations at shallow depths. Several phytoremediation plant species are well-studied crop plants, so there is an abundance of information on their application and management. One additional benefit of phytoremediation is the potential reduction or removal of secondary air or water-borne pollutants. This is achieved by the establishment of plant cover, which serves to stabilize the soil and mitigate the risk of wind or water erosion. The phenomenon of windblown dust and erosion gives rise to exposure pathways that involve the direct breathing of contaminated air and the consumption of food that has been contaminated through the deposition of suspended matter onto food plants. The utilization of hyperaccumulators enables the potential disposal of their biomass by incineration, hence resulting in a reduction in both the mass and volume of waste that would otherwise require deposition in landfills. Ratios as high as 200:1 have been documented in the literature for the evaluation of traditional remedial methods. Plant species possess genetic features and characteristics that categorize them into distinct groups, including indicators, species with poor tolerance, highly tolerant species, and hyperaccumulator species (Farraji et al., 2020). About 400 plant species have been identified as hyperaccumulators, with the plant family Brassicaceae demonstrating the highest prevalence of HA species.

   A problem with phytoremediation is that it is effective only if the plants used grow well and quickly. Because the soil and temperature in each site are different, the PR method used for a certain plant species in one site may not work in another site. So, it is only good for that place (Kafle et al., 2022). In addition to dirt and weather, insects, pests, and pathogens (living things that can hurt plants) may also have an effect on a plant's physiology. The potential coexistence of pollutants such as HMs, organic pollutants, radionuclides, and antibiotics, as well as the presence of insects, arthropods, and pathogens, may contribute to an increased susceptibility of plants to diseases, hence impeding the efficacy of phytoremediation endeavors. Moreover, it is worth noting that plants exhibit optimal growth exclusively in regions characterized by a specific level of pollution. Plants exhibit heightened sensitivity to elevated levels of pollutants, resulting in reduction in their growth rate (Huang et al., 2000). Consequently, this diminished growth capacity may impede their efficacy in mitigating pollution. In the context of remediating elevated levels of metal pollution, such as Pb, As, and U, it has been observed that the development rate of metal hyper-accumulators is notably sluggish, resulting in little biomass accumulation.

CONCLUSION

Phytoremediation has emerged as a promising approach in effectively tackling the pressing problem of HM-contaminated soil, as seen by recent advancements in this field. By using the inherent capacities of plants to take up, sequester, and remediate heavy metals, this environmentally conscious and enduring approach mitigates the ecological and physiological risks linked to these contaminants. The application of various techniques, including genetically modified plants, microbial-assisted phytoremediation, and the use of hyperaccumulators, has increased the scope and efficacy of phytoremediation techniques. Even though phytoremediation is an effective method, it still has some limitations. In the past few years, a large number of creative studies have been done to try to unlock phytoremediation's full potential and get around the problems. The efficacy of PR relies on a number of things, like choosing the right plant species, the type of pollution and how much of it is there, the state of the soil, and the weather. Also, the process of phytoremediation can take a long time. It might take many years to achieve considerable reductions in heavy metal concentrations, which means that it might not be a viable option in situations when there is an urgent need to clean up contamination. When plants reach their full maturity or die, there is a risk that they will leak heavy metals back into the surrounding environment, which raises certain worries. Maintenance and monitoring over an extended period of time are necessary in order to stop the contamination from happening again. The cost-effectiveness of phytoremediation, in particular for applications on a large scale, continues to be a substantial barrier to its widespread adoption.

REFERENCES

Adejube, A. A. H., A. Anteyi, F. H. Garba, O. A. Oyekunle, and F. O. Kudaisi. "Bioremediating activity of sunflower (*Helianthus annuus* L.) on contaminated soil from Challawa Industrial Area, Kano-State Nigeria." *International Journal of Agriculture and Earth Science* 3, no. 5 (2017): 1–11.

Al-Jobori, Kamil M., and Athar K. Kadhim. "Evaluation of sunflower (*Helianthus annuus* L.) for phytoremediation of lead contaminated soil." *Journal of Pharmaceutical Sciences and Research* 11, no. 3 (2019): 847–854.

Al-Khafaji, Mahmoud Saleh, Faris H. Al-Ani, and Alaa F. Ibrahim. "Removal of some heavy metals from industrial wastewater by Lemmna minor." *KSCE Journal of Civil Engineering* 22, no. 4 (2018): 1077–1082.

Antonkiewicz, Jacek, Barbara Kołodziej, and Elżbieta Jolanta Bielińska. "The use of reed canary grass and giant *Miscanthus* in the phytoremediation of municipal sewage sludge." *Environmental Science and Pollution Research* 23 (2016): 9505–9517.

Apori, O. S., E. Hanyabui, and Y. J. Asiamah. "Remediation technology for copper contaminated soil: A review." *Asian Soil Research Journal* 1, no. 3 (2018): 1–7.

Ariyakanon, Naiyanan, and Banchagan Winaipanich. "Phytoremediation of copper contaminated soil by *Brassica juncea* (L.) *Czern* and *Bidens alba* (L.) DC. var. radiata." *Journal of Scienctific Research Chula University* 31, no. 1 (2006): 49–56.

Babu, A. Giridhar, Patrick J. Shea, D. Sudhakar, Ik-Boo Jung, and Byung-Taek Oh. "Potential use of *Pseudomonas koreensis* AGB-1 in association with *Miscanthus sinensis* to remediate heavy metal (loid)-contaminated mining site soil." *Journal of Environmental Management* 151 (2015): 160–166.

Bafeel, Sameera O. "Contribution of mycorrhizae in phytoremediation of lead contaminated soils by *Eucalyptus rostrata* plants." *World Applied Science Journal* 5, no. 4 (2008): 490–498.

Bhat, Shakeel Ahmad, Omar Bashir, Syed Anam Ul Haq, Tawheed Amin, Asif Rafiq, Mudasir Ali, Juliana Heloisa Pinê Américo-Pinheiro, and Farooq Sher. "Phytoremediation of heavy metals in soil and water: An eco-friendly, sustainable and multidisciplinary approach." *Chemosphere* 303 (2022): 134788.

Bona, Elisa, Chiara Cattaneo, Patrizia Cesaro, Francesco Marsano, Guido Lingua, Maria Cavaletto, and Graziella Berta. "Proteomic analysis of *Pteris vittata* fronds: Two arbuscular mycorrhizal fungi differentially modulate protein expression under arsenic contamination." *Proteomics* 10, no. 21 (2010): 3811–3834.

Cantamessa, Simone, Nadia Massa, Elisa Gamalero, and Graziella Berta. "Phytoremediation of a highly arsenic polluted site, using *Pteris vittata* L. and arbuscular mycorrhizal fungi." *Plants* 9, no. 9 (2020): 1211.

Chamba, Irene, Daniel Rosado, Carolina Kalinhoff, Selvaraj Thangaswamy, Aminael Sánchez-Rodríguez, and Manuel Jesús Gazquez. "Erato polymnioides–A novel Hg hyperaccumulator plant in Ecuadorian rainforest acid soils with potential of microbe-associated phytoremediation." *Chemosphere* 188 (2017): 633–641.

Coelho, C. Livia, Ana Rosa R. Bastos, Paulo J. Pinho, Guilherme A. Souza, Janice G. Carvalho, Viviane A. T Coelho, Luiz Carlos A. Oliveira, Rimena R. Domingues, and Valdemar Faquin. "Marigold (*Tagetes erecta*): The potential value in the phytoremediation of chromium." *Pedosphere* 27, no. 3 (2017): 559–568.

Cummins, Ian, David P. Dixon, Stefanie Freitag-Pohl, Mark Skipsey, and Robert Edwards. "Multiple roles for plant glutathione transferases in xenobiotic detoxification." *Drug Metabolism Reviews* 43, no. 2 (2011): 266–280.

Dheri, G. S., M. S. Brar, and S. S. Malhi. "Comparative phytoremediation of chromium-contaminated soils by fenugreek, spinach, and raya." *Communications in Soil Science and Plant Analysis* 38, no. 11-12 (2007): 1655–1672.

Didier, Técher, Laval-Gilly Philippe, Henry Sonia, Bennasroune Amar, Martinez-Chois Claudia, and Falla Jairo. "Prospects of Miscanthus x giganteus for PAH phytoremediation: A microcosm study." *Industrial Crops and Products* 36, no. 1 (2012): 276–281.

Dogan, Muhammet, Mehmet Karatas, and Muhammad Aasim. "Cadmium and lead bioaccumulation potentials of an aquatic macrophyte *Ceratophyllum demersum* L.: A laboratory study." *Ecotoxicology and Environmental Safety* 148 (2018): 431–440.

Farraji, Hossein, Brett Robinson, Parsa Mohajeri, and Tayebeh Abedi. "Phytoremediation: Green technology for improving aquatic and terrestrial environments." *Nippon. Journal of Environmental Science* 1 (2020): 1–30.

Ghosh, M., and S. P. Singh. "Comparative uptake and phytoextraction study of soil induced chromium by accumulator and high biomass weed species." *Applied Ecology and Environmental Research* 3, no. 2 (2005): 67–79.

Huang, X. D., B. R. Glick, and B. M. Greenberg. "Combining remediation techniques increases kinetics for removal of persistent organic contaminants from soil." In *Environmental Toxicology and Risk Assessment: Science, Policy, and Standardization—Implications for Environmental Decisions: Tenth Volume*. ASTM International, 2000.

Ingole, Nitin W., and Vaibhav R. Dhawale. "Development of phytoremediation technology for arsenic removal-A state of art." *Development* 6, no. 1 (2021).

Jiang, X. J., Y. M. Luo, Q. Liu, S. L. Liu, and Q. G. Zhao. "Effects of cadmium on nutrient uptake and translocation by Indian mustard." *Environmental Geochemistry and Health* 26 (2004): 319–324.

Matheickal, Jose T., Qiming Yu, and Gavin M. Woodburn. "Biosorption of cadmium (II) from aqueous solutions by pre-treated biomass of marine alga *Durvillaea potatorum.*" *Water Research* 33, no. 2 (1999): 335–342.

Kafle, Arjun, Anil Timilsina, Asmita Gautam, Kaushik Adhikari, Anukul Bhattarai, and Niroj Aryal. "Phytoremediation: Mechanisms, plant selection and enhancement by natural and synthetic agents." *Environmental Advances* 8 (2022): 100203.

Kamran, M. A., J. H. Syed, S. A. M. A. S. Eqani, M. F. H. Munis, and H. J Chaudhary. "Effect of plant growth-promoting rhizobacteria inoculation on cadmium (Cd) uptake by *Eruca sativa.*" *Environmental Science and Pollution Research* 22 (2015): 9275–9283.

Kanwar, Mukesh Kumar, Renu Bhardwaj, Priya Arora, Sikandar Pal Chowdhary, Priyanka Sharma, and Subodh Kumar. "Plant steroid hormones produced under Ni stress are involved in the regulation of metal uptake and oxidative stress in *Brassica juncea* L." *Chemosphere* 86, no. 1 (2012): 41–49.

Karimi, Mahsa, Hamid Jalilvand, and Milad Pourahmad. "Spatial pattern *of Pistacia atlantica* desf. in Zagros forests of Iran." *Journal of Biodiversity and Environmental Sciences (JBES)* 5, no. 3 (2014): 299–307.

Khalid, Sana, Muhammad Shahid, Nabeel Khan Niazi, Behzad Murtaza, Irshad Bibi, and Camille Dumat. "A comparison of technologies for remediation of heavy metal contaminated soils." *Journal of Geochemical Exploration* 182 (2017): 247–268.

Khan, Sardar, Abd El-Latif Hesham, Min Qiao, Shafiqur Rehman, and Ji-Zheng He. "Effects of Cd and Pb on soil microbial community structure and activities." *Environmental Science and Pollution Research* 17 (2010): 288–296.

Koptsik, G. N. "Problems and prospects concerning the phytoremediation of heavy metal polluted soils: A review." *Eurasian Soil Science* 47 (2014): 923–939.

Leudo, Ana M., Yuby Cruz, Carolina Montoya-Ruiz, María del Pilar Delgado, and Juan F. Saldarriaga. "Mercury phytoremediation with *Lolium perenne*-Mycorrhizae in contaminated soils." *Sustainability* 12, no. 9 (2020): 3795.

Lewandowski, Iris, Jonathan MO Scurlock, Eva Lindvall, and Myrsini Christou. "The development and current status of perennial rhizomatous grasses as energy crops in the US and Europe." *Biomass and Bioenergy* 25, no. 4 (2003): 335–361.

Li, Hui, Na Luo, Li Jun Zhang, Hai Ming Zhao, Yan Wen Li, Quan Ying Cai, Ming Hung Wong, and Hui Ce Mo. "Do arbuscular mycorrhizal fungi affect cadmium uptake kinetics, subcellular distribution and chemical forms in rice?." *Science of the Total Environment* 571 (2016): 1183–1190.

Mangkoedihardjo, Sarwoko and S. Surahmaida. "*Jatropha curcas* L. for phytoremediation of lead and cadmium polluted soil." *World Applied Sciences Journal* 4, no. 4 (2008): 519–522.

Nsanganwimana, F., B. Pourrut, M. Mench, and F Douay. "Suitability of *Miscanthus* species for managing inorganic and organic contaminated land and restoring ecosystem services. A review." *Journal of Environmental Management* 143 (2014): 123–134.

Qian, Xiaoli, Yonggui Wu, Hongyun Zhou, Xiaohang Xu, Zhidong Xu, Lihai Shang, and Guangle Qiu. "Total mercury and methylmercury accumulation in wild plants grown at wastelands composed of mine tailings: Insights into potential candidates for phytoremediation." *Environmental Pollution* 239 (2018): 757–767.

Ramana, S., A. K. Biswas, Ajay, A. B. Singh, and Narendra K. Ahirwar. "Phytoremediation of chromium by tuberose." *National Academy Science Letters* 35 (2012): 71–73.

Rascio, Nicoletta, and Flavia Navari-Izzo. "Heavy metal hyperaccumulating plants: How and why do they do it? And what makes them so interesting?." *Plant Science* 180, no. 2 (2011): 169–181.

Ravanbakhsh, Mohammadhossein, Abdol-Majid Ronaghi, Seyed Mohsen Taghavi, and Alexandre Jousset. "Screening for the next generation heavy metal hyperaccumulators for dryland decontamination." *Journal of Environmental Chemical Engineering* 4, no. 2 (2016): 2350–2355.

Ronzan, M., L. Zanella, L. Fattorini, F. Della Rovere, D. Urgast, S. Cantamessa, and A. Nigro et al. "The morphogenic responses and phytochelatin complexes induced by arsenic in *Pteris vittata* change in the presence of cadmium." *Environmental and Experimental Botany* 133 (2017): 176–187.

Sheoran, V., A. S. Sheoran, and P. Poonia. "Phytomining: A review." *Minerals Engineering* 22, no. 12 (2009): 1007–1019.

Siahaan, B C., R S. Utami, and E Handayanto. "Phytoremediation of mercury polluted soil using *Lindernia crustacea*, *Digitaria radicosaa*, and *Cyperus rotundus* and their effects on growth and production of corn plants." *Journal of Land and Land Resources* 1, no. 2 (2014): 35–51.

Sidhu, Gagan Preet Singh, Aditi Shreeya Bali, Harminder Pal Singh, Daizy R. Batish, and Ravinder Kumar Kohli. "Phytoremediation of lead by a wild, non-edible Pb accumulator *Coronopus didymus* (L.) Brassicaceae." *International Journal of Phytoremediation* 20, no. 5 (2018): 483–489.

Sinha, Rajiv K., Sunil Herat, and P. K. Tandon. "Phytoremediation: role of plants in contaminated site management." In Singh, S. N., Tripathi, R. D. (eds) *Environmental Bioremediation Technologies* (pp. 315–330). Springer, 2007.

Suman, Jachym, Ondrej Uhlik, Jitka Viktorova, and Tomas Macek. "Phytoextraction of heavy metals: A promising tool for clean-up of polluted environment?." *Frontiers in Plant Science* 9 (2018): 1476.

Sun, Yuebing, Qixing Zhou, Yingming Xu, Lin Wang, and Xuefeng Liang. "Phytoremediation for co-contaminated soils of benzo [a] pyrene (B [a] P) and heavy metals using ornamental plant *Tagetes patula*." *Journal of Hazardous Materials* 186, no. 2–3 (2011): 2075–2082.

Sundaramoorthy, Perumal, Alagappan Chidambaram, Kaliyaperumal Sankar Ganesh, Pachikkaran Unnikannan, and Logalakshmanan Baskaran. "Chromium stress in paddy: (i) nutrient status of paddy under chromium stress; (ii) phytoremediation of chromium by aquatic and terrestrial weeds." *Comptes rendus biologies* 333, no. 8 (2010): 597–607.

Tang, K. H. D. "Phytoremediation of soil contaminated with petroleum hydrocarbons: A review of recent literature." *Global Journal of Civil and Environmental Engineering* 1, no. December (2019): 33–42.

Tangahu, Bieby Voijant, Siti Rozaimah Sheikh Abdullah, Hassan Basri, Mushrifah Idris, Nurina Anuar, and Muhammad Mukhlisin. "A review on heavy metals (As, Pb, and Hg) uptake by plants through phytoremediation." *International journal of chemical engineering* 2011, no. 1 (2011): 939161.

Techer, Didier, Claudia Martinez-Chois, Philippe Laval-Gilly, Sonia Henry, Amar Bennasroune, Marielle D'innocenzo, and Jairo Falla. "Assessment of *Miscanthus × giganteus* for rhizoremediation of long term PAH contaminated soils." *Applied Soil Ecology* 62 (2012): 42–49.

Török, Anamaria, Zsolt Gulyás, Gabriella Szalai, Gábor Kocsy, and Cornelia Majdik. "Phytoremediation capacity of aquatic plants is associated with the degree of phytochelatin polymerization." *Journal of Hazardous Materials* 299 (2015): 371–378.

Uddin Nizam, M., M. Mokhlesur Rahman, and Jang-Eok Kim. "Phytoremediation potential of kenaf (*Hibiscus cannabinus* L.), mesta (*Hibiscus sabdariffa* L.), and jute (*Corchorus capsularis* L.) in arsenic-contaminated soil." *Korean Journal of Environmental Agriculture* 35, no. 2 (2016): 111–120.

Vazquez, G., J. Gonzalez-Alvarez, S. Freire, M. López-Lorenzo, and G. Antorrena. "Removal of cadmium and mercury ions from aqueous solution by sorption on treated *Pinus pinaster* bark: Kinetics and isotherms." *Bioresource Technology* 82, no. 3 (2002): 247–251.

Vershinina, Z. R., D. R. Maslennikova, O. V. Chubukova, L. R. Khakimova, and V. V. Fedyaev. "Contribution of artificially synthetized phytochelatin encoded by the gene PPH6HIS to increase the phytoremediative qualities of tobacco plants." *Russian Journal of Plant Physiology* 69, no. 4 (2022): 71.

Vimal, Shobhit Raj, Jay Shankar Singh, Naveen Kumar Arora, and Surendra Singh. "Soil-plant-microbe interactions in stressed agriculture management: A review." *Pedosphere* 27, no. 2 (2017): 177–192.

Wanat, Nastasia, Annabelle Austruy, Emmanuel Joussein, Marilyne Soubrand, Adnane Hitmi, Cécile Gauthier-Moussard, Jean-François Lenain, Philippe Vernay, Jean Charles Munch, and Martin Pichon. "Potentials of *Miscanthus× giganteus* grown on highly contaminated technosols." *Journal of Geochemical Exploration* 126 (2013): 78–84.

Yadav, Krishna Kumar, Neha Gupta, Amit Kumar, Lisa M. Reece, Neeraja Singh, Shahabaldin Rezania, and Shakeel Ahmad Khan. "Mechanistic understanding and holistic approach of phytoremediation: A review on application and future prospects." *Ecological Engineering* 120 (2018): 274–298.

Yang, Yurong, Yan Liang, Xiaozhen Han, Tsan-Yu Chiu, Amit Ghosh, Hui Chen, and Ming Tang. "The roles of arbuscular mycorrhizal fungi (AMF) in phytoremediation and tree-herb interactions in Pb contaminated soil." *Scientific Reports* 6, no. 1 (2016): 20469.

Yannitsaros, Artemios. "New data on the naturalization and distribution of *Coronopus didymus* (Cruciferae) in Greece." *Willdenowia* (1986): 61–64.

Zhang, Xingfeng, Hanping Xia, Ping Zhuang, and Bo Gao. "Identification of a new potential Cd-hyperaccumulator *Solanum photeinocarpum* by soil seed bank-metal concentration gradient method." *Journal of Hazardous Materials* 189, no. 1–2 (2011): 414–419.

Zhao, Anqi, Lingyun Gao, Buqing Chen, and Liu Feng. "Phytoremediation potential of *Miscanthus sinensis* for mercury-polluted sites and its impacts on soil microbial community." *Environmental Science and Pollution Research* 26 (2019): 34818–34829.

Zhou, Chui-Fan, Yu-Jun Wang, Rui-Juan Sun, Cun Liu, Guang-Ping Fan, Wen-Xiu Qin, Cheng-Cheng Li, and Dong-Mei Zhou. "Inhibition effect of glyphosate on the acute and subacute toxicity of cadmium to earthworm *Eisenia fetida*." *Environmental Toxicology and Chemistry* 33, no. 10 (2014): 2351–2357.

4 CRISPR-Cas Technology for Enhanced Phytoremediation

A Promising Strategy to Combat Toxic Metals and Metalloids in Plants

*Madhusmita Barik, Animesh Pattnaik,
Swayamprabha Sahoo, Rukmini Mishra,
and Jatindra Nath Mohanty*

INTRODUCTION

The exponential growth of the human population and the continuous expansion of human activities, coupled with the beginning of the industrial revolution, have placed an overwhelming burden on the global environment (Gavrilescu et al., 2015). Following the industrial revolution, there has been a remarkable increase in the abundance of toxic pollutants, which have had a detrimental impact on various life forms existing on earth, including humans. These pollutants are present in air, soil, and water, posing serious environmental risks (Gavrilescu et al., 2015). Diverse substances such as hydrocarbons (like DDT), volatile organic compounds, chlorinated solvents, and explosives like trinitrotoluene are examples of organic pollutants (Gavrilescu et al., 2015). Nickel (Ni), zinc (Zn), molybdenum (Mo), copper (Cu), manganese (Mn), iron (Fe), and cobalt (Co) are inorganic pollutants that are required for plant and animal growth but can be harmful when found in excess amounts (Gavrilescu et al., 2015). Contrarily, toxic heavy metals and metalloids, which are known to have no good effects on living things—substances like cobalt (Co), uranium (U), lead (Pb), chromium (Cr), arsenic (As), and cadmium (Cd)—are harmful even at low doses (Clemens and Ma, 2016). Most specifically, because of their constant presence in nature and their tendency to bioaccumulate in the food chain due to their lipophilic nature, heavy metals have drawn substantial attention as environmental contaminants (Clemens and Ma, 2016). These contaminants have been shown to be extremely hazardous and to have carcinogenic consequences (Zhong et al., 2017).

Conventional approaches for remediating polluted sites, like excavation of soil and landfilling, immobilization and soil washing, and physicochemical extraction

DOI: 10.1201/9781003442295-4

methods, are often costly, labor-intensive, inefficient, and result in unsightly scars on the land after treatment (Khalid et al., 2020). However, plants offer a promising solution as they possess the ability to accumulate toxic heavy metals (HMs) within their cells and tissues. As a result, they can help with HM removal, transfer, and/or stability, effectively reducing pollution in the impacted areas (Khalid et al., 2020). Heavy metals-contaminated sites can be permanently cleaned up using phytoremediation, an economical and environmentally friendly approach (Fan et al., 2018; Pilon-Smits et al., 2019). This strategy harnesses the natural capabilities of plants in a solar-driven and energy-efficient manner. Phytoremediation encompasses various techniques and applications, each with distinct mechanisms for the accumulation, immobilization, removal, and degradation of HMs. An advanced approach to enhance the efficacy of phytoremediation plants involves the use of omics techniques, such as metabolomics, proteomics, genomics, and transcriptomics (Yadav et al., 2023). By applying genetic engineering methods in bioremediation, it becomes possible to develop plants with high biomass content, an extensive root system, and the ability to hyperaccumulate pollutants, even under challenging environmental conditions (Yadav et al., 2023). Researchers are particularly focused on characterizing and modifying genomic DNAs that are associated with accumulation and tolerance to pollutants. Therefore, the combination of omics approaches offers a promising avenue for genetically manipulating potent plants to enhance their efficiency in pollutant removal (Hassan et al., 2022). Broadly speaking, genome editing techniques allow researchers to change the regulation of gene expression at particular sites, providing significant insights into plant functional genomics (Nganso et al., 2022). Significant progress in this area has been achieved by phytoremediation based on clustered regularly interspaced palindromic repeats (CRISPR) (Kaur et al., 2019). Functional genomics holds great promise for CRISPR, a novel method for genome engineering that allows for the targeted modification of traits in plants (Hamim, 2022). Advanced plant phenotypes are now possible with the CRISPR-Cas system, which blends the usage of CRISPR and Cas proteins (Karmakar et al., 2022).

CRISPR technology offers a practical approach to achieve precise plant genome modifications (Naz et al., 2022). It has been successful in overcoming challenges associated with plant modifications that were previously difficult to achieve. By altering DNA sequences, CRISPR technology facilitates the modification of gene sequences in plants to mitigate the adverse effects of heavy metals and other soil pollutants. Through interactions with plant and microbe relationships, this strategy also increases nutrient absorption and bioavailability of metals to plants (Maurya et al., 2023). CRISPR-based phytoremediation offers a promising solution for soil purification and removal of toxic elements from the rhizosphere, making it suitable for agronomic sectors and other beneficial initiatives, including social and agroforestry (Thijs et al., 2016). To achieve successful phytoremediation, it is crucial to identify key genes involved in plant and microbial interactions and understand their roles. By manipulating the target genes and improving plants' phytoremediation abilities, gene editing strategies like CRISPR permit the usage of plants in a variety of difficult environmental situations (Agarwal and Rani, 2022). Here, we investigate the potential of CRISPR-Cas genome editing methods to alter plant genomes and produce plants with enhanced metal tolerance, absorption, and accumulation capabilities. We cover various preomics methods for phytoremediation. Extensively discussed are the candidate genes for these processes, highlighting their significance in enhancing the desired traits in plants like metal tolerance, uptake, and accumulation capacity.

PHYTOREMEDIATION: A POWERFUL POTENTIAL TO REOBTAIN VEGETATIVE LAND

The removal, degradation, detoxification, or immobilization of environmental pollutants in a growth matrix using various natural, biological, physical, or chemical processes of plants is known as phytoremediation. This method is emerging, environmentally benign, efficient, and advantageous. But according to Carolin et al. (2017), the use of plant-based technologies for the treatment of wastewater dates back about 300 years. Contamination is mainly caused by a number of organic and inorganic substances, which include precarious waste, rotting substances, heavy metals, combustible materials, and explosives, among others (Yadav and Kumar, 2019). These different pollutants disturb the ecosystem and ultimately create disorders in human health. In addition to reducing or preventing radioactive entities of site movement and filtration, the aim of the phytoremediation strategy is to partly rehabilitate waste regions so that local plants and animals can eventually use them (Poty et al., 2018). In different media around the world, multiple researchers have discussed various phytoremediation strategies (Favas et al., 2018; Muske et al., 2017; Yadav et al., 2018; Vidal et al., 2019). It involves the engineered use of green plants to get rid of, contain, and reduce the toxicity of environmental contaminants like organic molecules, heavy metals, trace elements, and radioactive substances (Hettiarachchi et al., 2019).

Pollutants are absorbed by plant roots, accumulate in tissues, degrade, and change into less toxic forms during this process. Due to their ability to reactivate the breakdown of organic compounds in the rhizosphere by releasing exudates, roots, and enzymes (Yaqoob et al., 2019), green plants play a crucial part in environmental planning. The plants' metal uptake process may allow them to function as "accumulators" or "excluders." The pollutants can be accumulated in the aerial tissues of the plants, where they can biodegrade or convert into active forms (Sinha et al., 2004). The procedure involves the buildup of contaminants in tissues, their absorption through plant roots, their decomposition, and their transformation into less dangerous forms.

In terms of the versatility of phytoremediation, some of the phytoremediation strategies for polluted environments include phytostabilization, phytoextraction, phytovolatilization, and rhizoremediation (Saleem et al., 2020). Since the Iron Age, human activities have caused environmental disruptions in one way or another, with the release of toxins from industrial effluents, coal mines, and agricultural wastes (Tarla et al., 2020). Different plant species have different abilities to accumulate, degrade, or immobilize specific pollutants, allowing for tailored approaches to different types of contamination. By using this remediation approach, tailings are given a definite landscape value, ecosystem services are restored, and the risk of human exposure to heavy metals is decreased (Zalesny et al., 2021). A process called phytoextraction involves plants absorbing pollutants from water, soil, or sediments through roots and transferring them to aboveground biomass where they accumulate, for example, in shoots or other beneficial parts of the plant (Singh and Santal, 2015; Sarwar et al., 2017; Yanitch et al., 2020). It is the most effective approach for the removal of metalloids and heavy metals from disturbed soils. It is cost-effective, with erosion control and soil stabilization properties (Chen et al., 2019). Phytoremediation can also support biodiversity and restore habitats as it involves the use of diverse plant species, including native and nonnative plants, to remediate contaminated sites. Plants used in phytoremediation perform a significant work in carbon sequestration, which is a process of absorbing and storing carbon dioxide from the atmosphere during photosynthesis, aiding in the mitigation of climate change (Lal, 2004; Basu et al., 2018a).

Overall, phytoremediation offers a sustainable and environmentally friendly approach to address pollution issues. By harnessing the natural capabilities of plants, it provides a range of benefits, including cost-effectiveness, biodiversity restoration, soil stabilization, public health improvement, and climate change mitigation. As we continue to face environmental challenges, phytoremediation plays a vital role in creating cleaner and healthier ecosystems.

UNLOCKING THE MOLECULAR STRATEGIES: HOW HIGHER PLANTS ADAPT TO TOXIC METALS

Toxic metals, including Cd, Cu, and Fe, exert damaging effects on plant cells due to their transitional properties. This disrupts oxidative potential and depletes essential biomolecules like glutathione (GSH) (Chirakkara et al., 2016; Chen et al., 2016). The interaction between these biomolecules and other transition metals might enhance the cellular redox balance. In addition, certain toxic metals can directly obliterate plant protein linkages and genetic components like RNA and DNA. Plants counteract the harmful effects of toxic metals by maintaining low concentrations of free metal ions within their cells (Maurya et al., 2022). Achieving this optimal ionic state involves processes such as metal accumulation, cellular translocation, interactions with proteins, and organic ligand formation. Transporter proteins play a key role in managing the initial steps: metal accumulation, translocation into cells, and protein interactions with metals. Essential transporter proteins include Zn and Fe-regulated transporters (ZIP), metal-binding proteins such as phytochelatins (PCs), metallothioneins (MTs), and Cu-chaperone ATX1 (Chaudhary et al., 2015). ZIP proteins are crucial for the absorption and transportation of divalent metal ions, maintaining equilibrium and homeostasis. In phytoremediation, toxic metal ATPase genes contribute to metal uptake, translocation, and sequestration. Cu binding domains, when coupled with protein molecules, assist in intracellular Cu balance due to their Cu-chelating abilities (Shin et al., 2012). Antioxidant proteins, including ATX1 and ATX2, exhibit significant sequence homology (Shin et al., 2012). In the third stage, organic ligands are made and then used to interact with plant genes through transcriptional and posttranslational processes. Molecular techniques using *A. thaliana* hypersensitive mutants are used to pinpoint genes responsible for producing organic ligands within plant tissues (Zhang et al., 2022)

NAVIGATION OF TOXIC METAL ACCUMULATION AND TRANSPORT IN PLANTS

Hyperaccumulator plants possess distinct qualities compared to regular plants, including their ability to absorb significant amounts of metals and transport efficiently these toxic metals from root to shoot part (Figure 4.1). They excel at binding or transforming toxic metals into less harmful chemical forms, reducing the presence of these metals in their ionic state. These traits are attributed to enhanced ion transport tissue in hyperaccumulators. For instance, *Thlaspi caerulescens* and *Arabidopsisrabidopsis halleri* contain genes associated with the ZIP family, encoding plasma membrane transporters like ZIP6, ZTN2 (*T. caerulescens*), ZIP9 (*A. halleri*), and ZTN1, resulting in increased Zn uptake compared to nonhyperaccumulator plants (Rascio and Navari-Izzo, 2011). While hyperaccumulators proficiently transfer toxic metals by xylem tissue, sensitive plants must detoxify these metals in vacuoles or root cytoplasm before the translocation and accumulation in shoots (Rascio and Navari-Izzo, 2011). Compared to metal-susceptible plants, *T. caerulescens*, a representative hyperaccumulator, exhibits nearly twice the Zn translocation rate from roots to shoots and significantly lower Zn concentrations in roots. To regulate metal accumulation, hyperaccumulators must maintain equilibrium within their plant tissues. They rely on various transporters, such as ATP-binding cassettes

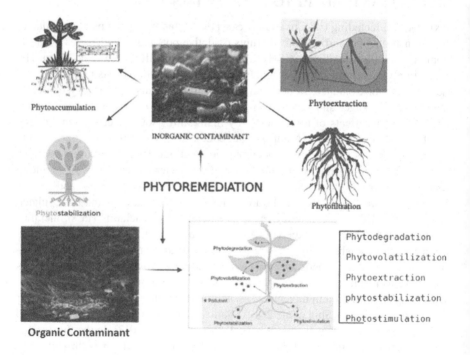

FIGURE 4.1 Utilizing phytoremediation and various filtration processes within plants effectively eliminates high levels of toxic metals and organic contaminants, promoting detoxification and enhancing plant stability against toxicity.

(ABCs), ATPases, copper transporters (COPTs), cation exchangers (CAXs), cation diffusion facilitators (CDFs), and ZIP proteins (Yaashikaa et al., 2022). For nonessential toxic metals like Cd, hyperaccumulators employ membranes or nonselective channels for translocation. Process transporters are mainly responsible for essential nutrient movement, such as Zn (Yaashikaa et al., 2022).

The P1B-ATPase subgroup of the HMA transporter family engages in the ATP-dependent transmembrane transport of both toxic and nontoxic metals in addition to metal detoxification. It has been suggested that members of the HMA family, such as HMA4 and HMA5, take part in long-distance metal transfer from root to shoot (Ma et al., 2021).

DETOXIFICATION OF TOXIC METALS

Hyperaccumulators possess the remarkable ability to detoxify various toxic metals while safeguarding their leaves and stems. Detoxification primarily occurs in the cuticle, epidermis, and trichomes (Asemaneh et al., 2006; Li et al., 2023). This enzyme-mediated process begins by removing organic ligands from the metabolic zone followed by the detoxification of reactive oxygen species (ROS). Crucially, thiol-producing biomolecules like MTs, PCs, and GSH efficiently counteract the negative effects of toxic metal on plants (Choppala et al., 2014). These complexes play a vital role in the tolerance of plants to metals. Plant tissue vacuoles are rich in phytochelatin and heavy metal (PC-HM) complexes, which the ABC family's HMT1 transporter, a member, facilitates across vacuolar membranes (Babu et al., 2021). Similarly, GSH plays a crucial role in detoxifying toxic metals and acts as an effective reducing agent, mostly for ROS. Elevated concentrations of toxic metals prompt the production of ROS in the sensitive organelles of plants (Choppala et al., 2014). GSHs also contribute to reducing H_2O_2 toxicity, mitigating xenobiotics' impact on flower growth, and facilitating salicylic acid formation. GSH's role in hazardous metal detoxification involves complexation with metals and sequestration in vacuoles, with potential release of these complexes into the apoplast (Thakur et al., 2022).

Also forming are MT-HM complexes, metallothiones comprised of chelating molecules with low-molecular-weight cysteine-rich protein. MTs fall into four groups on the basis of cysteine arrangement, varying in specificity of tissue and selectivity of metal. Among them, MT1 and MT2b are involved in Cd detoxification (Milner et al., 2014). The fourth category of MTs primarily detoxifies Zn and, compared to the other categories, accumulates higher amounts of Zn over time (Milner et al., 2014)

ORGANIC ACIDS: ENHANCING TOLERANCE TO TOXIC METALS

In plants, toxic metal sequestration is a genetically enhanced property orchestrated by cation diffusion facilitator (CDF) genes. According to Van De Mortel et al. (2008), metals are transported through the membranes (plasma membrane and tonoplast) by special CDF-shaped molecules known as metal transporter proteins (MTPs). Among these, MTP1, located in the tonoplast of leaves, significantly contributes to Zn/Ni hyperaccumulation in leaves of hyperaccumulator plants. In species like T.

goesingense, CDF gene overexpression results in the Zn/Ni hyperaccumulative trait of the vacuole (Skuza et al., 2022). However, higher-molecular-mass organic ligands like phytochelatins are limited in their ability to govern the process of detoxifying toxic metals due to the substantial metabolic costs and the requirement of excess sulfur for their synthesis (Skuza et al., 2022). In contrast, toxic metal detoxification is chiefly managed by antioxidant enzymes, enabling these enzymes to successfully combat stress caused by ROS and toxic metal poisoning in plants. In hyperaccumulator plants, both the increase in synthesis of GSH and antioxidant defense mechanisms, spurred by gene overexpression, work in tandem to detoxify toxic metals (Hu et al., 2019).

INVOLVEMENT OF TARGET GENES TO COMBAT METALLOIDS AND TOXIC METALS IN PLANTS

To combat metalloids and toxic metals in plants, a number of genes and their linked pathways are involved. The process involves multiple transgenic ways to obtain plants with improved phytoremediation traits. The methods for creating transgenic plants with improved phytoremediation traits for metal(loid) contamination involve manipulating metal transporter proteins to manage metal uptake and movement, boosting the production of metal-detoxifying compounds like phytochelatins, metallothioneins, glutathione, and organic acids, and undergoing metabolic transformations to convert toxic forms into safer ones (Luo et al., 2016). This approach can include both enhancing existing genes within a plant species and introducing new genes from different plants and bacteria (Dhankher et al., 2012).

GENE INVOLVEMENT IN METAL AND METALLOID TRANSPORT, TOLERANCE, UPTAKE, AND ACCUMULATION

Advanced genetic and molecular-based techniques like functional annotation, sequencing-based marker development, comparative analysis, and functional complementation study enable researchers to comprehend and characterize the genes involved in the transportation of heavy metals and metalloids in plants (Clemens et al., 2021). We find numerous gene families to be involved in the transport and uptake of different metal types, such as NRAMP, ZIP, HMAs, CDFs, and CPx-type ATPases, as well as copper transporters, MATE proteins, ABC transporters, and oligopeptide transporters. These gene families possess a wide range of metal specificity, indicating their potential to enhance the uptake of toxic metals for phytoremediation purposes.

A specific amino acid sequence characterizes the conserved family of transporters known as CPx-ATPases, commonly known as P-type ATPases. In order to move cations across cell membranes, they are essential. This family of heavy metal ATPases (HMAs) has a role in the localization, translocation, and distribution of metal. These ATPases utilize ATP to actively transport a variety of charged compounds across cell membranes, including essential and nonessential metals such as lead, zinc, copper, and cadmium (Guo et al., 2015). Transporters can be categorized into two groups on the basis of the metals they specialize in: one is the copper/silver group and the other is the zinc/cadmium/lead/cobalt group. According to the research model, in plants such as *Arabidopsis thaliana* and metal hyperaccumulators, HMAs play a key role in translocating and detoxifying metals like cadmium and zinc. In rice, nine HMA genes have been identified, with the OsHMA1-OsHMA3 subgroup being connected to the category Zn/Cd/Co/Pb. These genes are responsible for transporting and storing zinc and cadmium in various plant cells (Kaur et al., 2022). Increases in AtHMA4, P-1B, and ATPase expression from *Arabidopsis thaliana* led to improved growth in roots under stress caused by zinc, cadmium, and cobalt. Likewise, elevated expression of HMA4 resulted in enhanced tolerance to cadmium and zinc and greater accumulation of these metals in plants such as *T. caerulescens*, *A. halleri*, and *Nicotiana tabacum* (Grispen et al., 2011). OsHMA2, similar to OsHMA3, aids in the movement of cadmium from roots to shoots (Yamaji et al., 2013). It is most likely due to YCF1 transporting metal-glutathione conjugates to vacuoles that the transgenic plant *B. juncea* with yeast cadmium factor1 (YCF1) has shown better accumulation and tolerance to cadmium and lead. Compared to wild types, the increased expression of ATP-dependent ABC-type YCF1 in yeast vacuoles led to an increase in cadmium accumulation (Wei et al., 2014). NRAMP genes were initially discovered in rice for transporting the heavy metals along with three other genes recently identified. Using mutant yeast to express *Arabidopsis* genes AtNRAMP3, AtNRAMP1, and AtNRAMP4 demonstrated their role in transporting cadmium, manganese, and iron (Ehrnstorfer et al., 2014). Research on rice OsNRAMP5 revealed its involvement in the uptake and transportation of cadmium, manganese, and iron. According to the research by Oomen et al. (2009), the NRAMP3 and NRAMP4 help with tolerance to zinc and cadmium. Similarly, the CDF family of proteins participates in capturing heavy metals like zinc, cadmium, and cobalt. These proteins facilitate the movement of cations out of the cytoplasm, directing them through the cell membrane or into compartments like vacuoles (Chen et al., 2013).

Many genes have been characterized in rice and other important crops, as summarized here in detail, which are involved in the transport of heavy metal. Rice gene OsMTP1 enhances tolerance against various metals, including zinc, cadmium, copper, and iron. Transport of Mn, Cd, Fe, and Zn is associated with the ZIP family (zinc-iron permease), and greater Ni tolerance is associated with NtCBP4 transfer. IRT1 from *Arabidopsis* enhances the transport of divalent ions (Bhat et al., 2022). The copper transporter family (COPT) is linked to Cu transport in plants. *Hordeum vulgare*'s HvYS1 improves Fe uptake in rice. Laccases contribute to the formation of lignin and stress response. OsLAC10 overexpresses with copper exposure. NcTZN1 overexpression from *Neurospora crassa* enhances Zn accumulation in tobacco. Transgenic plants show increased accumulation of metal compared to wild-type (Yağız et al., 2022). Tobacco plants express BjCET3 and BjCET4 transporter genes from *B. juncea*, enhancing Cd tolerance and doubling Cd accumulation (Gu et al., 2021). A NiCoT family transporter from *Rhodopseudomonas palustris* raised the accumulation of Co by five times and Ni by two times in transgenic tobacco. The bacterium *Klebsiella aerogenes*' PPK gene, linked to Hg accumulation, was used in transgenic tobacco, which showed increased Hg accumulation when combined with bacterial merT genes. Tobacco overexpressing CAX2 and CAX4 from *Arabidopsis* demonstrated improved detoxification of Cd, Zn, and Mn. Similar results were seen with CAXcd overexpression in petunia, increasing growth and Cd accumulation in leaves (Liao et al., 2019).

THE PRE-CRISPR-CAS APPROACHES TO ENHANCE PHYTOREMEDIATION TO COMBAT TOXIC METALS AND METALLOIDS

The presence of several elements, such as biotic and abiotic stressors, climate change, population expansion, and the depletion of natural resources, is the main indication to the plant breeder to encourage the development of more tolerant varieties and increase agricultural output (Lateef, 2015). Various pregenomic approaches can be employed to enhance phytoremediation techniques, including nuclear genome transformation, marker-assisted selection (MAS), antisense silencing, plastome transformation, RNA interference, or non-tissue culture transformation.

MARKER-ASSISTED SELECTION

Molecular markers serve as genetic distinctions among species or organisms (Hayes et al., 2013), similar to chromosomal markers or flags that help the integration of economically important features (Collard et al., 2005). Employing the molecular marker method is a significant advance in modern crop improvement approaches, addressing a key constraint of conventional techniques (Devi et al., 2022). These markers consist of DNA sequences situated at particular genomic locations, passing down through generations following standard inheritance rules (Ruane and Sonnino, 2007). Genetic markers enable the direct selection of genes controlling desirable traits that are challenging and expensive to assess phenotypically (Jena and Mackill, 2008). Molecular marker-assisted selection (MAS) also allows for gene pyramiding. In cases where plants have low heritability, below 0.2, MAS proves particularly valuable. Techniques like MAS expedite the selection of crops resistant to heavy metal stress with heightened precision.

Moreover, these genetic markers remain unaffected by external cues and remain detectable throughout all growth phases, including seedling, vegetative, and reproductive, with the least amount of linkage drag. Single nucleotide polymorphism (SNP) markers have been associated with iron toxicity adaptation in *Oryza sativa*. Another use of the MAS method was found in *Silene paradoxa* L., where RAPD markers were used to detect genomic alterations brought on by heavy metal poisoning. RAPD analysis of crops grown in copper-contaminated soils close to mining areas showed polymorphism in soils with copper contamination. In Cu-tolerant plants, two distinct RAPD bands were identified (Mengoni et al., 2000).

NUCLEAR GENOME TRANSFORMATION

Three distinct genome types exist in plants—mitochondrial, plastid, and nuclear genomes—out of which the largest one is the nuclear genome, which is made up of billions of bases and nucleotide sequences. Nuclear genome transformation in plants involves introducing foreign genes into the nuclear genome (Figure 4.2). The majority

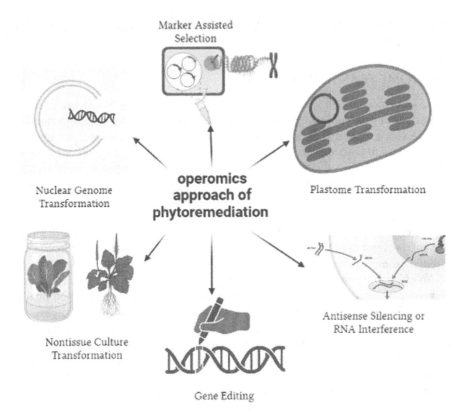

FIGURE 4.2 Various strategies are employed in the phytoremediation process to detoxify harmful metals and heavy ions effectively.

of genetically engineered crops have been produced by rDNA methods using nuclear genome transfer approaches like *Agrobacterium*-mediated or gene gun-regulated methods. While plants transfer a variety of genomes due to gene gun regulation, *Agrobacterium*-regulated gene transfer exclusively favors the nuclear genome (Kaur et al., 2019). This method allows the expression of genes encoding enzymes or transporters involved in the uptake of pollutants, transformation, or detoxification. By introducing specific genes, plants can be engineered to enhance their ability to remediate specific contaminants. Typically, dicotyledonous plants are altered using *Agrobacterium* vectors, while monocotyledonous plants are transformed through particle gun or electroporation techniques (Lateef et al., 2015). The bacterial genes *merA* (mercuric ion reductase) and merB (organomercurial lyase) were overexpressed to increase the tolerance of transgenic plants to mercury (Hg) stress (Lyyra et al., 2007). According to Yang et al. (2003), the procedure entails the protonolysis of Hg by lyase enzymes into Hg ions and its reduction (from Hg^{2+} to Hg^0) with the assistance of reductase enzymes. Furthermore, Hussein et al. (2007) demonstrated that increased reduction of Hg^{2+} to Hg^0 in response to the alteration of the chloroplast genome in the Hg-stressed tobacco plant was facilitated by overexpression of both the genes merA and merB (combined). The upshot of this quicker decrease was that mercury quickly volatilized.

PLASTOME TRANSFORMATION

Plastosome transformation techniques emerged as an outcome of the development of the particle or gene cannon approach (Svab et al., 1990). The plastome transformation method, characterized by the absence of a position effect, enables the precise targeting of transgenes to particular loci, thereby facilitating the high production of recombinant DNA (DeCosa et al., 2001). The goals of transgenic bioconfinement are accomplished through maternal plastid inheritance (Daniell, 2002; Jin and Daniell, 2015). The presence of numerous plastids per plant makes achieving homoplasmy—where all plastids within cells possess identical genetically engineered modifications— almost impractical (Maliga, 2003). Nonetheless, some reports detail the development of transgenic and homogenous cells, referred to as transplastomichomoplasmic, through the assistance of selectable markers and antibiotic combinations.

An effective approach of tissue culture is employed to eliminate untransformed plastomes or native chloroplasts. Recently, some species of the Solanaceae family like potato, tomato, and tobacco have undergone transplastomic profiling procedures (Maliga, 2003). Additionally, Ruiz et al. (2003) emphasized the importance of chloroplast genome engineering as a means to achieve high gene expression and to decrease the risk of gene alterations brought on by pollination. In another study, Ruiz et al. (2011) demonstrated the plastomic method, involving the insertion of DNA fragment or Increased expression of MT genes is the result of integrating a gene of interest into the chloroplast genome (mt1). This plastomic line showed heightened Hg sequestration levels and controlled chlorophyll content in transgenic stressed plants.

ANTISENSE SILENCING OR RNA INTERFERENCE

Antisense silencing and RNA interference (RNAi) techniques involve the introduction of specific RNA molecules that can inhibit the expression of target genes. By designing antisense or RNAi constructs against genes involved in pollutant tolerance or metabolism, the expression of these genes can be suppressed, potentially leading to enhanced phytoremediation capabilities (Kramer and Redenbaugh, 1994). Antisense silencing and RNA interference (RNAi) are powerful techniques used in molecular biology to selectively inhibit the expression of target genes (Figure 4.2). These techniques involve the introduction of specific RNA molecules that can bind to complementary messenger RNA (mRNA) molecules, thereby preventing their translation into proteins (Napoli et al., 1990). By targeting genes involved in pollutant tolerance or metabolism, antisense silencing and RNAi can be used to suppress the expression of these genes, potentially enhancing phytoremediation capabilities. In antisense silencing, a synthetic antisense RNA molecule is designed to be complementary to the target gene's mRNA sequence (Fire et al., 1998). For instance, *Brassica napus* seedlings exposed to cadmium and sulfate deficit conditions displayed different expression of the gene *miR395*, which in turn impacted the expression of sulfate transporter SULTR2;1 and ATP sulfurylase (APS). SULTR2;1

is responsible for sulfate transport, and the enzyme APS is involved in the assimilation of sulfur (Huang et al., 2010). Similarly, Zhou et al. (2012) reported comparable modulation of miR395 expression in response to mercury (Hg) stress.

In plants like *Arabidopsis thaliana*, the miR398 family is found at three different loci, miR398b, miR398a, and miR398c. Cu treatment led to the downregulation of miR398 expression in these plants at all three loci (Zhu et al., 2011). Other metal stress-regulating microRNAs (miRNAs) include miR408 and miR397 (Valdes-Lopez et al., 2010; Burklew et al., 2012). Both miRNAs exhibit laccase (LAC) activity, participating in biosynthesis of lignin and catalyzing the oxidation of several substrates like amines and phenols (Turlapati et al., 2011). This antisense RNA molecule binds to the target mRNA, forming a double-stranded RNA (dsRNA) complex. This dsRNA triggers the degradation of the target mRNA by cellular machinery, thereby reducing the availability of the mRNA for translation into proteins. As a result, the target gene's expression is downregulated. RNA interference (RNAi) operates through a similar mechanism, but with the use of small interfering RNA (siRNA) molecules. These siRNAs are typically short, double-stranded RNA molecules that are designed to specifically target and bind to the mRNA of the target gene. Once bound, the siRNA guides the RNA-induced silencing complex (RISC), an enzyme complex to the mRNA, leading to its degradation or inhibiting its translation. Both antisense silencing and RNAi techniques can be applied in phytoremediation, which uses plants to clean up the environment by removing contaminants. By designing antisense or RNAi constructs against genes involved in pollutant tolerance or metabolism, the expression of these genes can be effectively suppressed in transgenic plants. This targeted gene suppression can potentially enhance the plant's ability to remediate polluted sites by reducing or altering the expression of genes that are responsible for pollutant uptake, metabolism, or detoxification pathways. These techniques provide a means to precisely modulate gene expression, allowing researchers to study the functional roles of specific genes in pollutant tolerance or metabolism. By manipulating gene expression, scientists can optimize the phytoremediation potential of plants, improving their ability to absorb, metabolize, and detoxify environmental pollutants, thus contributing to the development of more efficient and sustainable remediation strategies.

NON-TISSUE CULTURE TRANSFORMATION

Non-tissue culture transformation methods are used to introduce foreign DNA into plants without the need for tissue culture or plant regeneration. These techniques, such as *Agrobacterium*-mediated transformation or particle bombardment, allow the direct transfer of genetic material into plant cells or tissues, facilitating the introduction of genes for phytoremediation purposes. The creation of genetically modified plants typically comprises the transformation of plant tissue in vitro, followed by regeneration. However, a limitation of crops regenerated through tissue culture is their susceptibility to variable phenotypes and fertility caused by somaclonal variations. The occurrence and severity of somaclonal variations depends on many factors, like the growth media, plant, tissue culture duration, and genotypes (Jiang et al., 2011; Stroud et al., 2013).

An alternative to tissue culture-based transformation is the non-tissue culture approach, exemplified by the floral dip process used by Liu et al. (2017) to develop transgenic plants like *Camelina sativa* and *A. thaliana*. Additionally, plant grafting can facilitate the protein transfer and RNA between rootstock and scions (Haroldsen et al., 2012). These genomic techniques offer a range of approaches to engineer plants with improved phytoremediation traits, contributing to more efficient and effective remediation of contaminated environments.

RECOMBINANT DNA TECHNOLOGY APPROACHES

Recombinant DNA technology or genetic engineering has become a viable tool for increasing plants' ability to detoxify heavy metal pollution through phytoremediation. This process involves the incorporation of foreign genes from various sources, including plant species, bacteria, or animals, into the target plant genome. Once the DNA recombination takes place, the introduced gene becomes a heritable trait that imparts specific characteristics to plants. When compared with conventional breeding methods, genetic engineering offers the advantage of swiftly imbuing plants with desired phytoremediation traits. Moreover, genetic engineering enables the transfer of beneficial genes from hyperaccumulator species to sexually incompatible plants, an outcome unattainable through conventional crossbreeding (Berken et al., 2002; Marques et al., 2009). Consequently, utilizing genetic engineering techniques to develop transgenic plants with desirable properties holds significant promise within the realm of phytoremediation. In practice, it is often more feasible to genetically modify high-biomass and fast-growing plants to enhance their resistance and accumulation capabilities for heavy metals, rather than focusing solely on engineering hyperaccumulators for high biomass production. As a result, in many applications, the emphasis is on engineering fast-growing, high-biomass plants to augment their resilience against heavy metals or to heighten their capacity for heavy metal accumulation, both of which are pivotal traits of hyperaccumulator plants. Consequently, in plants for genetic engineering the selection of genes should be grounded in a thorough understanding of the mechanisms driving heavy metal accumulation and tolerance.

Exposure to heavy metals in excess can trigger the overproduction of reactive oxygen species (ROS), leading to oxidative stress. Consequently, tolerance of heavy metal is often gauged by the efficacy of the defense system in plants against oxidative stress. Enhancing antioxidant activity has emerged as a prevalent approach to augment heavy metal tolerance (Wiszniewska et al., 2018). Achieving this can involve the overexpression of genes responsible for antioxidative machinery. Similarly, to promote the accumulation of heavy metals by genetic engineering techniques, a commonly used strategy involves the insertion and overexpression of genes linked to heavy metal absorption, sequestration, and translocation (Mani and Kumar, 2014; Das et al., 2023). For instance, genes producing metalloid/metal transporters like MTP, HMA, MATE, and ZIP family members are prime candidates for transfer and overexpression in target plants, as previously discussed. As metal chelators enhance heavy metal bioavailability, root-to-shoot translocation, uptake, and intracellular sequestration in organelles, stimulating metal chelator production through genetic engineering presents a promising avenue to amplify the accumulation of heavy metals. This can involve the overexpression of genes producing natural chelators, leading to improved absorption and translocation of heavy metals (Wu et al., 2010).

While genetic engineering techniques hold substantial promise for enhancing plant performance in heavy metal phytoremediation, certain challenges persist. The intricate and multifaceted nature of detoxification and accumulation mechanisms involving numerous genes makes manipulating those genes to achieve desired traits time-consuming and often unsuccessful. Moreover, gaining approval for field testing of genetically modified plants remains challenging in some regions due to concerns about food and ecosystem safety. Consequently, in plants, alternate approaches are needed to enhance their performance in phytoextraction when genetic engineering is not feasible.

Gene Editing Approaches

The gene editing approach involves introducing mutations into the targeted DNA sequence utilizing site-specific nucleases (SSNs). This method enables the addition, deletion, or modification of specific bases at predetermined locations, resulting in a double-stranded break or a nick. SSNs are divided into four groups that are used in gene editing: clustered regularly interspaced palindromic repeats (CRISPR) Cas9, transcription activator-like effector nucleases (TALENs), zinc finger nucleases (ZFNs), and meganucleases (Voytas and Gao, 2014; Foht, 2016). Meganucleases are single proteins that recognize a specific sequence of DNA with at least 12 nucleotides and cause a nick at the intended spot, making them the earliest SSNs discovered in bacteria and eukaryotes (Silva et al., 2011). This method has been applied to nick the genomes of *Nicotiana* sp. and *Zea mays* (Baltes and Voytas, 2015). ZFNs, on the other hand, are proteins that attach to a particular sequence of DNA and are present in many different plants. In some cases, they act as transcription factors (Urnov et al., 2010). TALENs, initially produced by bacteria, attach to plant gene promoters and reduce resistance to pathogens in plants. These bacterial-derived TALENs are encoded by specific codes, which are then engineered into proteins with designated target locations on the gene of interest (Moscou and Bogdanove, 2009). TALENs have been used in gene editing for plants like rice, maize, soybean, and wheat (Baltes and Voytas, 2015). CRISPR, a relatively recent gene editing approach, was identified in bacterial systems. Bacteria use CRISPR as a natural defense mechanism against viral stress and plasmids to control the movement of nucleases targeting foreign DNA sequences (Sander and Joung, 2014; Ledford, 2016).

Several iron and zinc regulated transporter (IRT and ZIP) proteins have been found to be upregulated in *Thlaspi caerulescens* and *Anemone halleri*, imparting tolerance to the toxicity of zinc and cadmium (Kramer, 2005). These transporters are characterized by the presence of zinc finger-like domains. Furthermore, the overexpression of ZAT, a zinc transporter in *Arabidopsis thaliana*, was found to increase the resistance of the plant *Arabidopsis* to Zn toxicity by twofold. In another study by Chen et al. (2012), the overexpression of zinc finger transcription factor 6 (ZAT6) in an *Arabidopsis thaliana* cadmium-tolerant mutant was reported. The upregulation of ZAT6 was found to be closely linked with glutathione concentrations and phytochelatin synthesis. Reduced expression of ZAT6 was associated with heightened transgenic *Arabidopsis* plants' susceptibility. Additionally, a novel *O. sativa* (indica) line was generated using gene editing through the CRISPR-Cas9 mechanism to knock out the metal transporter gene OsNramps, resulting in reduced Cd accumulation without the use of transgenes (Tang et al., 2017).

Researchers can now manipulate the expression of a gene at specific loci by the technique of genome editing, providing rapid insights into the plant's functional genomics (Wolt et al., 2016) (Figure 4.2). A significant advancement in this field is seen with CRISPR-assisted phytoremediation, which has improved the efficiency of the technique (Kour et al., 2019). CRISPR, a groundbreaking genome editing technique, has become pivotal for enhancing specific traits in plants and is a cornerstone of functional genomics. In the context of phytoremediation, Cas proteins and CRISPR combine to the form CRISPR-Cas system, a practical tool for generating advanced plant phenotypes (Bortesi and Fischer, 2015). This method seamlessly integrates with the plant genome and is particularly well-suited for GC-rich sequences, making it applicable to monocotyledonous crops. CRISPR technology has overcome some of the most challenging hurdles in plant modification, enabling transformations that were previously elusive. It enables precise DNA modifications, facilitating alterations in plant gene sequences to mitigate the adverse effects of heavy metals and other soil contaminants. This approach also leverages plant-microbe interactions to enhance nutrient uptake and increase the bioavailability of metals in plants (Abhilash et al., 2012). CRISPR-based phytoremediation has emerged as a potent approach for soil purification, effectively eliminating hazardous compounds derived from the rhizosphere. This allows the rejuvenated soil to be repurposed for use in agriculture, along with other ecologically beneficial activities such as forest preservation and agroforestry (Thijs et al., 2016). To achieve successful phytoremediation, it is imperative to pinpoint key genes within the intricate interactions between the plant and the microbe signaling network and to gain a comprehensive understanding of their functions. Gene editing technologies, for example CRISPR, offer the potential to edit these target genes, thereby enhancing the phytoremediation capabilities of plants across diverse environments (Basu et al., 2018b).

An illustrative instance involves the development of transgenic canola (*Brassica napus* L.) from its wild-variety counterpart using the gene OsMyb4 from rice, which codes the *A. thaliana* COR15a stress regulator (Raldugina et al., 2018). Remarkably, in the transgenic plant, spring canola cultivated under elevated copper levels (150 μM $CuSO_4$) and zinc (5000 μM $ZnSO_4$) exhibited prolonged survival (more than 15 days) compared to natural varieties. This underscores the pivotal role of the gene OsMyb4 in enhancing the process of survival that counteracts the effects of substantial copper and zinc exposure. Consequently, Raldugina et al. (2018) postulated that OsMyb4 might serve as a favorable regulator of the proline synthesis and phenylpropanoid pathway, thus holding promise for phytoremediation.

Boudet (2007) elucidated that noncellular polysaccharides, which fortify the structural integrity of plant cell walls, synergize with lignin to constitute the primary defense barrier against both biotic and abiotic stress. Subsequent research by Zhang et al. (2014) revealed that augmenting the expression of caffeoyl coenzyme A-3-O-methyltransferase (CCoAOMT) heightened the biosynthesis of lignin. Notably, under cadmium stress conditions in *Vicia sativa*, an increase in the production of lignin was observed. This insight served as the foundation for the development of a transgenic *Arabidopsis thaliana*. The Vs-CCoAOMT gene homolog obtained from *Vicia sativa* was cloned and transferred, aiming to bolster lignin production for potential phytoremediation applications (Xia et al., 2018). In this context, the Vs-CCoAOMT gene's products exhibited crucial roles in enhancing Cd stress tolerance within transgenic plants. The outcomes encompassed amplified biomass generation, improved Cd loading and transport mechanisms, and notably elevated Cd accumulation within cell wall structures. To enhance Cd resistance in transgenic *Nicotiana tabacum* (tobacco), Wang et al. (2018) harnessed the potential of heavy metal ATPases. These ATPases, categorized as P1B-type ATPase proteins, orchestrate the conversion of Cd between plant root and shoot (Takahashi et al., 2012). Such ATPases have been identified in various plant species, including *Hordeum vulgare* L. (Williams and Mills, 2005), *Thlaspi caerulescens* L., *Populus tomentosa* Carr. (Wang et al., 2018), and *O. sativa* L. (Takahashi et al., 2012). The strategy involved creating degenerate primer pairs derived from the *Populus trichocarpa* genome to facilitate the screening and cloning of heavy metal ATPase cDNA. The identified PtoHMA5, a heavy metal ATPase gene, was subsequently cloned and introduced into the tobacco plant (*Nicotiana tabacum*). Upon subjecting the transgenic tobacco to Cd stress, a substantial 25.05% increase in the accumulation of Cd within the leaves was observed. This study highlighted the potential utility of PtoHMA5 and PtPCs as valuable genetic resources for phytoremediation endeavors.

In the realm of environmental contamination, certain essential microelements like iron, copper, manganese, nickel, molybdenum, and zinc, indispensable for optimal growth and development, can result in toxicity when present in excessive amounts through water and soilborne routes (Kashyap et al., 2023). Each of these crucial components plays a different and pivotal role in the life cycle of a plant, necessitating their adequate provision to ensure proper developmental trajectories. The escalating accumulation of both toxic and nontoxic contaminants across several environmental strata like soil, air, and groundwater has emerged as a significant contemporary global concern (Padhye et al., 2023).

CRISPR-CAS APPROACH TO TACKLE TOXIC METALS AND METALLOIDS BY PLANTS

CRISPR technologies have emerged as a potent tool for genome editing in phytoremediation plants, holding substantial promise for improving the capacity for restoration of plants exposed to contaminants (Basharat et al., 2018). The integration of genome sequencing into the alteration process significantly enriches the investigative phase of phytoremediation. This technique facilitates the identification and classification of genetic components pivotal in various phytoremediation mechanisms, including phytostabilization, phytovolatilization, phytodesalination, and phytodegradation (Estrela and Cate, 2016). By deploying CRISPR techniques tailored to the processes of contaminant accumulation, volatilization, and degradation, targeted phytoremediation strategies can be engineered. Capitalizing on the insights from the sequencing data of genetically modified plants, CRISPR technology enables the direct insertion of precise instructions into the plant genome, fostering highly programmable genetic manipulation (Basu et al., 2018c). In contrast to earlier methods like ZFNs and TALENs, CRISPR offers a more straightforward and streamlined approach for delivering programmable instructions. This technology's adaptability has substantially broadened its applicability and facilitated the genetic engineering of an increased range of plant species. This expansion owes much to factors such as the accessibility of plant genome sequences and the availability of bioinformatics techniques and practices, such as flux balance analysis for evaluating metabolic pathways and omics technologies for assessing the integration of the genome (Flexas and Gago, 2018). Central to the CRISPR-mediated phytoremediation strategy is the modulation of gene expression to augment metal ligand synthesis (like phytochelatins and metallothioneins), metal transport proteins from ZIP, CDF, and HMA, families as well as plant growth regulators such as gibberellic acid. The potential of phytoremediation is further enriched through the transfer of genetic materials, as substantiated by the studies carried out since 2000. These studies encompass the transfer of particular bacterial and plant genes into target plants, demonstrating favorable outcomes for phytoremediation techniques (Pandey and Singh, 2019).

Arabidopsis and tobacco plants enhanced by the NAS1 gene exhibit significantly elevated tolerance of metals like Cd, Ni, Cu, Zn, and Fe, accompanied by heightened uptake of metals like manganese and nickel (Sebastian et al., 2019). Overexpression of genes that code for metallothionein (like MTA1, MT1, and MT2) led to an augmented ability in tobacco and *Arabidopsis* plants to tolerate and accumulate metals like Cu, Zn, and Cd (Sebastian et al., 2019). When MT2b (metallothionein gene) is active, *H. incana* exhibits a greater tolerance and the capacity to accumulate high levels of Pb (Huang et al., 2019). Introduction of APS and SMT genes into *Brassica juncea* conferred high selenium tolerance due to the fact that these genes produce selenocysteine methyltransferase and ATP sulfurylase. CRISPR technology offers the potential to significantly enhance these genes, elevating their effectiveness. Huang et al. (2019) explored these genes in metal-polluted soil reclamation through transgenically modified plants, with the potential for enhancement using CRISPR. Transfer of specific genes and their origins was investigated, resulting in outcomes ranging from increased resistance to toxic metal concentrations to augmented metal absorption capacity, and even instances of metal hyperaccumulation.

However, the introduction of specific genes that allow a certain metal to accumulate in the target plant might occasionally result in hypersensitivity to that element, causing plant deterioration. Overexpression of NtCBP4 (plasma membrane protein) in *Nicotiana tabacum* improved the capacity of Pb to accumulate while simultaneously heightening Pb sensitivity. Similarly, the MerC gene's expression in tobacco and *Arabidopsis* led to heightened accumulation of Hg metal, alongside hypersensitivity to mercury (Fasani et al., 2018). This method is applicable to various organic pollutants, including polycyclic aromatic hydrocarbons and polychlorinated biphenyls (Banerjee and Roychoudhury, 2019). It can also be applied to explosives like hexahydro-1,3,5-trinitro-1,3,5-triazine (RDX) and 2,4,6-trinitrotoluene (TNT). Jaiswal et al. (2019) laid the groundwork for possible CRISPR-based augmentation of enzyme systems in plants that removes and detoxifies organic pollutants by identifying genes in plants responsible for the detoxification of organic xenobiotics. Pollutants such as polychlorinated biphenyls and polycyclic aromatic hydrocarbons are suitable candidates for these methods. Thus, CRISPR-mediated advancement should be aimed toward improving enzyme systems in plants that remove pollutants (Pandey and Singh, 2019). CRISPR technology has also contributed to the advancement of growth-promoting rhizobacteria. Manipulating plant interactions with rhizobacteria, phytohormone levels, and nitrogen fixation capability underscores the significant role played by the interaction of plants and microbes, or the microbiome, in conferring resilience to adverse conditions. Altering phosphate ion solubility, alongside various direct and indirect approaches, offers the potential for upgrading facilities aimed at pollutant reclamation (Chinnaswamy et al., 2018). While CRISPR-Cas9 stands out as an exceptional genome editing tool, its outcomes in plants exhibit variability. Numerous factors, encompassing plant-specific target sites, delivery mechanisms, and plants' inherent genetic makeup, collectively influence the success response rate. Despite its promising potential in enhancing phytoremediation, the findings remain primarily confined to laboratory settings. Future strategies might use innovative approaches such as direct transfection of Cas9 into plant protoplasts using gRNAs, T-DNA-mediated gRNA-Cas9 delivery, and single-cell plant retrieval, presenting a futuristic path for phytoremediation enhancement (Morio et al., 2020). Although T-DNA-mediated gRNA-Cas9 has shown potential, its activity is limited to somatic tissues because of induction constraints.

CRISPR-Cas9 has huge potential for gRNA-guided gene regulation expression without the need for cloning. The Cas9 protein's ability to attach to transcriptional factors provides a means to control the activity of transcription by altering the transcriptional factor itself (Manna et al., 2021). This regulatory approach can operate effectively over a wide range, spanning up to a thousandfold. Recent strides have demonstrated that in rice, the CRISPR technique is capable of manipulating the metal transporter gene OsNramp5, opening up new horizons for employing CRISPR to regulate the transportation of metals (Peng et al., 2017). CRISPR-Cas9 has significantly developed the genomes of important plants such as poplar and maize, which were once considered unsuitable for phytoremediation due to their complex genome. Despite the intricate genome of maize being long considered challenging for CRISPR-Cas9 alteration, its high ploidy makes it an attractive candidate for addressing metal accumulation. Given its substantial root structure, capable of penetrating deep into the soil, maize holds the potential to effectively remediate a range of contaminants (Hussain et al., 2023). In a similar vein, other crop plants, such as those with extensive root systems, have also emerged as favored choices for phytoremediation (Gavrilescu, 2022).

CONCLUSION AND FUTURE OUTCOMES

In conclusion, the application of CRISPR-Cas technology into the field of phytoremediation holds immense promise for addressing the challenges posed by heavy metal and metalloid contamination in our environment. This chapter has underscored the significant potential of targeted gene modifications to enhance the metal tolerance, uptake, and accumulation capabilities of plants. By focusing on key candidate genes involved in metal transport, chelation, and antioxidative processes, researchers can develop plants that are better equipped to thrive in polluted environments while detoxifying and accumulating toxic substances effectively.

In sum, the integration of CRISPR-Cas technology to phytoremediation can revolutionize our approach to environmental cleanup. With continued research, responsible implementation, and collaborative efforts between scientists, policymakers, and the public, we can envision a future where contaminated sites are restored to health using genetically modified plants, contributing to a cleaner and more sustainable environment.

REFERENCES

Abhilash, P.C., et al. (2012). Plant–microbe interactions: Novel applications for exploitation in multipurpose remediation technologies. Trends in Biotechnology, 30(8), 416–420.

Agarwal, P., Rani, R. Strategic management of contaminated water bodies: Omics, genome-editing and other recent advances in phytoremediation. Environmental Technology & Innovation. 2022 Mar 9:102463.

Asemaneh, T., Ghaderian, S.M., Crawford, S.A., Marshall, A.T., Baker, A.J.M. (2006). Cellular and subcellular compartmentation of Ni in the Eurasian serpentine plants Alyssum bracteatum, Alyssum murale (Brassicaceae) and Cleome heratensis (Capparaceae). Planta 225, 193–202.

Babu, S.O., Hossain, M.B., Rahman, M.S., Rahman, M., Ahmed, A.S., Hasan, M.M., Rakib, A., Emran, T.B., Xiao, J., Simal-Gandara, J. (2021). Phytoremediation of toxic metals: A sustainable green solution for clean environment. Applied Sciences 11(21), 10348.

Baltes, N.J., Voytas, D.F. (2015). Enabling plant synthetic biology through genome engineering. Trends in Biotechnology 33(2), 120–31.

Banerjee, A., Roychoudhury, A. (2019). Genetic engineering in plants for enhancing arsenic tolerance. In: Transgenic Plant Technology for Remediation of Toxic Metals and Metalloids (pp. 463–475). Academic Press.

Basharat, A., et al. (2018). CRISPR interference for functional genomics in plants. Plant Cell Reports, 37(11), 1401–1411.

Basharat, Z., Novo, L., Yasmin, A. (2018). Genome editing weds CRISPR: What is in it for phytoremediation? Plants 7(3), 51.

Basu, S., et al. (2018a). An update on genetic engineering of plants for enhanced phytoremediation of heavy metals. World Journal of Microbiology and Biotechnology, 34(4), 54.

Basu, S., et al. (2018b). Emerging trends in the application of CRISPR-based gene editing in plant phytoremediation of toxic metals. Environmental Science and Pollution Research, 25(15), 14216–14227.

Basu, S., Rabara, R.C., Negi, S., Shukla, P. (2018c). Engineering PGPMOs through gene editing and systems biology: A solution for phytoremediation? Trends in Biotechnology 36(5), 499510.

Berken, A., Mulholland, M.M., LeDuc, D.L., Terry, N. (2002). Genetic engineering of plants to enhance selenium phytoremediation. Critical Reviews in Plant Sciences 21(6), 567–582.

Bhat, S.A., Bashir, O., Haq, S.A., Amin, T., Rafiq, A., Ali, M., Américo-Pinheiro, J.H., Sher, F. (2022). Phytoremediation of heavy metals in soil and water: An eco-friendly, sustainable and multidisciplinary approach. Chemosphere 303, 134788.

Bortesi, L., Fischer, R. (2015). The CRISPR/Cas9 system for plant genome editing and beyond. Biotechnology Advances 33(1), 41–52.

Boudet, A.M. (2007). Evolution and current status of research in phenolic compounds. Phytochemistry 68(22–24), 2722–2735.

Burklew, C.E., Ashlock, J., Winfrey, W.B., Zhang, B. (2012). Effects of aluminum oxide nanoparticles on the growth, development, and microRNA expression of tobacco (Nicotiana tabacum). PloS One 7(5), e34783.

Carolin, C.F., Kumar, P.S., Saravanan, A., Joshiba, G.J., Naushad, M. (2017). Efficient techniques for the removal of toxic heavy metals from aquatic environment: A review. Journal of Environmental Chemical Engineering 5, 2782–2799.

Chaudhary, K., Jan, S., Khan, S. (2015). Heavy metal ATPase (HMA2, HMA3, and HMA4) genes in hyperaccumulation mechanism of heavy metals. In: Plant Metal Interaction: Emerging Remediation Techniques (pp. 545–556). Elsevier: Amsterdam, The Netherlands.

Chen, F., Huber, C., May, R., Schröder, P. (2016). Metabolism of oxybenzone in a hairy root culture: Perspectives for phytoremediation of a widely used sunscreen agent. Journal of Hazardous Materials 306, 230–236.

Chen, Y., Li, X., Guo, J., Zhu, X., Yang, H., Ding, L. (2019). Comparative energy-saving analysis of different remediation technologies for crude oil-contaminated soil. Journal of Cleaner Production, 213, 49–59. https://doi.org/10.1016/j.jclepro.2018.12.270

Chen, Y.J., Lyngkjær, M.F., Collinge, D.B. (2012). Futureprospects for genetically engineering disease-resistantplants. In: Sessa, G. (Ed.), Molecular Plant Immunity.Wiley-Blackwell, Oxford, UK, 251-275.

Chen, Z., et al. (2013). Mn tolerance in rice is mediated by MTP8.1, a member of the cation diffusion facilitator family. Journal of Experimental Botany 64(14), 4375–4387.

Chinnaswamy, A., et al. (2018a). A nodule endophytic Bacillus megaterium strain isolated from Medicago polymorpha enhances growth, promotes nodulation by Ensifer medicae and alleviates salt stress in alfalfa plants. Annals of Applied Biology 172 (3), 295–308.

Chirakkara, R.A., Cameselle, C., Reddy, K.R. (2016). Assessing the applicability of phytoremediation of soils with mixed organic and heavy metal contaminants. Reviews in Environmental Science and Biotechnology 15, 299–326.

Choppala, G., Saifullah, Bolan, N., Bibi, S., Iqbal, M., Rengel, Z., Kunhikrishnan, A., Ashwath, N., Ok, Y.S. (2014). Cellular mechanisms in higher plants governing tolerance to cadmium toxicity. Critical Review in Plant Sciences 33, 374–391.

Clemens, S., Eroglu, S., Grillet, L., Nozoye, T. (2021). Metal transport in plants. Frontiers in Plant Science 12, 644960.

Clemens, S., Ma, J.F. (2016). Toxic heavy metal and metalloid accumulation in crop plants and foods. Annual Review of Plant Biology 67, 489–512.

Collard, B.C., Jahufer, M.Z.Z., Brouwer, J.B., Pang, E.C.K. (2005). An introduction to markers, quantitative trait loci (QTL) mapping and marker-assisted selection for crop improvement: The basic concepts. Euphytica 142, 169–196.

Cosa, B.D., Moar, W., Lee, S.B., Miller, M., Daniell, H. (2001). Overexpression of the Bt cry2Aa2 operon in chloroplasts leads to formation of insecticidal crystals. Nature Biotechnology 19(1), 71–74.

Das, S.K., Ghosh, G.K., Avasthe, R. (2023). Biochar application for environmental management and toxic pollutant remediation. Biomass Conversion and Biorefinery 13(1), 555–566.

Dhankher, O.P., Doty, S.L., Meagher, R.B., Pilon-Smits, E. (2012). Biotechnological approaches for phytoremediation. In: Altman, A., Hasegawa, P.M. (Eds.), Plant Biotechnology and Agriculture (pp. 309–328). Academic Press: Oxford, UK.

Devi, R., Chauhan, S. and Dhillon, T.S. (2022) Genome editing for vegetable crop improvement: Challenges and future prospects. Frontiers in Genetics, 13, 1037091.

Ehrnstorfer, I.A., Geertsma, E.R., Pardon, E., Steyaert, J., Dutzler, R. (2014). Crystal structure of a SLC11 (NRAMP) transporter reveals the basis for transitionmetal ion transport. Nature Structural & Molecular Biology 21(11), 990–996.

Estrela, R., Cate, J.H.D. (2016). Phytoremediation and phytomining: Using plants to remediate contaminated or mineralized environments. Elements 12(5), 325–331.

Estrela, R., Cate, J.H.D. (2016). Energy biotechnology in the CRISPR-Cas9 era. Current Opinion in Biotechnology 38, 79–84.

Fan, X., Liu, C., Wang, D., Li, D., Wang, Z. (2018). Phytoremediation of heavy metal-contaminated soils: Natural plants versus transgenic plants. International Journal of Environmental Research and Public Health 15(12), 2792.

Fasani, E., Manara, A., Martini, F., Furini, A., DalCorso, G. (2018). The potential of genetic engineering of plants for the remediation of soils contaminated with heavy metals. Plant, Cell & Environment 41(5), 1201–1232.

Favas, P.J.C., Pratas, J., Rodrigues, N., D'Souza, R., Varun, M., Paul, M.S. (2018). Metal (loid) accumulation in aquatic plants of a mining area: Potential for water quality bio-monitoring and biogeochemical prospecting. Chemosphere 194, 158–170.

Fire, A., Xu, S., Montgomery, M.K., Kostas, S.A., Driver, S.E., Mello, C.C. (1998). Potent and specific genetic interference by double-stranded RNA in Caenorhabditis elegans. Nature 391(6669), 806–811.

Flexas, J., Gago, J. (2018). Integrated processes determining photosynthetic performance under abiotic stress. Journal of Experimental Botany 69(13), 3107–3117.

Flexas, J., Gago, J. (2018). A role for ecophysiology in the 'omics' era. Plant Journal 96(2), 251–259.

Foht, H. (2016). Site-directed mutagenesis by Gene editing: A new approach towards improving plant stress tolerance. Frontiers in Plant Science 7, 1425.

Gavrilescu, M. (2022). Enhancing phytoremediation of soils polluted with heavy metals. Current Opinion in Biotechnology 74, 21–31.

Gavrilescu, M., Demnerová, K., Aamand, J. (2015). Emerging pollutants in the environment: Present and future challenges in biomonitoring, ecological risks and bioremediation. New Biotechnology 32(1), 147–156.

Grispen, V.M., Hakvoort, H.W., Bliek, T., Verkleij, J.A., Schat, H. (2011). Combined expression of the Arabidopsis metallothionein MT2b and the heavy metal transporting ATPase HMA4 enhances cadmium tolerance and the root to shoot translocation of cadmium and zinc in tobacco. Environmental and Experimental Botany 72(1), 71–76.

Gu, D., Zhou, X., Ma, Y., Xu, E., Yu, Y., Liu, Y., Chen, X., Zhang, W. (2021). Expression of a Brassica napus metal transport protein (BnMTP3) in Arabidopsis thaliana confers tolerance to Zn and Mn. Plant Science 304, 110754.

Guo, J., Green, B.R., Maldonado, M.T. (2015). Sequence analysis and gene expression of potential components of copper transport and homeostasis in Thalassiosira pseudonana. Protist 166(1), 58–77.

Hamim, I., Sekine, K.T., Komatsu, K. (2022). How do emerging long-read sequencing technologies function in transforming the plant pathology research landscape? Plant Molecular Biology 110(6), 469–484.

Haroldsen, V.M., Szczerba, M.W., Aktas, H., Lopez-Baltazar, J., Odias, M.J., Chi-Ham, C.L., Labavitch, J.M., Bennett, A.B., Powell, A.L. (2012). Mobility of transgenic nucleic acids and proteins within grafted rootstocks for agricultural improvement. Frontiers in Plant Science, 3, 39.

Hassan, S., Sabreena, Khurshid, Z., Bhat, S.A., Kumar, V., Ameen, F., Ganai, B.A. (2022). Marine bacteria and omic approaches: A novel and potential repository for bioremediation assessment. Journal of Applied Microbiology 133(4), 2299–2313.

Hayes, B.J., Lewin, H.A., Goddard, M.E. (2013). The future of livestock breeding: Genomic selection for efficiency, reduced emissions intensity, and adaptation. Trends in Genetics 29(4), 206–214.

Hettiarachchi, E., Paul, S., Cadol, D., Frey, B., Rubasinghege, G. (2019) Mineralogy controlled dissolution of uranium from airborne dust in simulated lung fuids (SLFs) and possible health implications. Environmental Science & Technology Letters 6(2), 62–67. https://doi.org/10.1021/acs.estlett.8b00557

Hu, Y., Lu, L., Tian, S., Li, S., Liu, X., Gao, X., Zhou, W., Lin, X. (2019). Cadmium-induced nitric oxide burst enhances Cd tolerance at early stage in roots of a hyperaccumulator Sedum alfredii partially by altering glutathione metabolism. Science of the Total Environment 650, 2761–2770.

Huang, X., Sang, T., Zhao, Q., Feng, Q., Zhao, Y., Li, C. (2010). Genome-wide association studies of 14agronomic traits in rice landraces. Nature Genetics 42(11), 961–967.

Huang, Y., et al. (2019). Effects of lead and cadmium on photosynthesis in Amaranthus spinosus and assessment of phytoremediation potential. International Journal of Phytoremediation 21, 1041–1049.

Hussein, H.S., Ruiz, O.N., Terry, N., Daniell, H. (2007). Phytoremediation of mercury and organomercurials in chloroplast transgenic plants: Enhanced root uptake, translocation to shoots, and volatilization. Environmental Science & Technology 41(24),8439–8446.

Hussain, I., Afzal, S., Ashraf, M.A., Rasheed, R., Saleem, M.H., Alatawi, A., Ameen, F., Fahad, S. (2023). Effect of metals or trace elements on wheat growth and its remediation in contaminated soil. Journal of Plant Growth Regulation 42(4), 2258–2282.

Jiang, C., Mithani, A., Gan, X., Belfield, E.J., Klingler, J.P., Zhu, J.K., et al. (2011). Regenerant Arabidopsis lineages display a distinct genome-wide spectrum of mutations conferring variant phenotypes. Current Biology 21(16), 1385–1390.

Jaiswal, S., Singh, D.K., Shukla, P., (2019). Gene editing and systems biology tools for pesticide bioremediation: A review. Frontiers in Microbiology 10, 87.

Jin, S., Daniell, H. (2015). The engineered chloroplast genome just got smarter. Trends in Plant Science 20(10), 622–640.

Jena, K.K., Mackill, D.J. (2008). Molecular markers and their use in marker-assisted selection in rice. Crop Science 48(4), 1266–1276.

Karmakar, S., Das, P., Panda, D., Xie, K., Baig, M.J., Molla, K.A. (2022). A detailed landscape of CRISPR-Cas-mediated plant disease and pest management. Plant Science 323, 111376.

Kashyap, S., Sharma, I., Dowarah, B., Barman, R., Gill, S.S., Agarwala, N. (2023). Plant and soil-associated microbiome dynamics determine the fate of bacterial wilt pathogen Ralstonia solanacearum. Planta 258(3), 57.

Kaur, H., Kohli, S.K., Khanna, K., Dhiman, S., Kour, J., Bhardwaj, T., Bhardwaj, R. (2022). Deciphering the role of metal binding proteins and metal transporters for remediation of toxic metals in plants. In Bioremediation of Toxic Metal (loid). CRC Press Boca Raton, United States, 257–272.

Kaur, R., Yadav, P., Kohli, S.K., Kumar, V., Bakshi, P., Mir B.A., Thukral A.K., Bhardwaj, R. (2019). Emerging trends and tools in transgenic plant technology for phytoremediation of toxic metals and metalloids. Transgenic plant technology for remediation of toxic metals and metalloids. 1, 63–88.

Khalid, S., Shahid, M., Niazi, N.K., Murtaza, B., Bibi, I., Dumat, C., Tripathi, R.D. (2020). Copper phytoextraction by Alternanthera bettzickiana: Physiological and molecular responses. Journal of Hazardous Materials 392, 122480.

Kour, G., et al. (2019). Recent progress in the CRISPR-Cas9 technology for genome editing applications. Journal of Applied Genetics, 60(3–4), 237–251.

Kramer, M.G., Redenbaugh, K. (1994). Commercialization of a tomato with an antisense polygalacturonase gene: The FLAVR SAVR™ tomato story. Euphytica 79, 293–297.

Kramer, U. (2005). Phytoremediation: Novel approaches to cleaning up polluted soils. Current Opinion in Biotechnology 16(2), 133–141.

Lal, R. (2004). Soil carbon sequestration to mitigate climate change. Geoderma, 123(1–2), 1–22. https://doi.org/10.1016/j.geoderma.2004.01.032

Lateef, A., Azeez, M.A., Asafa, T.B., Yekeen, T.A., Akinboro, A., Oladipo, I.C., Ajetomobi, F.E., Gueguim-Kana, E.B., Beukes, L.S. (2015). Cola nitida-mediated biogenic synthesis of silver nanoparticles using seed and seed shell extracts and evaluation of antibacterial activities. BioNanoScience 5, 196–205.

Ledford, H. (2016). CRISPR, the disruptor. Nature 531(7593), 276–279.

Li, C., Mo, Y., Wang, N., Xing, L., Qu, Y., Chen, Y., Yuan, Z., Ali, A., Qi, J., Fernández, V., Wang, Y. (2023). The overlooked functions of trichomes: Water absorption and metal detoxication. Plant, Cell & Environment 46(3), 669–687.

Liao, Q., Jian, S.F., Song, H.X., Guan, C.Y., Lepo, J.E., Ismail, A.M., Zhang, Z.H. (2019). Balance between nitrogen use efficiency and cadmium tolerance in Brassica napus and Arabidopsis thaliana. Plant Science 284, 57–66.

Liu, L., Huang, Z., Pan, F., Liu, Y., Zhang, X., Li, T. (2017). Effects of phytoremediation on soil erosion in a heavy metal-contaminated area. Environmental Science and Pollution Research, 24(31), 24479–24489.

Luo, Z.B., He, J., Polle, A., Rennenberg, H. (2016). Heavy metal accumulation and signal transduction in herbaceous and woody plants: Paving the way for enhancing phytoremediation efficiency. Biotechnology Advances 34(6), 1131–1148.

Lyyra, S., Meagher, R.B., Kim, T., Heaton, A., Montello, P., Balish, R.S., Merkle, S.A. (2007). Coupling two mercury resistance genes in Eastern cottonwood enhances the processing of organomercury. Plant Biotechnology Journal 5(2), 254–262.

Ma, Y., Wei, N., Wang, Q., Liu, Z., Liu, W. (2021). Genome-wide identification and characterization of the heavy metal ATPase (HMA) gene family in Medicago truncatula under copper stress. International Journal of Biological Macromolecules 193, 893–902.

Maliga, P. (2003). Progress towards commercialization of plastid transformation technology. Trends in Biotechnology 21(1), 20–28.

Mani, D., Kumar, C. (2014). Biotechnological advances in bioremediation of heavy metals contaminated ecosystems: An overview with special reference to phytoremediation. International Journal of Environmental Science and Technology 11, 843–872.

Manna, M., Thakur, T., Chirom, O., Mandlik, R., Deshmukh, R., Salvi, P. (2021). Transcription factors as key molecular target to strengthen the drought stress tolerance in plants. Physiologia Plantarum 172(2), 847–868.

Marques, A.P., Rangel, A.O., Castro, P.M. (2009). Remediation of heavy metal contaminated soils: Phytoremediation as a potentially promising clean-up technology. Critical Reviews in Environmental Science and Technology 39(8), 622–654.

Maurya, A., Sharma, D., Partap, M., Kumar, R., Bhargava, B. (2023). Microbially-assisted phytoremediation toward air pollutants: Current trends and future directions. Environmental Technology & Innovation 11, 103140.

Maurya, A.K., Sinha, D., Mukherjee, S. (2022). Plant response to heavy metals (at the cellular level). In: Heavy Metals in Plants (pp. 125–148). CRC Press.

Mengoni, A., Gonnelli, C., Galardi, F., Gabbrielli, R., Bazzicalupo, M. (2000). Genetic diversity and heavy metal tolerance in populations of Silene paradoxa L.(Caryophyllaceae): A random amplified polymorphic DNA analysis. Molecular Ecology 9(9), 1319–1324.

Milner, M.J., Mitani-Ueno, N., Yamaji, N., Yokosho, K., Craft, E., Fei, Z., Ebbs, S., Zambrano, M.C., Ma, J.F., Kochian, L.V. (2014). Root and shoot transcriptome analysis of two ecotypes of Noccaea caerulescens uncovers the role of NcNramp1 in Cd hyperaccumulation. Plant Journal 78, 398–410.

Morio, F., Lombardi, L., Butler, G. (2020). The CRISPR toolbox in medical mycology: State of the art and perspectives. PLoS Pathogens 16(1), e1008201.

Moscou, M.J., Bogdanove, A.J. (2009). A simple cipher governs DNA recognition by TAL effectors. Science 326(5959), 1501.

Muske, D.N., Gahukar, S.J., Akhare, A.A., Deshmukh, S.S. (2016). Phytoremediation: An environmentally sound technology for pollution prevention, control and remediation. Advances in Life Sciences 5(7), 2501–2509.

Napoli, C., Lemieux, C., Jorgensen, R. (1990). Introduction of a chimeric chalcone synthase gene into petunia results in reversible co-suppression of homologous genes in trans. The Plant Cell 2(4), 279–289.

Naz, M., Benavides-Mendoza, A., Tariq, M., Zhou, J., Wang, J., Qi, S., Dai, Z., Du, D. (2022). CRISPR/Cas9 technology as an innovative approach to enhancing the phytoremediation: Concepts and implications. Journal of Environmental Management 323, 116296.

Nganso, B.T., Pines, G., Soroker, V. (2022). Insights into gene manipulation techniques for Acari functional genomics. Insect Biochemistry and Molecular Biology 143, 103705.

Oomen, R.J., et al. (2009). Functional characterization of NRAMP3 and NRAMP4 from the metal hyperaccumulator Thlaspi caerulescens. New Phytology 181, 637–650.

Padhye, L.P., Srivastava, P., Jasemizad, T., Bolan, S., Hou, D., Sabry, S., Rinklebe, J., O'Connor, D., Lamb, D., Wang, H., Siddique, K.H. (2023). Contaminant containment for sustainable remediation of persistent contaminants in soil and groundwater. Journal of Hazardous Materials 6, 131575.

Pandey, S., Singh, S. (2019). Transgenic approach to phytoremediation: Recent advances and applications. Environmental Science and Pollution Research 26(22), 22515–22527.

Peng, J.S., Wang, Y.J., Ding, G., Ma, H.L., Zhang, Y.J., Gong, J.M. (2017). A pivotal role of cell wall in cadmium accumulation in the Crassulaceae hyperaccumulator Sedum plumbizincicola. Molecular Plant 10(5), 771–774.

Pilon-Smits, E.A. (2005). Phytoremediation. Annual Review of Plant Biology 56, 15–39.

Pilon-Smits, E.A.H., Quinn, C.F., Leduc, D.L., Terry, N. (2019). Phytoremediation: From timbre to forest. Trends in Plant Science 24(6), 497–507.

Poty, S., Francesconi, L.C., McDevitt, M.R., Morris, M.J., Lewis, J.S. (2018) α-Emitters for radiotherapy: From basic radiochemistry to clinical studies-part 2. Journal of Nuclear Medicine 59(7), 1020–1027. https://doi.org/10.2967/jnumed.117.204651

Raldugina, G.N., et al. (2018). Cloning of the OsMyb4 gene from rice cultivar Japonica and its expression analysis under the influence of copper and zinc ions. Journal of Genetics, 97(2), 429–441.

Rascio, N., Navari-Izzo, F. (2011). Heavy metal hyperaccumulating plants: How and why do they do it? And what makes them so interesting? Plant Science 180, 169–181.

Ruane, J., Sonnino, A. (2007). Marker-assisted selection as a tool for genetic improvement of crops, livestock, forestry and fish in developing countries: An overview of the issues. Marker-assisted Selection-Current Status and Future Perspectives in Crops, Livestock, Forestry and Fish (pp. 3–13). Food and Agriculture Organization of the United Nations (FAO): Rome.

Ruiz, J.F., Gómez-González, B., Aguilera, A. (2011). AID induces double-strand breaks at immunoglobulin switch regions and c-MYC causing chromosomal translocations in yeast THO mutants. PLoS Genetics 7(2), e1002009.

Ruiz-Lozano, J.M. (2003). Arbuscular mycorrhizal symbiosis and alleviation of osmotic stress. New perspectives for molecular studies. Mycorrhiza 13, 309–317.

Saleem, M.H., Ali, S., Rehman, M., Hasanuzzaman, M., Rizwan, M., Irshad, S., Shafiq, F., Iqbal, M., Alharbi, B.M., Alnusaire, T.S., Qari, S.H. (2020) Jute: A potential candidate for phytoremediation of metals—A review. Plants 9, 258. https://doi.org/10.3390/plants9020258

Sander, J.D., Joung, J.K. (2014). CRISPR-Cas systems for editing, regulating and targeting genomes. Nature Biotechnology 32(4), 347–355.

Sarwar, N., et al. (2017). Phytoremediation strategies for soils contaminated with heavy metals: Modifications and future perspectives. Chemosphere 171, 710–721. https://doi.org/10.1016/j.chemosphere.2016.12.116

Sebastian, A., Shukla, P., Nangia, A.K., Prasad, M.N.V. (2019). Transgenics in phytoremediation of metals and metalloids: From laboratory to field. In: Transgenic Plant Technology for Remediation of Toxic Metals and Metalloids (pp. 3–22). Academic Press.

Shin, L.J., Lo, J.C., Yeh, K.C. (2012). Copper chaperone antioxidant protein1 is essential for copper homeostasis. Plant Physiology 159, 1099–1110.

Silva, G.H., et al. (2011). The FNS2 proteins are required for meiosis in the protist Paramecium tetraurelia and have secondary functions in mitosis in conjunction with calmodulin. Eukaryotic Cell 10(3), 369–377.

Singh, N.P., Santal, A.R. (2015). Phytoremediation of heavy metals: The use of green approaches to clean the environment. In: Phytoremediation (pp. 115–129). Springer: Cham.

Sinha, R.K., Heart, S., Tandon, P.K. (2004). Phytoremediation: Role of plants in contaminated site management. In: Book of Environmental Bioremediation Technologies (pp. 315–330). Springer: Berlin, Germany.

Skuza, L., Szućko-Kociuba, I., Bożek, F.E. (2022). I. Natural molecular mechanisms of plant hyperaccumulation and hypertolerance towards heavy metals. International Journal of Molecular Sciences 23(16), 9335.

Stroud, H., Greenberg, M.V., Feng, S., Bernatavichute, Y.V., Jacobsen, S.E. (2013). Comprehensive analysis of silencing mutants reveals complex regulation of the Arabidopsis methylome. Cell 152(1), 352–364.

Svab, Z., Hajdukiewicz, P., Maliga, P. (1990). Stable transformation of plastids in higher plants. Proceedings of the National Academy of Sciences 87(21), 8526–8530.

Takahashi, R., et al. (2012). The role of P1B-type ATPases in the Zn/Co/Pb detoxification and accumulation in Noccaea caerulescens (J & C. Presl) F.K. Meyer. Plant & Cell Physiology, 53(2), 310–325.

Tang, L., Mao, B., Li, Y., Lv, Q., Zhang, L., Chen, C., He, H., Wang, W., Zeng, X., Shao, Y., Pan, Y. (2017). Knockout of OsNramp5 using the CRISPR/Cas9 system produces low Cd-accumulating indica rice without compromising yield. Scientific Reports 7(1), 14438.

Tang, X., et al. (2017). Knockout of OsNramp5 using the CRISPR/Cas9 system produces low Cd-accumulating indica rice without compromising yield. Scientific Reports, 7(1), 14438.

Tarla, D.N., Erickson, L.E., Hettiarachchi, G.M., Amadi, S.I., Galkaduwa, M., Davis, L.C., Nurzhanova, A., Pidlisnyuk, V. (2020) Phytoremediation and bioremediation of pesticide-contaminated soil. Applied Science 10(4), 1217. https://doi.org/10.3390/app10041217

Thakur, M., Praveen, S., Divte, P.R., Mitra, R., Kumar, M., Gupta, C.K., Kalidindi, U., Bansal, R., Roy, S., Anand, A., Singh, B. (2022). Metal tolerance in plants: Molecular and physicochemical interface determines the "not so heavy effect" of heavy metals. Chemosphere 287, 131957.

Thijs, S., et al. (2016). A review of selenate removal from aqueous solutions. Environmental Pollution, 208, 336–346.

Thijs, S., Sillen, W., Rineau, F., Weyens, N., Vangronsveld, J. (2016). Towards an enhanced understanding of plantmicrobiome interactions to improve phytoremediation: Engineering the metaorganism. Frontiers in Microbiology 7, 341.

Turlapati, P.V., Kim, K.W., Davin, L.B., Lewis, N.G. (2011). The laccase multigene family in Arabidopsis thaliana: Towards addressing the mystery of their gene function (s). Planta 233, 439–470.

Urnov, F.D., Rebar, E.J., Holmes, M.C., Zhang, H.S., Gregory, P.D. (2010). Genome editing with engineered zinc finger nucleases. Nature Reviews Genetics 11(9), 636–646.

Valdés-López, O., Yang, S.S., Aparicio-Fabre, R., Graham, P.H., Reyes, J.L., Vance, C.P., Hernández, G. (2010). MicroRNA expression profile in common bean (Phaseolus vulgaris) under nutrient deficiency stresses and manganese toxicity. New Phytologist 187(3), 805–818.

Van De Mortel, J.E., Schat, H., Moerland, P.D., Van Themaat, E.V.L., Van Der Ent, S., Blankestijn, H., Ghandilyan, A., Tsiatsiani, S., Aarts, M.G.M. (2008). Expression differences for genes involved in lignin, glutathione and sulphate metabolism in response to cadmium in Arabidopsis thaliana and the related Zn/Cd-hyperaccumulator Thlaspi caerulescens. Plant, Cell & Environment 31, 301–324.

Vidal, C.F., Oliveira, J.A., da Silva, A.A., Ribeiro, C., Farnese, F.D.S. (2019). Phytoremediation of arsenite-contaminated environments: Is Pistia stratiotes L. a useful tool? Ecological Indicators 104, 794–801.

Voytas, D.F., Gao, C. (2014). Precision genome engineering and agriculture: Opportunities and regulatory challenges. PLoS Biology 12(6), e1001877.

Wang, F., et al. (2018). Functional analysis of the heavy metal ATPase PtoHMA5 from Populustomentosa Carr. in tobacco (Nicotianatabacum L.). BMC Plant Biology 18(1), 362.

Wei, W., et al. (2014). YCF1-mediated cadmium resistance in yeast is dependent on copper metabolism and antioxidant enzymes. Antioxidants & Redox Signaling 21(10), 1475–1489.

Williams, L.E., Mills, R.F. (2005). P1B-ATPases–an ancient family of transition metal pumps with diverse functions in plants. Trends in Plant Science 10(10), 491–502.

Wiszniewska, A., Kamińska, I., Koźmińska, A., Hanus-Fajerska, E. (2018). Aspects of co-tolerance towards salt and heavy metal stresses in halophytic plant species. Plant Nutrients and Abiotic Stress Tolerance 477–498.

Wolt, J.D., Wang, K., Yang, B. (2016). The regulatory status of genome-edited crops. Plant Biotechnology Journal 14(2), 510–518.

Wu, G., Kang, H., Zhang, X., Shao, H., Chu, L., Ruan, C. (2010). A critical review on the bio-removal of hazardous heavy metals from contaminated soils: Issues, progress, eco-environmental concerns and opportunities. Journal of Hazardous Materials 174(1–3), 1–8.

Xia, X., et al. (2018). VsCOAOMT positively regulates cadmium tolerance and lignin biosynthesis in Viciasativa. International Journal of Molecular Sciences, 19(12), 3996.

Yaashikaa, P.R., Kumar, P.S., Jeevanantham, S., Saravanan, R. (2022). A review on bio-remediation approach for heavy metal detoxification and accumulation in plants. Environmental Pollution 301, 119035.

Yadav, D., Kumar, P. (2019). Phytoremediation of hazardous radioactive wastes. IntechOpen. https://doi.org/10.5772/intechopen.88055

Yadav, R., Singh, G., Santal, A.R., Singh, N.P. (2023). Omics approaches in effective selection and generation of potential plants for phytoremediation of heavy metal from contaminated resources. Journal of Environmental Management 336, 117730.

Yadav, K.K., Gupta, N., Kumar, A., Reecec, L.M., Singh, N., Rezania, S., Khan, S.A. (2018). Mechanistic understanding and holistic approach of phytoremediation: A review on application and future prospects. Ecological Engineering 120, 274–298.

Yağız, A.K., Yavuz, C., Naeem, M., Dangol, S.D., Aksoy, E. (2022). Genome editing for nutrient use efficiency in crops. In: Principles and Practices of OMICS and Genome Editing for Crop Improvement (pp. 347–383). Springer International Publishing: Cham.

Yamaji, N., Xia, J., Mitani-Ueno, N., Yokosho, K., Ma, J.F. (2013). Preferential delivery of zinc to developing tissues in rice is mediated by P-type heavy metal ATPase OsHMA2. Plant Physiology 162, 927–939.

Yang, J., Liu, L., He, D., Song, X., Liang, X., Zhao, Z.J. and Zhou, G.W. (2003). Crystal structure of human protein-tyrosine phosphatase SHP-1. Journal of Biological Chemistry 278(8),6516–6520.

Yanitch, A., Kadri, H., Frenette-Dussault, C., Joly, S., Pitre, F.E., Labrecque, M. (2020). A four-year phytoremediation trial to decontaminate soil polluted by wood preservatives: Phytoextraction of arsenic, chromium, copper, dioxins and furans. International Journal of Phytoremediation 22, 1505–1514. https://doi.org/10.1080/15226514.2020.1785387

Yaqoob, A., Nasim, F.U.H., Sumreen, A., Munawar, N., Zia, M.A., Choudhary, M.S., Ashraf, M. (2019) Current scenario of phytoremediation progresses and limitations. International Journal of Biosciences 14(3), 191–206.

Zalesny, R.S., Casler, M.D., Hallett, R.A., Lin, C.H., Pilipović, A. (2021). Bioremediation and soils. In Soils and landscape restoration (pp. 237–273). Academic Press.

Zhang, H., Zhu, J., Gong, Z., Zhu, J.K. (2022). Abiotic stress responses in plants. Nature Reviews Genetics 23(2), 104–19.

Zhang, W., et al. (2014). Overexpression of the WsSGTL1 gene of Withania somnifera enhances salt tolerance, heat tolerance and cold acclimation ability in transgenic Arabidopsis plants. PLoS One, 9(2), e93229.

Zhong, H., Lai, W., Wu, Y., Chen, B., Chen, Z., Hu, X. (2017). Heavy metals in soil, dust, and aerosol from electronic workshop in Guangzhou, South China: Indicators of air quality. Ecotoxicology and Environmental Safety 144, 479–486.

Zhou, Z.S., Zeng, H.Q., Liu, Z.P., Yang, Z.M. (2012). Genome-wide identification of Medicago truncatula microRNAs and their targets reveals their differential regulation by heavy metal. Plant, Cell & Environment 35(1), 86–99.

Zhu, H., et al. (2011). Arabidopsis Argonaute10 specifically sequesters miR166/165 to regulate shoot apical meristem development. Cell 145(2), 242–256.

5 Unraveling the Mechanism of Phytoremediation Prospects of Oilseed and Aromatic Plants

Updesh Chauhan, Durgesh Singh, Sandhya Maurya, and Garima Gupta

INTRODUCTION

Phytoremediation is a magnificent technology for the reduction of environmental contaminants and xenobiotic compounds present in soil and water. Here, phytoremediation has been done using a wide array of crops and wild plants, but they may cause biomagnification, affecting the food chain, food web, and entire ecosystems and causing severe repercussions. Therefore, the use of oilseed and aromatic plants may serve as a sustainable approach for phytoremediation of heavy metals, as these plants have the tremendous potential to tolerate and accumulate an upsurge in trace metal concentrations in different plant organs. A recent report claimed that contaminated soil, various kinds of pollution, and deterioration have contributed to 20% of ground pollution or around 100 million acres [1]. Remarkably, industrialization, urbanization, and various agriculture practices have contaminated the environment with hazardous chemicals and toxic compounds, and specifically heavy metals, that can pose severe threats to humankind and nature. Living organisms and their surrounding environment are being continuously exposed to heavy metals, both those required for growth and those that retard growth like arsenic, lead, nickel, and cadmium. Generally, growth-enhancing heavy metals can cause toxicity at higher concentrations, whereas growth-retarding metals act as poisons even when low in concentration [2]. They can cause adverse impacts on human health due to bioaccumulation, and there is great risk in human consumption. This environmental challenge can be handled by using phytoremediation as it can serve as a strong, cost-effective, eco-friendly alternative, and it also shows practical utility in remediation, industrial/environmental waste stabilization, and decontamination.

DOI: 10.1201/9781003442295-5

PHYTOREMEDIATION MECHANISMS OF OILSEED AND AROMATIC PLANTS

There are several defense mechanisms employed by plants to reduce the effect of heavy metals either by taking up, storing, or translocating them (Table 5.1).

These plants most commonly trap heavy metals by releasing some plant exudates, such as organic acids or anionic groups of cell wall, to create obstacles for the entry of heavy metals into them through the root system. Nevertheless metal contaminants may enter the cell through the plasma membrane.

1. **Root uptake:** Heavy metals can be taken up by the roots of oilseed and aromatic plants through the same mechanisms as water and nutrients. A wide variety of proteins, including aquaporins, are involved in transporting water as well as small molecules across the cell membrane. Then detoxification of heavy metals takes place through amino acids, organic acids, peptides present in the cell, or by accumulating inside vacuoles or being converted into free ionic forms to limit metals in the below-ground part [3].

2. **Phytotolerance:** These plants can undergo extraction, accumulation, transformation, stabilization, and storing huge amounts of heavy metals without increasing phytotoxicity. Hypertolerant plants can detoxify or accumulate heavy metals, mostly in roots, and limiting their transfer to shoots, and hypertolerant plants can also take a high quantity of trace elements to further translocate to aboveground plant tissues. Further, the hypertolerant biomass might be incinerated for the reduction of metals [4].

3. **Phytoextraction:** Metals may be bio-unavailable and in soilbound form when plants secrete organic acids and alter soil pH to enhance the bioavailability of metals for extraction. Plant-released hydrogen ions replace the heavy metals and render them bioavailable [5].

TABLE 5.1
Various Phytoremediation Techniques and Their Mechanisms Involved

Techniques Involved	Possible Mechanism of Phytoremediation
Phytodegradation	Root exudates and respective enzymes convert toxic contaminants into less toxic forms.
Phytovolatilization	Toxins volatilize via stomata into environment.
Phytofiltration	Filter polluted water (groundwater, moisture, surface) and accumulate in roots; released water can be used for irrigation purposes.
Phytostabilization	Root accumulation of contaminants and root exudates immobilize in soil matrix.
Rhizodegradation	Plants associate with microbes to enhance toxin degradation by microbial enzymes.
Phytoextraction	Plant roots take up toxins and transfer them to shoots.

4. **Chelating agents:** Some chelating agents are artificially added to soil, like elemental sulfur, citric acid, ammonium sulfate, and EDTA, to enhance the bioavailability of metals, although at higher concentrations chelating agents can also cause toxicity. These plants can produce certain chelating agents, such as phytohormones, which can quench the heavy metals and increase their solubility. Most known chelating agents synthesized by plants include phytochelatins and metallothioneins, which can quench the heavy metals [6].

5. **Transport proteins:** Some heavy metal-specific proteins do active transport inside the plant cell and play a role in regulating heavy metal concentrations in the plant tissues [6].

6. **Lipid-rich seeds in immobilizing organic pollutants:** Oilseed plants are rich in lipid content and have a high affinity for other organic pollutants like pesticides and polycyclic aromatic hydrocarbons (PAHs) rich in heavy metals. Certain pesticides have an affinity for lipid-rich environments. These pollutants are either adsorbed on the seed surface or attached to the seed matrix. This type of immobilization increases under hydrophobic conditions and inhibits the release of contaminants again in the environment and decreases bioavailability [7].

7. **Rhizofiltration:** Plants acts as natural filters as they take up contaminants through roots and accumulate them in root tissues. It is beneficial for the removal of organic compounds, ions, and heavy metals [8].

8. **Rhizodegradation:** In general, like other plants, oilseed plants produce chemoattractant exudates through roots and attract microorganisms for associating with host plants, and these associated microorganisms help in bioaugmentation of heavy metals. Additionally, oilseed and aromatic plants can accumulate heavy metals in the leaves, depending on the heavy metal type, concentration, and the plant species. The exact mechanism of heavy metal removal by oilseed and aromatic plants is quite complicated and unknown. However, researchers are trying to get a clear-cut understanding of the mechanism of phytoremediation of heavy metals by oilseed and aromatic plants [9]. (Figure 5.1)

PHYTOREMEDIATION BY OILSEED PLANTS

Oilseed plants can be most promising but have been less explored for trace metal phytoremediation. Oilseed crops are generally cultivated to obtain the oil present in the seed. Compared to the oil content present in grains such as wheat, which is approximately 1–2%, oilseed crops contain 20–40% oil [10]. Oilseed crops include safflower, sunflower, olive plant, brassica, soybean, and rapeseed [11]. Environmental stresses caused by reduced rainfall, the use of wastewater for irrigation, solid waste dumping on agriculture lands, and the inappropriate use of agrochemicals may lead to heavy metal contamination. Oilseed plants have great tolerance and accumulation capacity for metals. As mentioned earlier, there are mechanisms to remove heavy metals such as phytofiltration, phytovolatilization, phytoextraction, phytostabilization and phytodegradation [8].

Oil-bearing plants can be classified into two types, those that produce edible oils like corn oil, sesame oil, cottonseed, safflower, mustard, sunflower, and groundnut

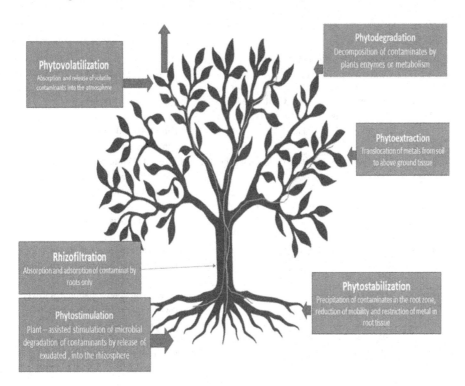

FIGURE 5.1 Phytoremediation mechanisms used by oilseed and aromatic plants.

oil and those that produce nonedible oils such as jatropha, rapeseed, and castor. Among them, a few plants like flax and hemp are low oil-yielding plants [12]. After the removal of oils, a huge amount of remnant biomass is left as waste, which can serve as fuel either by pyrolysis or direct burning. And oilseeds have immense capacity to remediate, although the absorption rate, mechanism, and metal distribution in various plant organs may vary [13]. Various oilseed plants which are being exploited for their phytoremediation capacity are listed in Table 5.2.

Previously a study concluded that castor has the potential to trap heavy metals in the roots rather than in stem and leaf. The bioconcentration factor (BCF) of castor root indicated its good heavy metal accumulation capacity from fly ash-contaminated sites [14]. So *R. communis* L. through phytostabilization and revegetation remediated the polluted site. Huang et al. (2011) unveiled the extraordinary capacity of castor plant in the accumulation of DDT, which was mostly retained in the root part and slightly transferred to shoots. This led to approximately 95.6–99.4% and 82.1–93.7% accumulation of DDT and Cd, respectively, of total plant uptake.

In another study, *Brassica napus* line DH O-120 was investigated for phytoremediation potential by modulating the phytochelatin system involving glutathione (GSH) and phytochelatins (PCs) sequestering heavy metals for the removal of lead. Generally, like *B. juncea*, it showed increased biomass production, heavy metals accumulation, and tolerance, implying its possible use as a phytoremediator. Genetic engineering approaches can be utilized to enhance phytoremediation capacity by targeting the

TABLE 5.2
Oilseed Plants with Phytoremediation Potential

Oilseed Crop Type	Toxin Type	Soil/ Other Contaminants	Remarks
Sunflower	Heavy metals	Cu, Cr, Ni, Fe, Zn, Mn, Pb, U, Pu, As, Cd	Has good phytoremediation, but analyze before oil production and food consumption
Indian mustard	Heavy metals	Zn, Pb, Cd, Cr, Cu, Fe, Ni, Mn	Has good phytoremediation, but safety check required before oil production and food consumption
Sesame	Heavy metals/ Metals	Cr, Mn, Cu, Pb, K, Na, Fe, Zn, Ni and Cd Fe, Mn, Zn, Cu, Co, Ni, Cd, Pb	Has good phytoremediation, but analyze before oil production and food consumption
Safflower	Heavy metals	Cd, Pb, As	Has good phytoremediation, but safety check required before oil production and food consumption
Rapeseed	Heavy metals	Zn and Cd	Has good phytoremediation, but safety check required before oil production and food consumption
Castor bean	Ricin (toxin)/ heavy metals	Ricin, Cd	Has good phytoremediation, but analyze before oil production and food consumption
Hemp	Heavy metals, pesticides	Absorbs contaminants	Has good phytoremediation, but safety check required before oil production and food consumption

phytochelatin system of *Brassica napus*. Recently, *Brassica campestris* L., along with endophyte *Aspergillus niger* CSR3 (chromate tolerant), was challenged for phytorhizoremediation of chromium toxicity, and the study inferred that host growth parameters were significantly enhanced and chromate stress was highly reduced [15].

Previous studies have reported on the phytoremediation potential of *M. pinnata* for remediation of coal mine dumps along with other phytoremediator species [16]. A study suggested a 50–60% decrement in heavy metals, and plant growth parameters were found to be enhanced in coal waste-containing soil compared to control soil [9]. Apart from that, [17] explored the phytoremediation potential of various bioenergy crops such as *Brassica, Glycine, Ricinus, Arachis, Helianthus, Cannabis, Carthamus,* and *Linum* for cadmium accumulation and found the highest Cd phytostabilization by hemp (*Cannabis sativa*). In another study, hemp was also studied for accumulating Cr, Ni, and Zn in the leaf area rather in roots, suggesting that such highly contaminated hemp (>1050 k Bq/m^2) cannot be used for food or cloth manufacturing. It has also been used for bioremediating polycyclic aromatic hydrocarbons, and it in return increased plant growth [9]. Synergistic approaches such as mycorrhizoremediation by using biocrops like vetiver grass and hemp plants along with rhizobacteria were also used for phytoremediation of uranium-contaminated mine sites [18].

A few years back, oilseed rape and peanuts were reported to be high accumulators of metals, and a study concluded that heavy metals (Pb and Cd) concentration were reduced after Soxhlet extraction and extracting agents like potassium citrate, potassium tartrate, and diammonium ethylene diamine tetraacetate further enhanced the removal efficiency [19]. Ghnaya et al. (2009) reported a cultivar effect in phytoextraction capacity as certain cultivars (cossair and pactol) decreased growth parameters during metallic stress, while jumbo and drakkar were resistant and significantly accumulated Zn and Cd in different plant parts [20]. Chen et al. (2015) made a combinatorial approach of intercropping oilseed rape (*Brassica napus* L.) and *Sedum alfredii*, and then inoculating endophytic bacterium *Acinetobacter calcoaceticus* Sasm3 on an intercropped system for removal of nitrate-cadmium compounds [21]. This combined phytoremediation and rhizoremediation approach enhanced not only cadmium clearance efficiency but also the crop quality.

Lin et al. (2003) explored *Helianthus annuus* L. (sunflower) for bioaccumulating copper by various plant parts like roots, hypocotyls, cotyledons, and leaves at different concentrations. There might be several mechanisms involved, like detoxifying proteins such as phytochelatins and metallathionein, bioaccumulation, and enhanced tolerance by restricted transportation from root to shoot. Apart from this, sunflower has also been utilized for remediating nickel ion [22] and tannery sludge (other metals). Varying magnitude and relative distribution in different parts indicate the involvement of various bioaccumulation and translocation mechanisms. Gupta and Sinha (2006) conducted a pot study of *Sesamum indicum* (L.) and soil+tannery sludge, and bioaccumulation was directly proportional to increased sludge except for Pb ion. Abdel-Sabour and El-Seoud (1996) deduced that organic waste compost reduced the heavy metal content in sesame seeds below the phytotoxicity level. Water treatment can be conducted using *Moringa oleifera* (MO) due to their flocculation capacity, cost-effectiveness, and toxin-free nature, and they can be used for *Chlorella* collection. Apart from that, it has superseded chemical flocculants, being an excellent flocculating agent (>95% at 20 min sedimentation), ecofriendly, with decreased cost of harvesting microalgae, with microalgae culture expansion serving as an effective green technology [23]. Recently Poursattari and Hadi (2022) used plant growth regulators (PGE) in canola and EDTA to enhance plant tolerance to lead metal stress and recommended the use of canola with safe PGE instead of EDTA as a cost-cutting phytoremediator for lead. Sunflowers are able to accumulate high concentrations of lead, cadmium, and zinc. Several varieties of sunflower seeds in a field study were assessed for yield, growth, lignocellulose material for bioenergy production, and metal accumulation for remediating arsenic and mercury-contaminated soil. The study suggested that sunflower can be utilized for oil production as well as for As-/Hg-contaminated soils remediation [24].

PHYTOREMEDIATION BY AROMATIC PLANTS

Like crop species, aromatic plants are propagated to extract essential oils mostly used in cosmetics, perfumery, and the medicinal industry [25]. These can also be promising candidates for phytoremediation as they are also not a part of the food chain. Recently, many aromatic plants' phytoremediation capacity has been tested for heavy metals belonging to a wide array of families like Asteraceae, Geraniaceae, Lamiaceae, and Poaceae. The aromatic grasses have a significant amount of biomass and are in high demand for essential oil; examples include palmarosa, vetiver, lemongrass, and citronella. Other than grasses, plant species like geranium, lavender, *Ocimum*, rosemary, *Salvia*, *Mentha*, and chamomile are being explored for their response to phytoremediation challenges. Economic aromatic crops can be a great choice for phytoremediation as they can tolerate toxic environments, enhance soil nutrients, reduce soil erosion, lack any contamination in essential oil, and alteration in oil composition is observed and removed after harvesting.

Citronella (*Cymbopogon winterianus* Jowitt.): This can be used for phytoremediation as previously it has been reported to accumulate Cd, mostly in roots, although higher accumulations may lead to death [6]. Recently a pot study was conducted to evaluate the phytoremediation capacity of citronella (*C. nardus* Rendle) for Cu–Ni mine tailings. Various amendments such as cow manure, FA, and compost increased the survival and growth of citronella. The increased biomass enhanced the phytoremediation capacity of citronella for heavy metals through phytostabilization [26]. It has been less explored for its phytoremediation capacity, so it can be explored for other heavy metals as well.

Palmarosa (*Cymbopogon martinii*): Palmarosa has been used for the treatment of tannery sludge, and this phytostabilization mechanism was also used for heavy metals removal; uptake was found to be highest for Cr and least for Cd [27]. Recently an annual study in a plastic sack was done to assess the sewage sludge phytoremediation potential of *C. martinii*. Bioconcentration and bioaccumulation analysis and other advanced techniques revealed the increment in concentrations of micronutrients (Fe, Zn, Cu, Mn), toxic metals (Pb, Cd, Ni, Cr), and biomass of *C. martinii* with upsurges in sewage sludge concentration and palmarosa acting as a hyperaccumulator [28].

Vetiver (*Chrysopogon zizanioides* or *Vetiveria zizanoides*): Vetiver has been found to be tolerant of extreme conditions such as sodicity, acidity, pH, salinity, contaminated soil, harsh weather, and heavy metals. Another report indicates that vetiver has a long lifespan, which makes it an ideal candidate for phytoremediation [29]. Previous study infers that metal uptake is enhanced by regularly cutting shoots of vetiver, so that plant growth and in turn biomass can increase the contaminant uptake capacity [30]. Vetiver can be utilized to treat landfills for the removal of heavy metal and to further soil reclamation [31]. Andra et al. (2009) utilized a combined approach for chelation of heavy metals by phytochelatins of vetiver and EDTA for the removal of Pb. It has been observed that chelated Pb is less toxic and can easily be taken up by plants. The hyperaccumulation capacity of vetiver was exploited for the removal of Cd [32]. Further, Xia (2004) found that the phytostabilization mechanism involved in the quenching of Pb through vetiver roots and even Pb can enhance vetiver oil production. Complex formations such as chelant-bound lead are less harmful and are easily taken up by plants, and thus they can enhance phytoremediation capacity. The major advantage of vetiver is that its large biomass after reclamation is nontoxic and can be exploited for bioenergy production, artistic craft works, or compost. Earlier investigation suggests that Cr and Pb reduction from lab-synthesized wastewater samples containing Cr and Pb can be achieved using vetiver [33]. There could be several possible mechanisms for heavy metal reduction, such as chelation by glutathione S-transferase, thiols, and phenolics in the cell wall [34]. Recently a review was done to assess the synergistic approach of a nanotechnology, mycorrhizoremediation, phytoremediation (NMPR) strategy for decontamination of the environment and further use of the bioremediated crop for essential oil production. It also emphasized building up the strategy for the large-scale environmental restoration technology using various remediation approaches specifically involving vetiver and cannabis [35].

Lemongrass (*Cymbopogon flexuosous* or *Cymbopogon citratus*): Sobh et al. (2014) found that *C. citratus* stem has good capacity to adsorb Pb(II) ions, and being less expensive, environmentally friendly, and with good adsorption can perform wastewater treatment efficiently. Das and Maiti (2009) achieved reclamation of toxic Cu tailings by using chicken fertilizer (@5% rate) with lemongrass. They even proved irrigation using wastewater can enhance the biomass of lemongrass and its oil yield [36]. Lemongrass has been used for the treatment of industrial effluents and found to be effective for Zn(II), Cd(II), and Pb(III) (Hassan et al. 2016), and recently lemongrass along with chicken manure gave promising results for Pb removal [37]. Similar to vetiver, the oil content of lemongrass increases in the presence of heavy metals and using fungi [38]. Lemongrass also uses the phytostabilization approach for Cu, Fe, and Mn in roots and the bioaccumulation approach for Ni, Pb, Al, Zn, As, Cd, Cr, and Pb [36]. A few reports suggested that *C. flexuosus* can accumulate Cr and As (Jha and Kumar 2017), and *C. citratus* can achieve Ni^{2+} removal using $-R$, $-OH$, and $-CO$ groups (Lee et al., 2014). Comparative study of lemongrass and vetiver was done to understand the potential for decontaminating Ni–Cd battery EW (electrolyte waste); contaminated soil and vetiver grass was reported to be more effective [39].

Geranium (*Pelargonium* species): It uses the accumulation and detoxification approach for the sequestration of cadmium and nickel and interestingly does not affect the metabolically active sites of the plant [40]. It can tolerate metal stress by using photosystem II for plant metabolic activities and avoiding any damage to the photosynthetic apparatus. A few studies have suggested that geranium can be used for the treatment of various metals acting as a hyperaccumulator and by utilizing a phytoextraction mechanism for the removal [41]. Even though it is sensitive to metals and certain metals like Cd, Ni, and Pb can affect its oil content and plant biomass by accumulating inside the plant [42]. In a previous study, a field study was conducted to utilize the high biomass of geranium using the bioaccumulation approach to accumulate Pb, and in another study organic compounds in roots were found to make complexes with Pb [43]. Apart from that, plant-based chelating agents such as phytochelatins and metallothioneins are associated with amino acids like cysteine and glutathione to accumulate heavy metals in plants. The phytoextraction potential of geranium plants for contaminated sludge impacted the quality of essential oil, its growth, and its yield [44]. Geraniums have been proposed as the best candidate for the treatment of tannery sludge, and also enhancement in oil yield was observed [45].

Mint (*Mentha* species): Like other aromatic plants, *Mentha* also uses a phytostabilization approach for accumulating heavy metals in roots and debarring them from entering the aboveground parts. Accumulation of Cr and Pb causes changes in the composition of *Mentha* oil [46]. *Mentha* is considered a hyperaccumulator and can be cultivated in soil along with vermicompost and organic matter without showing any signs of toxicity. Another study claimed that Pb and Cd do not affect the essential oil [47], but Sa et al. (2015) also supported the earlier findings by reporting the changes in carvone concentration on Pb treatment. *Mentha aquatica* (Cd, Pb) and *Mentha piperata* along with citric acid have been used recently for phytoremediation [48].

Chamomile (*Matricaria* species): Chamomile has been observed to be a metal remover rather than a hyperaccumulator. Chamomile has been reported to be Cd tolerant due to production of antioxidative enzymes, and certain molecules like phenol are responsible for controlling the movement of heavy metals like Cd (help in root to shoot translocation) and as barriers for Ni. Similarly, in chamomile Cr(VI) is reduced to Cr(III) and is deposited in its roots, and nickel stress is efficiently reduced [49]. Recently, chamomile (*Matricaria chamomilla* L.) was use to detoxify metals like Cd, Ni, and Pb, whereas results revealed that plants are acting as Cd accumulators and showed no phytotoxic effect, which makes them ideal candidates for phytoremediation [50].

Basil (*Ocimum* species): Basil also tolerates heavy metal concentration by a phytostabilization mechanism, although being edible, phytoremediated basil plants should not be consumed as food to avoid food chain contamination. As in other aromatic plants, the heavy metals Cr, Cd, and Pb changed basil oil composition and were found to reduce the concentration of linalool content and to increase methyl chavicol when compared to controls [51]. Like this study, a few others reported the increment in essential oil yield under heavy metal stress of Cd and Pb and found increment in the concentration of commercially important components [52].

Rosemary (*Rosmarinus officinalis* L.): Heavy metal accumulation depends on the part of the plant, and different metals are accumulated with different rates in different plant parts [53]. Previous studies show that rosemary has a great potential tolerance for heavy metal stress, and generally less heavy metal is transferred to aboveground parts via roots [54]. It also acts as a phytostabilizer and hyperaccumulator of heavy metal such as nickel. Recently rosemary along with *Glomus mossea* was used to decontaminate cadmium and lead concentrations, and their symbiotic capacity enhanced the phytoremediation ability of the rosemary plant acting as a phytoextractor, and its efficiency was enhanced in the presence of citric acid and humic acid-like chelators [55].

Sage (*Salvia* species): It works as a hyperaccumulator of metals, and after remediation its essential oil can be obtained for utilization in cosmetics, perfumery, and the tobacco industry [56]. Chand et al. (2015) reported the use of clary sage in treating sludge containing high amounts of iron, lead, and nickel. Recently, a study proved that clary sage plants treated with excessive amount of Zn can help in mitigating Cd toxicity in sage plants, and sage was also used for the decontamination of uranium contaminants [57].

Lavender (*Lavandula* species): Lavender also has a hyperaccumulator mechanism for bioaccumulation of heavy metals [58]. Previous studies revealed that under heavy metal stress, Cd (more in leaves than in roots) and Pb (more in stem than in leaves) get accumulated in the plant, but the essential oil was free from any kind of contamination [59]. A recent study suggests that AMF inoculation enhanced the growth responses and oil yield of lavender in heavy metal-contaminated environments, and it helped in mitigating the stress in Pb- and Ni-contaminated soil [60]. Another study confirmed the remediation capacity of lavender for heavy metals like Pb, Cu, Cd, and Zn [61].

CHALLENGES AND LIMITATIONS IN PHYTOREMEDIATION

Although efficacy, low cost, and the eco-friendly approach of removing heavy metals by using oilseed and aromatic plants lure scientists to use these plants for phytoremediation purposes, there are still certain challenges and limitations to be addressed. Disposal, safe reuse, and containment of harvested biomass after phytoremediation has to be well planned and conducted to avoid the chances of secondary pollution. Certain other points may affect the remediation of heavy metals, such as the type of heavy metal, contamination level, soil type, climatic conditions, and cost of oilseed and aromatic plants used. Still, they are easy to grow and maintain, and they remove a broad range of pollutants. Further research is needed to exploit the capacity of oilseed and aromatic plants for phytoremediation [61]. (Table 5.3)

TABLE 5.3
Phytoremediation Potential of Various Aromatic Plants

Plant Types	Toxin Type	Soil Type	Usage Considerations	Reference
Citronella	Cd, Cr	Adaptable to most of the soil types	Essential oil, insect repellent, can be used after safety considerations as not involved in food chain.	[63]
Palmarosa	Fe, Zn, Cr, Cd, Pb, Ni, Cu	In drained soil with high OC	Aromatherapy, perfumery and can be used after safety considerations as not involved in food chain.	[27]
Vetiver	Removal of Cd	Wide array of soils types (wet or waterlogged preferably)	Soil erosion eradication, enhance water quality, can be used after safety considerations as not involved in food chain.	[32]
Lemongrass	Cd, Hg, Pb, Cu, Ni, Cr	Preferably well-drained and good fertility	Lemon aroma for culinary and medicines, cannot be used for foods.	[64]
Geranium	Cd, Hg, Pb, Cu, Ni, Cr	Well-drained soils with good OC	Cosmetics and can be used after safety considerations as not involved in food chain.	[45]
Mint	Cr, Pb, Ni, Cd, Cu, Zn	Fertile and well-drained soil	Medicines, culinary, herbal formulations. Cannot be used for foods.	[65]
Chamomile	Cd, Zn, Cu, Pb, Ni	Well-drained, loamy	Beverages, cosmetics, and cannot be used for foods.	[66]
Basil	Cr, Cd, Ni, Pb, As, Zn	Well-drained, good fertility	Cooking, aroma, cannot be used for foods.	[67]
Sage	Cd, Pb, Cr	Sandy soils, well-drained	Medicines, culinary, cannot be used for foods.	[57]
Lavender (*Lavandula*)	Cd, Pb, Cu, Zn, Fe	Alkaline, well-drained	Fragrance, essential oil, cosmetics and can be used after safety considerations as not involved in food chain.	[60]

ADVANTAGES OF OILSEED PLANTS IN PHYTOREMEDIATION

Aromatic and oilseed plants have certain advantages for use in phytoremediation, adding great value for addressing environmental contamination and offering additional advantages as well. The most significant outcome of using these plants for remediation is that after remediation, valuable oils can be harvested from oilseeds and can be used for the production of biofuels. So this kind of multipurpose approach provides economic incentives by treating contaminated sites and obtaining biofuels. In addition to this, in crop rotation oilseed plants can improve soil health as they are deep-rooting, improving soil structure and the recycling of nutrients. They can reduce contaminants, inhibit soil compaction, and most importantly enhance soil organic matter, aeration, and microbial activity, ultimately increasing soil fertility and restoring ecosystem health to favor the growth of the next crop [68]. Aromatic plants' unique fragrances and essential oil content make them suitable candidates for reducing heavy metals present in volatile organic compound (VOC) pollution and thus, in stress mitigation. VOCs are organic chemicals which readily evaporate into the air and are emitted by industries, vehicles, and house chimneys. VOCs can cause respiratory trouble and neuro-related disorders, and they lead to smog and ground-level ozone formation. Aromatic plants produce essential oils containing volatile compounds, and they inhibit the volatilization of VOCs; they also have large surface areas in trichomes and glands and can quench VOCs from the air. They also have special biochemical pathways to metabolize and degrade VOCs accumulated in their tissues for detoxification. Apart from this, aromatic plants can use their symbiotic relations with rhizospheric microorganisms like mycorrhizal fungi and rhizobacteria for the degradation and removal of VOCs [66].

ECONOMIC AND POLICY IMPLICATIONS

The successful implementation of phytoremediation using oilseed and aromatic plants goes beyond scientific considerations. Economic feasibility and supportive policy frameworks play essential roles in determining the viability and widespread adoption of these innovative strategies [69].

CONCLUSION

Exploitation of oilseed and aromatic plants can be a sustainable, eco-friendly, and innovative approach to reduce heavy metal contamination. Apart from phytoremediation, these crops may make a significant contribution to bioenergy production, ecosystem health and sustainability, and VOC remediation, making these plants extremely valuable for broad areas. An interdisciplinary effort involving exhaustive research and technology development is needed to understand and apply oilseed and aromatic plants for phytoremediation purposes. In the future, genetic engineering interventions can be utilized to upgrade the phytoremediation ability of these plants. Gene alteration technology can be used to improve metal uptake, translocation, bioaccumulation capacity, and VOC degradation within plant tissues [70]. Integrated approaches like bioaugmentation (using microorganisms), biochar for improving soil quality, physical methods such as electrokinetics or *in situ* chemical oxidation, along with phytoremediation strategies can be used to develop comprehensive and sustainable solutions for ecosytem restoration and protection [71]. The proper execution of such remediation strategies also requires site selection, plant species selection, monitoring, project budgets, attractive government incentive programs, clear and flexible regulatory guidelines, public–private partnerships, environmental impact assessment, and long-term monitoring.

The ongoing advancement of scientific understanding and technology opens up exciting avenues for the application of oilseed and aromatic plants in phytoremediation. As research progresses, further exploration into the genetic basis of plant-remediation interactions could lead to the development of designer plants optimized for specific contaminants and conditions. Additionally, interdisciplinary collaboration between ecologists, geneticists, engineers, and policymakers will play a pivotal role in translating laboratory findings into scalable, real-world solutions.

REFERENCES

1. N. Rodríguez Eugenio, M. J. McLaughlin and D. J. Pennock, *Soil Pollution: a Hidden Reality*. Rome: Food and Agriculture Organization of the United Nations, 2018.
2. B. Mishra and M. Chandra, "Evaluation of phytoremediation potential of aromatic plants: A systematic review," *J. Appl. Res. Med. Aromat. Plants*, vol. 31, p. 100405, Dec. 2022, doi: 10.1016/j.jarmap.2022.100405.
3. V. Page and U. Feller, "Heavy metals in crop plants: Transport and redistribution processes on the whole plant level," *Agronomy*, vol. 5, no. 3, pp. 447–463, Sep. 2015, doi: 10.3390/agronomy5030447.
4. M. N. V. Prasad and H. M. De Oliveira Freitas, "Metal hyperaccumulation in plants - Biodiversity prospecting for phytoremediation technology," *Electron. J. Biotechnol.*, vol. 6, no. 3, pp. 0–0, Dec. 2003, doi: 10.2225/vol6-issue3-fulltext-6.
5. H. Ali, E. Khan and M. A. Sajad, "Phytoremediation of heavy metals—Concepts and applications," *Chemosphere*, vol. 91, no. 7, pp. 869–881, May 2013, doi: 10.1016/j.chemosphere.2013.01.075.
6. M. I. Lone, Z. He, P. J. Stoffella and X. Yang, "Phytoremediation of heavy metal polluted soils and water: Progresses and perspectives," *J. Zhejiang Univ. Sci. B*, vol. 9, no. 3, pp. 210–220, Mar. 2008, doi: 10.1631/jzus.B0710633.
7. A. T. Lawal, "Polycyclic aromatic hydrocarbons. A review," *Cogent Environ. Sci.*, vol. 3, no. 1, p. 1339841, 2017, doi: 10.1080/23311843.2017.1339841.
8. M. Ashfaq et al., "Robust late twenty-first century shift in the regional monsoons in RegCM-CORDEX simulations," *Clim. Dyn.*, vol. 57, no. 5–6, pp. 1463–1488, Sep. 2021, doi: 10.1007/s00382-020-05306-2.
9. S. Kumar, D. Machiwal, D. Dayal and A. K. Mishra, "Enhanced quality fodder production through grass-legume intercropping under arid eco-system of Kachchh, Gujarat," *LEGUME Res. - Int. J.*, Oct. 2016, doi: 10.18805/lr.v0i0.7596.
10. M. Farooq, M. Hussain, A. Wahid and K. H. M. Siddique, "Drought Stress in Plants: An Overview," in *Plant Responses to Drought Stress*, R. Aroca, Ed., Berlin, Heidelberg: Springer Berlin Heidelberg, 2012, pp. 1–33. doi: 10.1007/978-3-642-32653-0_1.
11. H. K. Woodfield and J. L. Harwood, "Oilseed Crops: Linseed, Rapeseed, Soybean, and Sunflower," in *Encyclopedia of Applied Plant Sciences*, Elsevier, 2017, pp. 34–38. doi: 10.1016/B978-0-12-394807-6.00212-4.
12. A. Ohifuemen, E. A. Fagbemi and K. A. Fasina, "Bio-lubricants: A renewable alternative to mineral oils," *Int. J. Agric. Environ. Bioresearch*, vol. 05, no. 02, pp. 22–32, 2020, doi: 10.35410/IJAEB.2020.5487.

13. J. Zhou, L. H. Chen, L. Peng, S. Luo and Q. R. Zeng, "Phytoremediation of heavy metals under an oil crop rotation and treatment of biochar from contaminated biomass for safe use," *Chemosphere*, vol. 247, p. 125856, May 2020, doi: 10.1016/j.chemosphere.2020.125856.

14. D. Panda, L. Mandal, J. Barik, B. Padhan and S. S. Bisoi, "Physiological response of metal tolerance and detoxification in castor (*Ricinus communis* L.) under fly ash-amended soil," *Heliyon*, vol. 6, no. 8, p. e04567, Aug. 2020, doi: 10.1016/j.heliyon.2020.e04567.

15. M. Qadir, A. Hussain, M. Shah, M. Hamayun, A. Iqbal and Nadia, "Enhancement of chromate phytoremediation and soil reclamation potential of *Brassica campestris* L. by *Aspergillus niger*," *Environ. Sci. Pollut. Res.*, vol. 30, no. 4, pp. 9471–9482, 2022, doi: 10.1007/s11356-022-22678-6.

16. A. A. Juwarkar, R. R. Misra and J. K. Sharma, "Recent Trends in Bioremediation," in *Geomicrobiology and Biogeochemistry*, vol. 39, N. Parmar and A. Singh, Eds., Berlin, Heidelberg: Springer Berlin Heidelberg, 2014, pp. 81–100. doi: 10.1007/978-3-642-41837-2_5.

17. G. Shi and Q. Cai, "Cadmium tolerance and accumulation in eight potential energy crops," *Biotechnol. Adv.*, vol. 27, 555–561, 2009, doi: 10.1016/j.biotechadv.2009.04.006.

18. A. G. Khan, "In Situ Phytoremediation of Uranium Contaminated Soils," in *Phytoremediation: Concepts and Strategies in Plant Sciences*, B. R. Shmaefsky, Ed., Cham: Springer International Publishing, 2020, pp. 123–151. doi: 10.1007/978-3-030-00099-8_5.

19. Y. Yang, H. Li, L. Peng, Z. Chen and Q. Zeng, "Assessment of Pb and Cd in seed oils and meals and methodology of their extraction," *Food Chem.*, vol. 197, pp. 482–488, 2016, doi: 10.1016/j.foodchem.2015.10.143.

20. M. Tandon, P. Vasudevan, S. N. Naik and P. Davies, "Oil bearing seasonal crops in India: Energy and phytoremediation potential," *Int. J. Energy Sect. Manag.*, vol. 7, no. 3, pp. 338–354, Sep. 2013, doi: 10.1108/IJESM-02-2013-0005.

21. B. Chen *et al.*, "An endophytic bacterium *Acinetobacter calcoaceticus* Sasm3-enhanced phytoremediation of nitrate–cadmium compound polluted soil by intercropping *Sedum alfredii* with oilseed rape," *Environ. Sci. Pollut. Res.*, vol. 22, no. 22, pp. 17625–17635, Nov. 2015, doi: 10.1007/s11356-015-4933-5.

22. Y. Tadayon, M. E. Bahrololoom and S. Javadpour, "An experimental study of sunflower seed husk and zeolite as adsorbents of Ni(II) ion from industrial wastewater," *Water Resour. Ind.*, vol. 30, p. 100214, 2023, doi: 10.1016/j.wri.2023.100214.

23. S. H. Abdul Hamid *et al.*, "Harvesting microalgae, *Chlorella* sp. by bio-flocculation of *Moringa oleifera* seed derivatives from aquaculture wastewater phytoremediation," *Int. Biodeterior. Biodegrad.*, vol. 95, pp. 270–275, Nov. 2014, doi: 10.1016/j.ibiod.2014.06.021.

24. A. Ruiz Olivares, R. Carrillo-González, Ma. D. C. A. González-Chávez and R. M. Soto Hernández, "Potential of castor bean (*Ricinus communis* L.) for phytoremediation of mine tailings and oil production," *J. Environ. Manage.*, vol. 114, pp. 316–323, Jan. 2013, doi: 10.1016/j.jenvman.2012.10.023.

25. A. Lubbe and R. Verpoorte, "Cultivation of medicinal and aromatic plants for specialty industrial materials," *Ind. Crops Prod.*, vol. 34, no. 1, pp. 785–801, Jul. 2011, doi: 10.1016/j.indcrop.2011.01.019.

26. Y. Shi *et al.*, "Biochar enhanced phytostabilization of heavy metal contaminated mine tailings: A review," *Front. Environ. Sci.*, vol. 10, p. 1044921, Nov. 2022, doi: 10.3389/fenvs.2022.1044921.

27. P. Pandey, V. Ramegowda and M. Senthil-Kumar, "Shared and unique responses of plants to multiple individual stresses and stress combinations: Physiological and molecular mechanisms," *Front. Plant Sci.*, vol. 6, Sep. 2015, doi: 10.3389/fpls.2015.00723.

28. G. Singh, U. Pankaj, P. V. Ajayakumar and R. K. Verma, "Phytoremediation of sewage sludge by *Cymbopogon martinii* (Roxb.) Wats. var. *motia* Burk. grown under soil

amended with varying levels of sewage sludge," *Int. J. Phytoremediation*, vol. 22, no. 5, pp. 540–550, Apr. 2020, doi: 10.1080/15226514.2019.1687422.

29. A. RoyChowdhury, P. Mukherjee, S. Panja, R. Datta, C. Christodoulatos and D. Sarkar, "Evidence for phytoremediation and phytoexcretion of NTO from industrial wastewater by vetiver grass," *Mol. Basel Switz.*, vol. 26, no. 1, p. 74, Dec. 2020, doi: 10.3390/molecules 26010074.

30. J. F. Sabouang, R. N. Mbongko and L. L. Mohamadou, "Mineral uptake of heavy metals by some marine organisms along the Limbe Coastline in Cameroon and health risk assessment," *J. Geosci. Environ. Prot.*, vol. 10, no. 06, pp. 106–120, 2022, doi: 10.4236/gep. 2022.106007.

31. N. Roongtanakiat, S. Tangruangkiat and R. Meesat, "Utilization of vetiver grass (*Vetiveria zizanioides*) for removal of heavy metals from industrial wastewaters," *ScienceAsia*, vol. 33, no. 4, p. 397, 2007, https://www.thaiscience.info/journals/Article/SCAS/10460374.pdf.

32. J. Chen, A. D. Del Genio, B. E. Carlson and M. G. Bosilovich, "The spatiotemporal structure of twentieth-century climate variations in observations and reanalyses. Part I: Long-term trend," *J. Clim.*, vol. 21, no. 11, pp. 2611–2633, Jun. 2008, doi: 10.1175/2007JCLI2011.1.

33. S. Singh, S. Inamdar, M. Mitchell and P. McHale, "Seasonal pattern of dissolved organic matter (DOM) in watershed sources: Influence of hydrologic flow paths and autumn leaf fall," *Biogeochemistry*, vol. 118, no. 1–3, pp. 321–337, Apr. 2014, doi: 10.1007/s10533-013-9934-1.

34. T. Nyenda, S. M. Jacobs, W. Gwenzi and J. Muvengwi, "Biological crusts enhance fertility and texture of gold mine tailings," *Ecol. Eng.*, vol. 135, pp. 54–60, Sep. 2019, doi: 10.1016/j.ecoleng.2019.03.007.

35. A. G. Khan, "Promises and potential of *in situ* nano-phytoremediation strategy to mycorrhizo-remediate heavy metal contaminated soils using non-food bioenergy crops (*Vetiver zizinoides & Cannabis sativa*)," *Int. J. Phytoremediation*, vol. 22, no. 9, pp. 900–915, Jul. 2020, doi: 10.1080/15226514.2020.1774504.

36. R. Lal, "Intensive agriculture and the soil carbon pool," *J. Crop Improv.*, vol. 27, no. 6, pp. 735–751, Nov. 2013, doi: 10.1080/15427528.2013.845053.

37. Y. Yoshii *et al.*, "Evaluation of phytoremediation effects of chicken manure, urea and lemongrass on remediating a lead contaminated soil in Kabwe, Zambia," *South Afr. J. Plant Soil*, vol. 37, no. 5, pp. 351–360, Nov. 2020, doi: 10.1080/02571862.2020.1772386.

38. C. Lermen *et al.*, "Essential oil content and chemical composition of *Cymbopogon citratus* inoculated with arbuscular mycorrhizal fungi under different levels of lead," *Ind. Crops Prod.*, vol. 76, pp. 734–738, Dec. 2015, doi: 10.1016/j.indcrop.2015.07.009.

39. Kriti *et al.*, "Nickel and cadmium phytoextraction efficiencies of vetiver and lemongrass grown on Ni–Cd battery waste contaminated soil: A comparative study of linear and nonlinear models," *J. Environ. Manage.*, vol. 295, p. 113144, Oct. 2021, doi: 10.1016/j.jenvman.2021.113144.

40. T. Amari, T. Ghnaya and C. Abdelly, "Nickel, cadmium and lead phytotoxicity and potential of halophytic plants in heavy metal extraction," *South Afr. J. Bot.*, vol. 111, pp. 99–110, Jul. 2017, doi: 10.1016/j.sajb.2017.03.011.

41. M. Shahid, E. Pinelli and C. Dumat, "Review of Pb availability and toxicity to plants in relation with metal speciation; role of synthetic and natural organic ligands," *J. Hazard. Mater.*, vol. 219–220, pp. 1–12, Jun. 2012, doi: 10.1016/j.jhazmat.2012.01.060.

42. E. Goyal, S. K. Amit, R. S. Singh, A. K. Mahato, S. Chand and K. Kanika, "Transcriptome profiling of the salt-stress response in *Triticum aestivum* cv. Kharchia Local," *Sci. Rep.*, vol. 6, no. 1, p. 27752, 2016, doi: 10.1038/srep27752.

43. M. Arshad *et al.*, "Phosphorus amendment decreased cadmium (Cd) uptake and ameliorates chlorophyll contents, gas exchange attributes, antioxidants, and mineral nutrients in wheat (*Triticum aestivum* L.) under Cd stress," *Arch. Agron. Soil Sci.*, vol. 62, no. 4, pp. 533–546, Apr. 2016, doi: 10.1080/03650340.2015.1064903.

44. A. Mazeed, *et al.*, "Evaluation of phytoaccumulation potential of toxic metals from sewage sludge by high- value aromatic plant geranium," *J. Environ. Biol.*, vol. 41, no. 4, pp. 761–769, Jul. 2020, doi: 10.22438/jeb/41/4/MRN-1210.

45. V. Pandey, A. Patel and D. D. Patra, "Amelioration of mineral nutrition, productivity, antioxidant activity and aroma profile in marigold (*Tagetes minuta* L.) with organic and chemical fertilization," *Ind. Crops Prod.*, vol. 76, pp. 378–385, Dec. 2015, doi: 10.1016/j.indcrop.2015.07.023.

46. M. Rajkumar, N. Ae, M. N. V. Prasad and H. Freitas, "Potential of siderophore-producing bacteria for improving heavy metal phytoextraction," *Trends Biotechnol.*, vol. 28, no. 3, pp. 142–149, Mar. 2010, doi: 10.1016/j.tibtech.2009.12.002.

47. S. Amirmoradi, P. R. Moghaddam, A. Koocheki, S. Danesh and A. Fotovat, "Effect of cadmium and lead on quantitative and essential oil traits of peppermint (*Mentha piperita* L.)," *Not. Sci. Biol.*, vol. 4, no. 4, pp. 101–109, Nov. 2012, doi: 10.15835/nsb448185.

48. N. Alioghli, S. A. A. Fathi, J. Razmjou and M. Hassanpour, "Does intercropping patterns of potato and safflower affect the density of *Leptinotarsa decemlineata* (Say), predators, and the yield of crops?" *Biol. Control*, vol. 175, p. 105051, 2022, doi: 10.1016/j.biocontrol.2022.105051.

49. F. Ghanavatifard, A. Mohtadi and A. Masoumiasl, "Investigation of tolerance to different nickel concentrations in two species *Matricaria chamomilla* and *Matricaria aurea*," *Int. J. Environ. Sci. Technol.*, vol. 15, no. 5, pp. 949–956, May 2018, doi: 10.1007/s13762-017-1435-7.

50. E. A. Serban, G. G. Vasile, S. Gheorghe and C. Ene, "Effects of toxic metals Cd, Ni and Pb on *Matricaria Chamomilla* L. growth in a laboratory study," *Rev. Chim.*, vol. 71, no. 4, pp. 325–335, May 2020, doi: 10.37358/RC.20.4.8072.

51. Y. Ma, M. N. V. Prasad, M. Rajkumar and H. Freitas, "Plant growth promoting rhizobacteria and endophytes accelerate phytoremediation of metalliferous soils," *Biotechnol. Adv.*, vol. 29, no. 2, pp. 248–258, Mar. 2011, doi: 10.1016/j.biotechadv.2010.12.001.

52. N. A. Youssef, "Changes in the morphological traits and the essential oil content of sweet basil (*Ocimum basilicum* L.) as induced by cadmium and lead treatments," *Int. J. Phytoremediation*, vol. 23, no. 3, pp. 291–299, Feb. 2021, doi: 10.1080/15226514.2020.1812508.

53. S. A. Abdul-Wahab and B. Yaghi, "Total suspended dust and heavy metal levels emitted from a workplace compared with nearby residential houses," *Atmos. Environ.*, vol. 38, no. 5, pp. 745–750, Feb. 2004, doi: 10.1016/j.atmosenv.2003.10.017.

54. E. Bozdogan Sert, M. Turkmen and M. Cetin, "Heavy metal accumulation in rosemary leaves and stems exposed to traffic-related pollution near Adana-İskenderun Highway (Hatay, Turkey)," *Environ. Monit. Assess*, vol. 191, no. 9, p. 553, Sep. 2019, doi: 10.1007/s10661-019-7714-7.

55. A. Eren, "Phytoextraction of nickel contaminated soil with citric acid and humic acid treatments using Rosemary (*Rosmarinus officinalis*) plant," *Int. J. Environ. Sci. Nat. Resour.*, vol. 19, no. 4, May 2019, doi: 10.19080/IJESNR.2019.19.556016.

56. Z. Angelova, S. Georgiev and W. Roos, "Elicitation of plants," *Biotechnol. Biotechnol. Equip.*, vol. 20, no. 2, pp. 72–83, Jan. 2006, doi: 10.1080/13102818.2006.10817345.

57. A. Dobrikova, E. Apostolova, I.-D. S. Adamakis, A. Hanć, I. Sperdouli and M. Moustakas, "Combined impact of excess Zinc and Cadmium on elemental uptake, leaf anatomy and pigments, antioxidant capacity, and function of photosynthetic apparatus in clary Sage (*Salvia sclarea* L.)," *Plants*, vol. 11, no. 18, p. 2407, Sep. 2022, doi: 10.3390/plants11182407.

58. R. M. Hlihor, M. Roşca, L. Hagiu-Zaleschi, I. M. Simion, G. M. Daraban and V. Stoleru, "Medicinal plant growth in heavy metals contaminated soils: Responses to metal stress and induced risks to human health," *Toxics*, vol. 10, no. 9, p. 499, Aug. 2022, doi: 10.3390/toxics10090499.

59. J. M. Al-Khayri, A. Banadka, R. Rasheavy metalsi, P. Nagella, F. M. Alessa and M. I. Almaghasla, "Cadmium toxicity in medicinal plants: An overview of the tolerance strategies, biotechnological and omics approaches to alleviate metal stress," *Front. Plant Sci.*, vol. 13, p. 1047410, 2022, doi: 10.3389/fpls.2022.1047410.

60. F. Rasouli *et al.*, "Improvements in the biochemical responses and Pb and Ni phytoremediation of lavender (*Lavandula angustifolia* L.) plants through *Funneliformis mosseae* inoculation," *BMC Plant Biol.*, vol. 23, no. 1, p. 252, May 2023, doi: 10.1186/s12870-023-04265-0.

61. R. Vidican, T. haiescu, A. D. Plesa, A. Malinas and B. Pop, "The phytoremediation potential of *Lavandula Angustifolia* Mill. grown in soils historically polluted with heavy metals: A case study from Baia Mare, Romania," *J. Appl. Life Sci. Environ.*, vol. 55, no. 4(192), pp. 495–504, May 2023, doi: 10.46909/alse-554078.

62. A. Kafle, A. Timilsina, A. Gautam, K. Adhikari, A. Bhattarai and N. Aryal, "Phytoremediation: Mechanisms, plant selection and enhancement by natural and synthetic agents," *Environ. Adv.*, vol. 8, p. 100203, Jul. 2022, doi: 10.1016/j.envadv.2022.100203.

63. A. Boruah and S. Ganapathi, "Organic richness and gas generation potential of Permian Barren Measures from Raniganj field, West Bengal, India," *J. Earth Syst. Sci.*, vol. 124, no. 5, pp. 1063–1074, Jul. 2015, doi: 10.1007/s12040-015-0596-3.

64. V. Rana, S. Bandyopadhyay and S. K. Maiti, "Potential and Prospects of Weed Plants in Phytoremediation and Eco-Restoration of Heavy Metals Polluted Sites," in *Phytoremediation Technology for the Removal of Heavy Metals and Other Contaminants from Soil and Water*, Elsevier, 2022, pp. 187–205. doi: 10.1016/B978-0-323-85763-5.00015-5.

65. P. V. V. Prasad, R. Bheemanahalli and S. V. K. Jagadish, "Field crops and the fear of heat stress—Opportunities, challenges and future directions," *Field Crops Res.*, vol. 200, pp. 114–121, Jan. 2017, doi: 10.1016/j.fcr.2016.09.024.

66. J. Pandey, R. K. Verma and S. Singh, "Suitability of aromatic plants for phytoremediation of heavy metal contaminated areas: A review," *Int. J. Phytoremediation*, vol. 21, no. 5, pp. 405–418, Apr. 2019, doi: 10.1080/15226514.2018.1540546.

67. S. Prasad and B. B. Aggarwal, "Turmeric, the Golden Spice: from Traditional Medicine to Modern Medicine," in *Herbal Medicine: Biomolecular and Clinical Aspects*, 2nd ed., I. F. F. Benzie and S. Wachtel-Galor, Eds., Boca Raton (FL): CRC Press/Taylor & Francis, 2011. Accessed: Oct. 19, 2023. [Online]. Available: http://www.ncbi.nlm.nih.gov/books/NBK92752/

68. H. Ashfaq *et al.*, "Phytoremediation Potential of Oilseed Crops for Lead- and Nickel-Contaminated Soil," in *Plant Ecophysiology and Adaptation under Climate Change: Mechanisms and Perspectives II*, M. Hasanuzzaman, Ed., Singapore: Springer Singapore, 2020, pp. 801–820. doi: 10.1007/978-981-15-2172-0_31.

69. A. K. Priya, M. Muruganandam, S. S. Ali and M. Kornaros, "Clean-up of heavy metals from contaminated soil by phytoremediation: A multidisciplinary and eco-friendly approach," *Toxics*, vol. 11, no. 5, p. 422, May 2023, doi: 10.3390/toxics11050422.

70. A. Yan, Y. Wang, S. N. Tan, M. L. Mohd Yusof, S. Ghosh and Z. Chen, "Phytoremediation: A promising approach for revegetation of heavy metal-polluted land," *Front. Plant Sci.*, vol. 11, p. 359, Apr. 2020, doi: 10.3389/fpls.2020.00359.

71. L. Xiang *et al.*, "Integrating biochar, bacteria, and plants for sustainable remediation of soils contaminated with organic pollutants," *Environ. Sci. Technol.*, vol. 56, no. 23, pp. 16546–16566, Dec. 2022, doi: 10.1021/acs.est.2c02976.

6 Exploring the Potential of Grasses in the Phytoremediation of Various Xenobiotic Compounds

*Durgesh Singh, Updesh Chauhan,
Sandhya Maurya, and Garima Gupta*

INTRODUCTION

Phytoremediation has been recognized as a cutting-edge environmental restoration technology for addressing environmental pollution, rejuvenating and enhancing the aesthetic features of damaged lands [1]. This technology uses the immense capacity of plant diversity to take up, detoxify, decompose, or immobilize toxic contaminants present in the environment. This technique may serve as a sustainable and eco-friendly alternative to traditional remediation methodologies such as leaching, vitrification, electrokinetic treatment, off-site treatment, excavation, or chemical treatment. In phytoremediation, plants which are tolerant and can easily accumulate the contaminants without harm are generally chosen and used. Phytoremediation technology includes certain mechanisms like hyperaccumulation in plant tissues, contaminant detoxification, rhizofiltration, phytoextraction, phytodegradation, and phytostabilization [1]. Phytoremediation has other advantages over traditional methods, such as cost-effectiveness, lower maintenance, long-term usage, practical applicability for large land areas, specifically for urban landscapes like buildings, parks, and roadsides, and it improves aesthetic value. It can further be improved by adding other remediation methods, such as rhizoremediation and mycoremediation, which encourages biodiversity and improves soil fertility [2]. Although phytoremediation can be a most promising technology, exhaustive studies and extensive field trials are still required to select the suitable plant for a particular contaminant, to understand the underlying phytoremediation mechanisms, and to check on its sustainability, eco-friendliness, and efficiency.

Grasses can be a promising candidate for phytoremediation as they do not accumulate metals in their aboveground parts, which reduces the exposure of animals to toxic compounds. They densely cover the land, help prevent erosion, produce huge amounts of dry biomass, are less flammable, increase soil fertility, and improve biodiversity. Grasses like rye, Bermuda grass, sorghum, and fescue are generally used

DOI: 10.1201/9781003442295-6

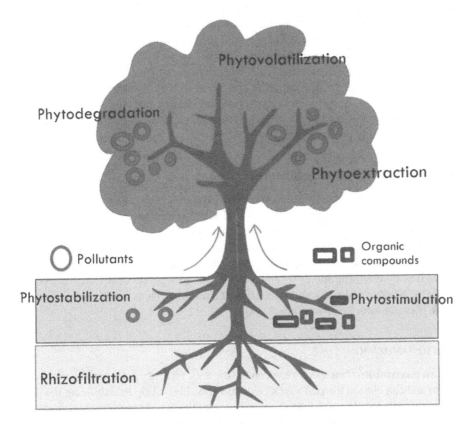

FIGURE 6.1   Phytoremediation mechanisms used by grasses.

for remediation as their fibrous root systems with long roots cover a large surface area per unit volume of soil, and they are even better for rhizospheric colonization than taproots [3]. (Figure 6.1)

MECHANISTIC INSIGHTS OF PHYTOREMEDIATION

Different plant species have certain strategies to conduct phytoremediation, which may vary depending on the kind of element, chemical properties, and bioavailability of the potential toxic elements. This includes phytoremediation strategies such as phytoextraction, phytofiltration or rhizofiltration, phytovolatilization, phytostabilization, and phytodegradation.

PHYTOEXTRACTION

This can also be called phytoabsorption or phytoaccumulation or phytosequestration. It is the process in which the plants remove toxic elements from the soil by accumulating them in the harvested plant parts. Further, these accumulator plants transport and accumulate various toxic materials from soil to shoot. It includes extrusion, concentration, and transfer of material to the aboveground part. It is a highly preferable technique because toxins part can be obtained from the harvested parts, generally the shoot [4]. It effectively removes from the soil heavy metals required for growth, like Fe, Mn, Zn, Cu, Mg, Mo, and Ni, as well as hazardous metals like Co, As, Cd, Cr, Pb, Se, and Hg. Phytoextraction can be done effectively only by choosing the correct plant having good phytoextraction capacity, higher bioaccumulation capacity, and large biomass [5].

PHYTOFILTRATION OR RHIZOFILTRATION

In this process, toxic materials are absorbed through roots and mostly by aquatic plants through absorption, precipitation, and accumulation in the plant part [6]. Then phytofiltration stops the transfer of contaminants to the various parts of the ecosystem and thus prevents environmental contamination. Terrestrial plant species associated with rhizospheric bacteria may enhance the filtration process by phytofiltration as well as by microfiltration [7]. This can be used to separate heavy metals like Pb, Cd, Cu, Ni, Zn, and Cr from contaminated sites, and plants can then be removed once they get saturated by pollutants [8].

PHYTOSTABILIZATION

In this mechanism, plant species can convert the toxic form of a pollutant into a nontoxic form and can change the pollutant's bioavailability. Plants help in stabilizing the toxic form present in soil to avoid the chances of contaminants causing severe groundwater and air pollution by reducing leaching of metals, erosion, entry into the food chain, and surface runoff [9]. But this method immobilizes metals for a limited time only, and therefore it is less useful than phytoextraction methods [10], although it enhances the structural stability and immobilization rate of heavy metals through roots [11].

PHYTOTRANSFORMATION OR PHYTODEGRADATION

Like phytostabilization, in this process transformation of toxic elements into inactive forms takes place, and plants release certain enzymes (phosphatase, nitroreductase, nitrilase oxidase, etc.) into the surroundings while plant-associated bacteria also reduce the metal toxicity in the soil. The degradation occurs as soil microorganisms consider plant-released exudates as food sources, so they release enzymes to degrade and utilize them and do the same for hazardous chemicals [12]. Phytodegradation requires lots of labor and soil amendments, and is reported to be unreliable [13]. It gives good results for organic pollutants but is found to be ineffective for inorganic contaminants.

PHYTOVOLATILIZATION

This basically converts heavy metal contaminants into gaseous form and releases them into the atmosphere, so the transfer of toxic material from environment to atmosphere makes it less popular than other techniques [14].

PHYTOREMEDIATION OF PCBs AND PAHs BY GRASSES

Environmental contaminants like polychlorinated biphenyls (PCBs) and polycyclic aromatic hydrocarbons (PAHs) can be a dreadful threat to human health and the ecosystem. Instead of using traditional remediation procedures like chemical and physical treatments, which are costly and time-consuming, grasses can be explored. They grow rapidly, have dense and fibrous roots, generate large biomass, and can be used in the degradation of PAH and PCB [15]. In a lab study, alfalfa (*Medicago sativa*), bluestem grass (*Schizachyrium scoparium*), and switchgrass (*Panicum virgatum*) were used for the decontamination of PAH-contaminated soil and achieved a 57% reduction within six months of treatment. Therefore, PAH removal can be efficiently achieved economically for environmental cleanup to restore PAH-contaminated lands [16]. Ryegrass (*Lolium annual*) was used for a field study on a site contaminated by a crude oil spill, revealing a 42% reduction in total PHC concentration and 50% on treating with St. Augustine grass (*Stenotaphrum secundatum*) after 21 months. In an another study, a crude oil spill-contaminated site phytoremediated with grasses and fertilizer led to the removal of PAH and other recalcitrant components. Similarly, tall fescue was used continuously for three years for decontaminating soil containing aged creosote and caused degradation of the fluoranthene, pyrene, acenaphthene, fluorene, and chrysene [17].

A greenhouse study was done on annual beard grass (*Polypogon monspeliensis*) to remediate oil-contaminated soils and inferred that it can significantly reduce TPH concentration and has huge potential for the transfer of hydrocarbons through roots [18]. In a greenhouse study, the phytoremediation potential of four grasses, *Leymus cinereus* Scribn. & Merr., *Agropyron spicatum* (Pursh) Scribn. & Smith, *Agropyron cristatum* (L.) Gaertn., and *Bromus tectorum* L., was evaluated for Cs-contaminated soils, and no Cs toxicity was reported. Still, the study concluded that these grasses were not efficient to use for phytoremediation purposes due to low transfer factors of Cs for Cs-contaminated soils [19].

PHYTOREMEDIATION OF OIL AND DIESEL BY GRASSES

Cyperus rotundus (nut grass) was examined for decontaminating soil contaminated with crude oil. The study reported that *C. rotundus* did reduce contamination in 180 days in varied amounts in different treatments and proved the effectiveness of nut grass in phytoremediation [19, 20]. Grass pea (*Lathyrus sativus* L.) and tall fescue (*Festuca arundinacea*) phytoremediation capacity was monitored on the basis of diesel concentration and irrigation water type, and it was found that tall fescue removed diesel by 47.9–62.4% and grass pea by 44.2–63.0% by phytoextraction and phytodegradation-rhizodegradation mechanisms. No significant difference was observed for a few different diesel concentrations, while a significant difference was observed in certain irrigation water types. The study showed that plant species, diesel level, and irrigation water may have a significant impact on diesel reduction, although it may have different effectiveness in different treatments [21]. The combinatorial approach of phytoremediation along with rhizoremediation was used in which two grasses (*Festuca arundinacea* Schreb. and *Festuca pratensis* Huds.), infected (E+) and noninfected (E−) by endophytic fungi (*Neotyphodium coenophialum* and *Neotyphodium uncinatum*, respectively), were investigated for petroleum hydrocarbon degradation. The results revealed that higher degradation was achieved in grasses infected with endophytic fungi of TPH in oil-contaminated soils [22]. Vetiver grass (*Vetiveria zizanioides*), bahia grass (*Paspalum notatum*), St. Augustine grass (*Stenotaphrum secundatum*), and bana grass (*Pennisetum glaucum* × *P. purpureum*) were compared to restore degraded oil-mined land in China. In this study, vetiver showed the best survival rate, followed by bahia and St. Augustine; bana showed the lowest survival rate of 62%. Vetiver was suggested as the best option for vegetation rehabilitation in oil-contaminated sites [23]. Recently *Panicum maximum* (phytoremediation), *Pontoscolex corethrurus* (vermiremediation), and encapsulated bacterial consortium (bioaugmentation) were together used for the removal of alkane and PAH achieved in 28 days; after 112 days, degradation of alkane was 78.5–94.5%, and PAH 54.5–77.2% in unsterilized soil. The combination of different remediation methods enhanced the effects, and better TPH removal was obtained in an earthworm and bacteria-treated sample than by all the three strategies together (phyto+vermi+rhizoremediation) [24]. (Table 6.1)

PHYTOREMEDIATION OF HEAVY METALS BY GRASSES

Metals can be categorized into different categories: essential micronutrients (heavy metals like Fe, Mn, Cu, Zn, and Ni) required for plant development and nonessential metals (like Cd, Pb, As, Hg, and Cr) for plant growth. Both are toxic in higher concentrations, so grasses can be a good option as they have extensive root systems; being pioneers, they generate large biomass in a short time and can be harvested yearly or seasonally [37].

A few decades back, certain soil amendments like EDTA were being used to chelate heavy metals, which could form complexes with heavy metals and get leaked into groundwater, causing pollution and damage to flora, fauna, microflora, and entire environments [38]. So in a study, vetiver grass (*Vetiveria zizanioides*) along with EDTA were used for uptake and transport from Pb-contaminated soils [39]. The study found that a soil matrix having vetiver can absorb Pb, Cd, Cu, and Zn (98%, 88%, 54%, 41% respectively) and decrease the chances of heavy metal entry into the groundwater [40]. Mine wastes containing Cu, Pb, and Zn were restored and remediated by grasses like *Agrostis tenuis* for copper mines and in calcareous lead and zinc mine wastes by *Festuca rubra* [41]. In another study, Shu et al. (2004) inferred that *V. zizanioides* was the best metal extruder among grasses like *Paspalum notatum*, *Cynodon dactylon*, and *Imperata cylindrica* var. major in the field study. It accumulated a high concentration of metals in roots and shoots through phytostabilization and phytoextraction. Among Poaceae family, the accumulation of different metals was higher in various grass species such as *Paspalum* sp., *Eriochloa ramosa* for Cu, *Holcus lanatus* and *Pennisetum clandestinum* for Ag, *Paspalum racemosum* for Cu and Zn, *Bidens humilis* for Pb, Cu in *Mullinum spinosum* and in *B. cynapiifolia*, Zn in *Baccharis amdatensis*, and *Rumex crispus* for Ag. *Sorghum sudanense* (Piper) Stapf. (Sudan grass), *Lolium perenne* L. (ryegrass), and *Cynodon dactylon* (Bermuda grass), using soil conditioner (TH-LZ01) and straw, were utilized for the ecological

TABLE 6.1
Trace Metal Phytoremediation by Various Grasses

Trace Element	Grass Species	Growth Medium	Change Induced by Toxicity	Reference
As	*Brachiaria decumbens*	Soil	Reduction in total biomass	[25]
	Pennisetum clandestinum	Nutrient solution	Less phosphorous absorption and root elongation	[26]
	Spartina patens	Hydroponics	Reduction in total biomass	[27]
Cd	*Brachiaria decumbens*	Soil	Affected nitrogen metabolism, size and starch percent in grains, oxidative stress in source tissues, less nutrient uptake	[28]
	Lolium perenne	Nutrient solution	Affected cytokinin production, fewer tillers and auxiliary buds	[29]
	Panicum maximum	Soil	Lack in nutrient acquisition, oxidative stress in source tissues, affected size and percent of starch grains	[28]
Cu	*Arundo donax*	Nutrient solution	Decrement in root and shoot length	
	Pennisetum purpureum	Soil	Affected photosynthetic efficiency and total biomass	[30]
	Phragmites australis	Soil	Affected photosynthetic efficiency and total biomass	[30]
Hg	*Avena sativa*	Soil	Less proline production, stomata conductance, photosynthetic rate	[31]
	Chloris barbata	Soil	Reduced root length	[32]
	Paspalum distichum	Soil	Reduced root length, biomass, and oxidative stress	[33]
Ni	*Agrostis gigantea*	Nutrient solution	Growth retardation	[34]
	Deschampsia cespitosa	Nutrient solution	Growth retardation	[34]
	Phalaris arundinacea	Soil	Reduction in total biomass	[35]
Zn	*Agrostis gigantea*	Nutrient solution	Growth retardation	[34]
	Chloris barbata	Soil	Reduced root length	[32]
	Deschampsia cespitosa	Nutrient solution	Growth retardation	[34]
	Phalaris arundinacea	Soil	Reduction in total biomass	[36]
	Poa compressa	Nutrient solution	Growth retardation	[34]

restoration of a copper tailings field. Among all these, ryegrass was found to be more efficient (best for Cd followed by Zn, Pb, and Cu) than Sudan grass and Bermuda, and amendments were increasing the stability of metals deposited in roots more than leaves [42]. *Chara canescens* (musk-grass) has shown the capacity to absorb contaminants like selenium and mercury and further turn them into gaseous form inside the plant and then release them into the atmosphere [3]. The genus *Lolium* from the Poaceae family, acting as a hyperaccumulator, can absorb 1–2% of potentially toxic compounds and can be an ideal candidate due to its fast growth, root structure, metal sensitivity, and degradation capacity.

Cruz tested its phytoremediation ability against heavy metals like Cd and Hg, and more tolerance was observed during germination and increment in biomass of plant parts. *Festuca rubra* L. is a perennial grass having several advantages, like an extensive rhizosystem, rapid growth, higher biomass, and stress tolerance and can remediate heavy metals such as Mo, B, Zn, Cu, and Mn. *Poa pratensis* L. prevents mechanical damage, forms a dense turf on new surfaces, specifically on fertile and moist soils, and has great potential to remediate Cd, and its seeds are resistant to copper stress. *L. perenne* and *T. repens* form a plant cover for soil contaminated with Cd, Pb, and Zn [5]. Vetiver grass enhanced the accumulation of Cr ion in wastewater through phytostabilization and phytoextraction, and it accumulated metal in its leaves, indicating that it can be used for the reuse of ash [43]. A comparative study was carried out between cultivated grasses and local wild grass for Zn-contaminated soil. In this study, four species of cultivated grasses, i.e., *Poa pratensis*, *Lolium perenne*, *Festuca rubra*, and *Festuca pratensis*, were compared with wild native grass: *Deschampsia cespitosa*. Phytoremediation capacity was highest in wild grass for Zn, and the least transfer from roots to aboveground parts occurred. So these grasses can be used for phytostabilization of Zn in the contaminated sites. Zn tolerance was in the following order among studied grass species: *D. cespitosa>L. perenne>F. rubra>F. pratensis>P. pratensis*. In addition, there were differences in the accumulation and distribution of Zn between the roots and shoots, which is related to the different defense mechanisms of the studied grasses against Zn phytotoxicity [35]. Recently, King grass was investigated for phytoremediation capacity as having high biomass and heavy metals resistance, and microbial diversity associated with it might play a significant role in decontaminating heavy metals and enhancing plant performance [44]. *Solanum nigrum* L. (black nightshade), an annual grass, has been reported as a Cd hyperaccumulator, and this study reported a phytoremediation index enhancement by natural biostimulant (yeast extract). This may be due to increased protection of photosynthetic pigments, enhanced antioxidant defenses, and nutrient uptake [45].

PHYTOREMEDIATION OF WASTEWATER

Wetland systems can serve as great wastewater treatment technology. Many plant types (grasses like canna indica, typha grass, para grass) for phytoremediation, hydraulic retention time (HRT), and bed material (gravels and sand) are being used for the creation of a wide variety of wetlands, to remove pollutants using microphytes for making the water parameters suitable as potable water [46]. Recently, a nanophytoremediation approach was reviewed using several plant species (Bermuda grass, *Eichhornia crassipes*, *Helianthus annus*) and by increasing the pollutant adsorption on nanoparticles, remediation was done on wastewater and organic wastes [47]. Per- and polyfluoroalkyl substances (PFASs) are released through industries and domestic waste in different water sources. Pilot plant along with reed grasses reduced PFASs by 50% through phytodepuration, and thus commended the use of highly sustainable constructed wetlands in the remediation of wastewater [48].

PHYTOREMEDIATION OF DYES

Elephant grass (EG) has been investigated, and mathematical calculation using batch degradation of methylene blue (MB) dye in the wastewater contaminated by a local dyeing industry in Nigeria found it good for the remediation of dyes as well [49].

PHYTOREMEDIATION OF SALTY AREAS

Different grasses such as *Lepironia articulata*, *Eleocharis dulcis*, *Typha orientalis*, *Scirpus littoralis*, *Brachiaria mutica*, *Paspalum atratum*, and *Setaria sphacelata* were studied for salt phytoremediation and sodium ion bioaccumulation at varied salinity levels. *S. littoralis* was found to be a salt tolerant plant, and Na$^+$ accumulation was highest in *S. sphacelata* followed by *S. littoralis* and *T. orientalis* acting as good hyperaccumulators; these can be used for salt reclamation purposes [50]. In Pakistan, Kallar grass (*Leptochloa fusca* [L.] Kunth), and sesbania (*Sesbania bispinosa* [Jacq.] W. Wight) were used to replace the costlier strategy of using chemical gypsum for treating sodicity and soil salinity, and the study indicated that the stress amelioration capacity of sesbania and Kallar grass was almost same for moderately saline sodic calcareous soils [51].

MISCELLANEOUS PHYTOREMEDIATION BY GRASSES

Vetiver grass apart from heavy metals has the immense potential to absorb and biodegrade organic wastes such atrazine, phenol, 2,4,6-trinitrotoluene, benzo [a]pyrene, and ethidium bromide and is considered a good choice for the phytoremediation of organic wastes [52]. It has been observed that aromatic grasses like *Cymbopogon flexuosus, Chrysopogon zizanioides* (L.) Nash., and *Cymbopogon martini* are also used for restoration of metal and pesticide-contaminated soils, dry land and mine-affected areas. This study reviewed the capacity of aromatic plants for ecological restoration and carbon sequestration in various soil conditions and agro-based regions specifically [53]. With the advancement of electronics and the technology industry, there is a huge generation of e-waste containing hazardous heavy metals, so in one study coco grass (*Cyperus rotundus* L.) as a hyperaccumulator of metals like Cd and Pb was used for the restoration of an environment contaminated with hazardous e-waste material [45]. Recently, grass species such as *Lolium multiflorum* Lam., *Lolium perenne* L., and *Dactylis glomerata* L., were treated with different bacterial consortia and used to treat uranium-contaminated soil. The results indicated that inoculation of consortia enhanced the phytoextraction by all grasses as well as their health and total biomass. The highest extraction efficiency was observed in *L. perenne* for uranium compared to the other two grasses, and mowing grass along with these treatments further encouraged plant growth and increased biomass to increase U-uptake compared to other treatments [54]. Municipal sewage sludge containing heavy metals was phytoremediated by two energy crops, i.e., reed canary grass (*Phalaris arundinacea* L.) and giant miscanthus (*Miscanthus* × *giganteus* GREEF et DEU), and with increment in sewage sludge heavy metals concentration was significantly increased in these energy crops. In the presence of high concentrations of heavy metals, the yield of energy plants was increased, leading to more bioaccumulation and increased metal uptake, and miscanthus was found to be more efficient for sewage treatment [55].

PHYTOREMEDIATION ALONG WITH BIOGAS PRODUCTION

Apart from phytoremediation, biogas production can also be carried out after using grasses for phytoremediation. Water hyacinth (*Eichhornia crassipes*) and channel grass (*Vallisneria spiralis*) were used for lignin, metal-rich pulp, and acidic distillery effluent remediation. These grasses showed excellent growth in the presence of diluted effluent and could take metals and toxic compounds from wastewater. Later on, after phytoremediation, the slurry of these two grasses was used for biogas production, and a significantly higher amount of biogas was produced from phytoremediated plants than from control plants. The effect was found to be greater in channel grass and faster than water hyacinth, which may be due to variation in C, N, and C/N ratio of slurry caused by phytoremediation [56].

PHYTOREMEDIATION ALONG WITH BIOFUEL PRODUCTION

M. pinnata along with biofertilizers (VAM and diazotrophic bacteria) and sugar factory sludge were employed for the remediation of coal mine dumps. Researchers reported a decrement in heavy metal concentration (Cd, Zn, Fe, Pb, Ni, Cr, Mn, and Cu), and the degradation was even higher in the presence of biofertilizer. Similarly, *M. pinnata* gave the same kind of results in copper-contaminated soil after adding amendments like biofertilizers and zeolite. Other than its phytoremediation capacity, *M. pinnata* has also become a highly popular second-generation biodiesel plant cell as it has high oil content, fatty acid presence, performance, and abiotic stress tolerance [57]. Common reed (*Phragmites australis*) has been distributed around the globe and can remediate wastewater, sediment, and soils. The present reviewed work gave an extensive overview on heavy metals removal from the environment using common reed. Additionally, the other uses of common reed for bioenergy and animal feedstock production have also been mentioned [58]. Similarly, reed canary grass (*Phalaris arundinacea* L.) is a rapidly growing, perennial, C3 plant of family Poaceae. It has great tolerance and can be used for bioenergy, pulp, biogas, ethanol, and paper production [59]. Trace metals were analyzed by taking samples from water, bottom sediments, and different organs of *Phalaris arundinacea* L. obtained from the Bystrzyca River (Lower Silesia) and to understand the metal accumulation abilities in various organs of *P. arundinacea*. This study revealed that trace metals were present in the following order: root > leaf > stem in reed canary grass, which showed high accumulation of Pb and Zn due to vehicle release and less in dam and forests, suggesting the great potential of *P. arundinacea*. Likewise, *Pennisetum purpureum* (Napier grass) has the required features of a phytoremediator like a high growth rate and high tolerance level to survive in most contaminated soils. This study analyzed the various applications of Napier grass in a comprehensive manner and different mechanisms of phytoremediation involved in soil and wastewater. Further, the phytoremediation was coupled with bioenergy production like ethanol from Napier grass, which can develop a zero-waste system by using *P. purpureum* after phytoremediation to produce biofuel using its high lignocellulosic content [60].

TECHNIQUES FOR INCREASING PHYTOREMEDIATION'S EFFICACY

Grasses used in phytoremediation may have certain limitations and restrictions, such as slow growth, environmental challenges in plant growth, and lack of understanding of phytoremediation mechanisms. Here genetic selection and recombinant DNA technology-based interventions can be used to handle such challenges [5]. Genetic engineering can be employed and successfully used for selecting grass species for the selected place and contaminant in a more feasible manner than conventional methods. Using this approach, even hyperaccumulation genes can be introduced into the desired plant species [61].

Before genetic engineering approaches are used, certain challenges have to be tackled, such as detoxification strategies of plants. These might be quite complex and may involve a large number of genes, and further choosing the genes for modification to obtain the desired results might be time-consuming, and it might be difficult to obtain positive outcomes. Recently, scientists are trying to modify and overexpress the genes involved in producing the detoxifying enzymes for degrading hazardous chemical compounds or producing certain molecules for capturing, transferring, and acquisition of metals [62]. Apart from that, with the increased knowledge of the adverse effects of genetically modified crops on ecosystems, it has become quite difficult to obtain the permissions for field trials of these GM phytoremediating grasses [63, 64]. At present, several approaches of using phytoremediation along with rhizoremediation may boost plant growth, resistance to heavy metals, and phytoremediation by grasses [53].

Axonopus affinis, a carpet grass, was inoculated with species like *Pseudomonas* sp. ITRH25, *Pantoea* sp. BTRH79, and *Burkholderia* sp. PsJN, and the study observed that bacterial inoculation enhanced the hydrocarbon-degrading bacteria in the rhizosphere, increasing grass biomass and hydrocarbon degradation [65].

In the past half century, there have been several studies confirming the role of arbuscular mycorrhizal fungi in photosynthesis, increased root area for absorption of water, and increased mineral nutrient uptake for essential metals like phosphorus [66]. They enhance the uptake of more lipid and carbohydrates, increase resilience to stress, and generate antioxidant defenses; they also help plants in immobilization, detoxification, and amelioration of heavy metal stress, increase bioabsorption of heavy metals, and decrease metal transfer to shoot parts [27]. The adsorption and stabilization capacity was enhanced by AMF hyphae, and they entangle metal ions on soil particles, reducing their availability and movement [67]. In connection to this, a greenhouse study was done to prove the effectiveness of arbuscular mycorrhizal (AM) fungi (*Rhizophagus intraradices*) along with phytoremediation (vetiver grass) of lead-contaminated soil [68]. Iron concentration and uptake of Fe and Mn was higher in vetiver shoots treated with *Rhizophagus intraradices* at different concentrations of Pb than control treatment [69]. Fungal treatment enhanced the efficiency of Pb metal extraction, its uptake and transfer [57]. Sheoran et al. claim that the ability of rhizosphere bacteria to release enzymes greatly improves the availability of heavy metals [69]. Some endophytes release siderophores and biosurfactants that cause heavy metals in the soil to move around. By chelating with heavy metals, siderophores boost their bioavailability for rhizobacteria and plants [64].

Nanobioremediation, in which nanotechnology is used along with bioremediation, has been a trend now. Nanoparticles, microorganisms, and plants have shown immense potential in adsorbing, detoxifying, or removing pollutants. These nanoparticles can work as catalysts and increase pollutant breakdown [70, 71].

REFERENCES

1. Boorboori, M. R., and Zhang, H.-Y., "Arbuscular mycorrhizal fungi are an influential factor in improving the phytoremediation of arsenic, cadmium, lead, and chromium," *J. Fungi Basel Switz.*, vol. 8, no. 2, p. 176, Feb. 2022, doi: 10.3390/jof8020176.
2. Asante-Badu, B., Kgorutla, L. E., Li, S. S., Danso, P., Xue, Z., and Qiang, G., "Phytoremediation of organic and inorganic compounds in a natural and an agricultural environment: A review," *Appl. Ecol. Environ. Res.*, vol. 18, pp. 6875–6904, Jan. 2020, doi: 10.15666/aeer/1805_68756904.
3. Muthusaravanan, S. *et al.*, "Phytoremediation of heavy metals: Mechanisms, methods and enhancements," *Environ. Chem. Lett.*, vol. 16, no. 4, pp. 1339–1359, Dec. 2018, doi: 10.1007/s10311-018-0762-3.
4. Usman, M. *et al.*, "Nanotechnology in agriculture: Current status, challenges and future opportunities," *Sci. Total Environ.*, vol. 721, p. 137778, Jun. 2020, doi: 10.1016/j.scitotenv.2020.137778.
5. Sladkovska, T., Wolski, K., Bujak, H., Radkowski, A., and Sobol, Ł., "A review of research on the use of selected grass species in removal of heavy metals," *Agronomy*, vol. 12, no. 10, p. 2587, Oct. 2022, doi: 10.3390/agronomy12102587.
6. Singh, S., Inamdar, S., Mitchell, M., and McHale, P., "Seasonal pattern of dissolved organic matter (DOM) in watershed sources: Influence of hydrologic flow paths and autumn leaf fall," *Biogeochemistry*, vol. 118, no. 1–3, pp. 321–337, Apr. 2014, doi: 10.1007/s10533-013-9934-1.
7. Wei, Y. *et al.*, "A systematic review and meta-analysis reveals long and dispersive incubation period of COVID-19," Infectious Diseases (except HIV/AIDS), preprint, Jun. 2020. doi: 10.1101/2020.06.20.20134387.
8. Sharma, J., "Introduction to phytoremediation – A green clean technology." Rochester, NY, May 12, 2018. doi: 10.2139/ssrn.3177321.
9. Yan, R., Zhang, Y., Li, Y., Xia, L., Guo, Y., and Zhou, Q., "Structural basis for the recognition of SARS-CoV-2 by full-length human ACE2," *Science*, vol. 367, no. 6485, pp. 1444–1448, Mar. 2020, doi: 10.1126/science.abb2762.
10. Radziemska, M., Bilgin, A., and Vaverková, M. D., "Application of mineral-based amendments for enhancing phytostabilization in *Lolium perenne* L. cultivation," *CLEAN – Soil Air Water*, vol. 46, no. 1, p. 1600679, Jan. 2018, doi: 10.1002/clen.201600679.
11. Marques, A., Rangel, A., and Castro, P., "Remediation of heavy metal contaminated soils: Phytoremediation as a potentially promising clean-up technology," *Crit. Rev. Environ. Sci. Technol.*, vol. 39, Aug. 2009, doi: 10.1080/10643380701798272.
12. Sladkovska, T., Wolski, K., Bujak, H., Radkowski, A., and Sobol, Ł., "A review of research on the use of selected grass species in removal of heavy metals," *Agronomy*, vol. 12, no. 10, Art. no. 10, Oct. 2022, doi: 10.3390/agronomy12102587.

13. Mishra, L., Gupta, T., and Shree, A., "Online teaching-learning in higher education during lockdown period of COVID-19 pandemic," *Int. J. Educ. Res. Open*, vol. 1, p. 100012, 2020, doi: 10.1016/j.ijedro.2020.100012.

14. Suman, J., Uhlik, O., Viktorova, J., and Macek, T., "Phytoextraction of heavy metals: A promising tool for clean-up of polluted environment?" *Front. Plant Sci.*, vol. 9, 2018, Accessed: May 31, 2023. [Online]. Available: https://www.frontiersin.org/articles/10.3389/fpls.2018.01476

15. Shahsavari, E., Aburto-Medina, A., Taha, M., and Ball, A. S., "Phytoremediation of PCBs and PAHs by grasses: A critical perspective," in *Phytoremediation: Management of Environmental Contaminants, Volume 4*, A. A. Ansari, S. S. Gill, R. Gill, G. R. Lanza, and L. Newman, Eds., Cham: Springer International Publishing, 2016, pp. 3–19. doi: 10.1007/978-3-319-41811-7_1.

16. Pradhan, A.K., Pradhan, N., "Microbial Biosurfactant for Hydrocarbons and Heavy Metals Bioremediation," in: Sukla, L.B., Pradhan, N., Panda, S., Mishra, B.K. (Eds.), *Environmental Microbial Biotechnology, Soil Biology*. Springer International Publishing, Cham, 2015, pp. 91–104, doi: 10.1007/978-3-319-19018-1_5.

17. Gerhardt, K. E., Huang, X.-D., Glick, B. R., and Greenberg, B. M., "Phytoremediation and rhizoremediation of organic soil contaminants: Potential and challenges," *Plant Sci.*, vol. 176, no. 1, pp. 20–30, Jan. 2009, doi: 10.1016/j.plantsci.2008.09.014.

18. Azeez, N., Al-Abbawy, D., and Al- Nabhan, E., "The use of annual beard grass in phytoremediation of petroleum-contaminated soils," *J. Sustainability Sci. and Management*, vol. 18, pp. 87–97, Aug. 2023, doi: 10.46754/jssm.2023.08.006.

19. Hatfield, J. L., and Prueger, J. H., "Temperature extremes: Effect on plant growth and development," *Weather Clim. Extrem.*, vol. 10, pp. 4–10, Dec. 2015, doi: 10.1016/j.wace.2015.08.001.

20. Basumatary, B., Saikia, R., and Bordoloi, S., "Phytoremediation of crude oil contaminated soil using nut grass, *Cyperus rotundus*," *J. Environ. Biol.*, vol. 33, no. 5, pp. 891–896, Sep. 2012.

21. Mottaghi, S., Bahmani, O., and Pak, V.A., "Phytoremediation of Diesel Contaminated Soil Using Urban Wastewater and Its Effect on Soil Concentration and Plant Growth," *Water Supply* vol.11, 8104–19, Aug. 2022, doi: 10.2166/ws.2022.312.

22. Soleimani, B. *et al.*, "Genome wide association study of frost tolerance in wheat," *Sci. Rep.*, vol. 12, no. 1, Art. no. 1, Mar. 2022, doi: 10.1038/s41598-022-08706-y.

23. Xia, H. P., "Ecological rehabilitation and phytoremediation with four grasses in oil shale mined land," *Chemosphere*, vol. 54, no. 3, pp. 345–353, Jan. 2004, doi: 10.1016/S0045-6535(03)00763-X.

24. Rodriguez-Campos, J. *et al.*, "Bioremediation of soil contaminated by hydrocarbons with the combination of three technologies: Bioaugmentation, phytoremediation, and vermiremediation," *J. Soils Sediments*, vol. 19, no. 4, pp. 1981–1994, Apr. 2019, doi: 10.1007/s11368-018-2213-y.

25. Araújo, M. S., Bolnick, D. I., and Layman, C. A., "The ecological causes of individual specialisation," *Ecol. Lett.*, vol. 14, no. 9, pp. 948–958, Sep. 2011, doi: 10.1111/j.1461-0248.2011.01662.x.

26. Okem, A., Kulkarni, M. G., and Staden, J. V., "Enhancing phytoremediation potential of *Pennisetum clandestinum*hochst in cadmium-contaminated soil using smoke-water and smoke-isolated karrikinolide," *Int. J. Phytoremediation*, vol.17, no. 11, pp.1046–1052, 2015, doi: 10.1080/15226514.2014.981245.

27. Carbonell-Barrachina, A. A., Burlo, F., Lopez, E., and Martinez-Sanchez, F., "Arsenic toxicity and accumulation in radish as affected by arsenic chemical speciation," *J. Environ. Sci. Health Part B*, vol. 34, no. 4, pp. 661–679, Jul. 1999, doi: 10.1080/03601239909373220.

28. Rabelo, M. M., Paula-Moraes, S. V., Pereira, E. J. G., and Siegfried, B. D., "Contrasting susceptibility of lepidopteran pests to diamide and pyrethroid insecticides in a region

of overwintering and migratory intersection," *Pest Manag. Sci.*, vol. 76, no. 12, pp. 4240–4247, Dec. 2020, doi: 10.1002/ps.5984.

29. Niu, K., Zhang, R., Zhu, R., Wang, Y., Zhang, D., and Ma, H., "Cadmium stress suppresses the tillering of perennial ryegrass and is associated with the transcriptional regulation of genes controlling axillary bud outgrowth," *Ecotoxicol. Environ. Saf.*, vol. 212, p. 112002, Apr. 2021, doi: 10.1016/j.ecoenv.2021.112002.

30. Liu, Z. *et al.*, "Liu et al 2009 Science EOT dataset," 2009, doi: 10.13140/RG.2.1. 3045.5127.

31. De Lima, C. Z., Buzan, J. R., Moore, F. C., Baldos, U. L. C., Huber, M., and Hertel, T. W., "Heat stress on agricultural workers exacerbates crop impacts of climate change," *Environ. Res. Lett.*, vol. 16, no. 4, p. 044020, Apr. 2021, doi: 10.1088/1748-9326/abeb9f.

32. Patra, J., Lenka, M., and Panda, B. B., "Tolerance and co-tolerance of the grass *Chloris barbata* Sw. to mercury, cadmium and zinc," *New Phytol.*, vol. 128, no. 1, pp. 165–171, 1994, doi: 10.1111/j.1469-8137.1994.tb03999.x.

33. Ding, Y., Shi, Y., and Yang, S., "Advances and challenges in uncovering cold tolerance regulatory mechanisms in plants," *New Phytol.*, vol. 222, no. 4, pp. 1690–1704, Jun. 2019, doi: 10.1111/nph.15696.

34. Rauser, W. E., and Winterhalder, E. K., "Evaluation of copper, nickel, and zinc tolerances in four grass species," *Can. J. Bot.*, vol. 63, no. 1, pp. 58–63, Jan. 1985, doi: 10.1139/b85-009.

35. Korzeniowska, J., and Stanislawska-Glubiak, E., "The phytoremediation potential of local wild grass versus cultivated grass species for zinc-contaminated soil," *Agronomy*, vol. 13, no. 1, p. 160, Jan. 2023, doi: 10.3390/agronomy13010160.

36. Gołda, S., and Korzeniowska, J., "Comparison of phytoremediation potential of three grass species in soil contaminated with cadmium," *Ochr. Srodowiska Zasobów Nat.*, vol. 27, no. 1, pp. 8–14, Mar. 2016, doi: 10.1515/oszn-2016-0003.

37. Laghlimi, M., Baghdad, B., Hadi, H. E., and Bouabdli, A., "Phytoremediation mechanisms of heavy metal contaminated soils: A review," *Open J. Ecol.*, vol. 05, no. 08, pp. 375–388, 2015, doi: 10.4236/oje.2015.58031.

38. Barona, A., Aranguiz, I., and Elías, A., "Metal associations in soils before and after EDTA extractive decontamination: Implications for the effectiveness of further cleanup procedures," *Environ. Pollut.*, vol. 113, no. 1, pp. 79–85, Jun. 2001, doi: 10.1016/S0269-7491(00)00158-5.

39. Chen, Y., Shen, Z., and Li, X., "The use of vetiver grass (*Vetiveria zizanioides*) in the phytoremediation of soils contaminated with heavy metals," *Appl. Geochem.*, vol. 19, no. 10, pp. 1553–1565, Oct. 2004, doi: 10.1016/j.apgeochem.2004.02.003.

40. Parvin, S., Van Geel, M., Yeasmin, T., Lievens, B., and Honnay, O., "Variation in arbuscular mycorrhizal fungal communities associated with lowland rice (*Oryza sativa*) along a gradient of soil salinity and arsenic contamination in Bangladesh," *Sci. Total Environ.*, vol. 686, pp. 546–554, Oct. 2019, doi: 10.1016/j.scitotenv.2019.05.450.

41. Smith, R. A. H., and Bradshaw, A. D., "The use of metal tolerant plant populations for the reclamation of metalliferous wastes," *J. Appl. Ecol.*, vol. 16, no. 2, p. 595, Aug. 1979, doi: 10.2307/2402534.

42. Wang, W., Xue, J., You, J., Zhang, Z., Qi, H., and Zhang, X., "Phytoremediation of multi-metal contaminated copper tailings with herbaceous plant and composite amendments," In Review, preprint, Aug. 2023. doi: 10.21203/rs.3.rs-3041107/v1.

43. Masinire, F., Adenuga, D. O., Tichapondwa, S. M., and Chirwa, E. M. N., "Phytoremediation of Cr(VI) in wastewater using the vetiver grass (*Chrysopogon zizanioides*)," *Miner. Eng.*, vol. 172, p. 107141, Oct. 2021, doi: 10.1016/j.mineng.2021.107141.

44. Khalid, M. *et al.*, "Responses of microbial communities in rhizocompartments of king grass to phytoremediation of cadmium-contaminated soil," *Sci. Total Environ.*, vol. 904, p. 167226, Dec. 2023, doi: 10.1016/j.scitotenv.2023.167226.

45. AL-Huqail, A. A., "Stimulating the efficiency of Cd-phytoremediation from contaminated soils by *Solanum nigrum* L.: Effect of foliar and soil application of yeast extract," *South Afr. J. Bot.*, vol. 161, pp. 512–518, Oct. 2023, doi: 10.1016/j.sajb.2023.08.053.

46. Bilgaiyan, P., Shivhare, N., and Gowripathi Rao, N. R. N. V., "Phytoremediation of wastewater through implemented wetland – A review," *E3S Web Conf.*, vol. 405, p. 04026, 2023, doi: 10.1051/e3sconf/202340504026.

47. Bharadvaja, N. Garima, "Phytonanoremediation of metals and organic waste in wastewater treatment," in *Advanced Application of Nanotechnology to Industrial Wastewater*, M. P. Shah, Ed., Singapore: Springer Nature, 2023, pp. 241–261. doi: 10.1007/978-981-99-3292-4_12.

48. Ferrario, C. *et al.*, "Assessment of reed grasses (*Phragmites australis*) performance in pfas removal from water: A phytoremediation pilot plant study," *Water*, vol. 14, no. 6, p. 946, Mar. 2022, doi: 10.3390/w14060946.

49. Mustapha, O. R., Osobamiro, T. M., Sanyaolu, N. O., and Alabi, O. M., "Adsorption study of methylene blue dye: An effluents from local textile industry using *Pennisteum pupureum* (elephant grass)," *Int. J. Phytoremediation*, vol. 25, no. 10, pp. 1348–1358, Aug. 2023, doi: 10.1080/15226514.2022.2158781.

50. Trang, T. D., Tung, N. C. T., Han, P. T., and Viet, V. H., "Screening wetland and forage plants for phytoremediation of salt-affected soils in the Vietnamese Mekong Delta," *Bull. Environ. Contam. Toxicol.*, vol. 110, no. 1, p. 29, Jan. 2023, doi: 10.1007/s00128-022-03667-4.

51. Qadir, M., Qureshi, R. H., and Aheavy metalsad, N., "Amelioration of calcareous saline sodic soils through phytoremediation and chemical strategies," *Soil Use Manag.*, vol. 18, no. 4, pp. 381–385, Jan. 2006, doi: 10.1111/j.1475-2743.2002.tb00256.x.

52. Danh, L. T., Truong, P., Mammucari, R., Tran, T., and Foster, N., "Vetiver grass, *Vetiveria zizanioides* : A choice plant for phytoremediation of heavy metals and organic wastes," *Int. J. Phytoremediation*, vol. 11, no. 8, pp. 664–691, Oct. 2009, doi: 10.1080/15226510902787302.

53. Yadav, D., Yadav, A., Singh, M., and Khare, P., "Cultivation of aromatic plant for nature-based sustainable solutions for the management of degraded/marginal lands: Techno-economics and carbon dynamic," *Carbon Res.*, vol. 2, no. 1, p. 27, Jul. 2023, doi: 10.1007/s44246-023-00055-3.

54. Qi, X. *et al.*, "Integrated phytoremediation system for uranium-contaminated soils by adding a plant growth promoting bacterial mixture and mowing grass," *J. Soils Sediments*, vol. 19, no. 4, pp. 1799–1808, Apr. 2019, doi: 10.1007/s11368-018-2182-1.

55. Antonkiewicz, J., Kołodziej, B., and Bielińska, E. J., "The use of reed canary grass and giant miscanthus in the phytoremediation of municipal sewage sludge," *Environ. Sci. Pollut. Res.*, vol. 23, no. 10, pp. 9505–9517, May 2016, doi: 10.1007/s11356-016-6175-6.

56. Singhal, V., and Rai, J. P. N., "Biogas production from water hyacinth and channel grass used for phytoremediation of industrial effluents," *Bioresour. Technol.*, vol. 86, no. 3, pp. 221–225, Feb. 2003, doi: 10.1016/S0960-8524(02)00178-5.

57. Kumar, D., Singh, B., and Sharma, Y. C., "Bioenergy and phytoremediation potential of *Millettia pinnata*," in *Phytoremediation Potential of Bioenergy Plants*, K. Bauddh, B. Singh, and J. Korstad, Eds., Singapore: Springer Singapore, 2017, pp. 169–188. doi: 10.1007/978-981-10-3084-0_6.

58. Rezania, S., Park, J., Rupani, P. F., Darajeh, N., Xu, X., and Shahrokhishahraki, R., "Phytoremediation potential and control of *Phragmites australis* as a green phytomass: An overview," *Environ. Sci. Pollut. Res.*, vol. 26, no. 8, pp. 7428–7441, Mar. 2019, doi: 10.1007/s11356-019-04300-4.

59. Wrobel, C., Coulman, B. E., and Smith, D. L., "The potential use of reed canarygrass (*Phalaris arundinacea* L.) as a biofuel crop," *Acta Agric. Scand. Sect. B - Plant Soil Sci.*, vol. 59, no. 1, pp. 1–18, Jan. 2009, doi: 10.1080/09064710801920230.

60. Osman, N. A., Roslan, A. M., Ibrahim, M. F., and Hassan, M. A., "Potential use of *Pennisetum purpureum* for phytoremediation and bioenergy production: A mini review," *Asia Pac. J. Mol. Biol. Biotechnol.*, vol. 28, no. 1, pp. 14–26, Jan. 2020, doi: 10.35118/apjmbb.2020.028.1.02.

61. Wu, F. *et al.*, "Unraveling cadmium toxicity in *Trifolium repens* L. seedling: Insight into regulatory mechanisms using comparative transcriptomics combined with physiological analyses," *Int. J. Mol. Sci.*, vol. 23, no. 9, p. 4612, Apr. 2022, doi: 10.3390/ijms23094612.

62. Pathak, B., Khan, R., Fulekar, J., and Fulekar, M., "Biotechnological strategies for enhancing phytoremediation," *Biotechnol. Crucif.*, pp. 63–90, Apr. 2013, https://doi.org/10.1007/978-1-4614-7795-2_5.

63. Raklami, A., Meddich, A., Oufdou, K., and Baslam, M., "Plants-microorganisms-based bioremediation for heavy metal cleanup: Recent developments, phytoremediation techniques, regulation mechanisms, and molecular responses," *Int. J. Mol. Sci.*, vol. 23, no. 9, p. 5031, May 2022, doi: 10.3390/ijms23095031.

64. Yan, A., Wang, Y., Tan, S. N., Mohd Yusof, M. L., Ghosh, S., and Chen, Z., "Phytoremediation: A promising approach for revegetation of heavy metal-polluted land," *Front. Plant Sci.*, vol. 11, p. 359, Apr. 2020, doi: 10.3389/fpls.2020.00359.

65. Tara, N., Afzal, M., Ansari, T. M., Tahseen, R., Iqbal, S., and Khan, Q. M., "Combined use of alkane-degrading and plant growth-promoting bacteria enhanced phytoremediation of diesel contaminated soil," *Int. J. Phytoremediation*, vol. 16, no. 12, pp. 1268–1277, 2014, doi: 10.1080/15226514.2013.828013.

66. Parniske, M., "Arbuscular mycorrhiza: The mother of plant root endosymbioses," *Nat. Rev. Microbiol.*, vol. 6, no. 10, pp. 763–775, Oct. 2008, doi: 10.1038/nrmicro1987.

67. Wang, F. Y., Wang, L., Shi, Z. Y., Li, Y. J., and Song, Z. M., "Effects of AM inoculation and organic amendment, alone or in combination, on growth, P nutrition, and heavy-metal uptake of tobacco in Pb-Cd-contaminated soil," *J. Plant Growth Regul.*, vol. 31, no. 4, pp. 549–559, Dec. 2012, doi: 10.1007/s00344-012-9265-9.

68. Bahraminia, M., Zarei, M., Ronaghi, A., and Ghasemi-Fasaei, R., "Effectiveness of arbuscular mycorrhizal fungi in phytoremediation of lead-contaminated soil by vetiver grass," *Int. J. Phytoremediation*, vol. 18, no. 7, pp. 730–737, Jul. 2016, doi: 10.1080/15226514.2015.1131242

69. Banerjee, R., Goswami, P., Lavania, S., Mukherjee, A., and Lavania, U. C., "Vetiver grass is a potential candidate for phytoremediation of iron ore mine spoil dumps,"*Ecol. Eng.*, vol. 132, pp. 120–136, Jul. 2019, doi: 10.1016/j.ecoleng.2018.10.012.

70. Sheoran, V., Sheoran, A. S., and Poonia, P., "Role of hyperaccumulators in phytoextraction of metals from contaminated mining sites: A review," *Crit. Rev. Environ. Sci. Technol.*, vol. 41, no. 2, pp. 168–214, Dec. 2010, doi: 10.1080/10643380902718418.

71. Rajput, V. D. *et al.*, "Nanotechnology in the restoration of polluted soil," *Nanomaterials*, vol. 12, no. 5, p. 769, Feb. 2022, doi: 10.3390/nano12050769.

7 A Review of Phytoremediation Techniques Used for Removal of Arsenic from Groundwater

Akansha Singh, Sandhya Maurya, Abhishek Saxena, and Garima Gupta

INTRODUCTION

Arsenic pollution in drinking water presents a significant worldwide challenge, impacting both the environment and public health on a global scale. Current estimates indicate that a considerable number of people, ranging from 95 to 230 million, are affected by this issue. About 105 countries face the risk of exposure to elevated levels of arsenic in groundwater, with the majority (94%) of these cases concentrated in Asia (M. U. Khan et al., 2023; Rajendran et al., 2021). Addressing arsenic-contaminated water is a significant concern in various regions across Asia. These regions include the Ganga-Brahmaputra-Meghna river system spanning India and Bangladesh, the Red River in Vietnam, the Mekong River that flows through Laos and Cambodia, the Indus River in Pakistan, the Irrawaddy River in Myanmar, and the Yellow River in China.

In the past, arsenic had various industrial applications, although its use has significantly diminished due to its toxicity. It was utilized in the production of pesticides, herbicides, wood preservatives, alloys, and certain pharmaceuticals. However, due to concerns about health and the environment, numerous countries have either restricted or completely banned its use in consumer products. To address the severe consequences of arsenic exposure on human health and the ecosystem, leading health organizations like the World Health Organization (WHO) and the International Agency for Research on Cancer (IARC) have established strict guidelines for arsenic (0.01 mg/L) in drinking water.

Arsenic, a naturally occurring element within the earth's crust, exists in various forms and concentrations across different environmental contexts (Ambrožič-Dolinšek et al., 2023). Its presence can arise from either natural processes or human activities, with potential exposure pathways encompassing air, drinking water, and food. Groundwater contamination with arsenic can result in decreased agricultural

DOI: 10.1201/9781003442295-7

135

productivity and compromised food quality, as arsenic infiltrates the food chain, leading to disruptions (A. K. Patel et al., 2022). Long-term consumption of arsenic-contaminated drinking water is closely linked to adverse effects on bodily systems and organs, including the cardiovascular, respiratory, nervous, and endocrine systems, as well as sensitive tissues and the skin (A. K. Patel et al., 2022). Various methods have been implemented worldwide to remediate arsenic contamination, each carrying its own advantages and disadvantages. A comprehensive analysis of existing scholarly literature has provided insights into these methods and their implications. Arsenic can exist in different forms.

CHEMISTRY OF ARSENIC

Arsenic is a chemical element with the symbol As and atomic number 33. It occurs naturally in the earth's crust and can be found in various forms and concentrations in different environments.

Arsenic is highly toxic to humans and animals. Arsenic forms organic as well as inorganic compounds, and its compounds can pose serious health risks. Prolonged exposure to arsenic can lead to such health problems as skin lesions, cardiovascular diseases, and an increased risk of cancer.

Organic Arsenic Compounds

In organic compounds, arsenic is bonded to carbon atoms. These compounds are primarily found with biological processes in nature. Arsanilic acid, roxarsone, and arsphenamine (Salvarsan), which was historically used as an antibiotic, are a few examples.

Inorganic Arsenic Compounds

a. Arsenic trioxide (As_2O_3) is a white compound commonly used in the production of glass and pesticides. It is also a by-product of the roasting of certain metal ores. Arsenic trioxide is toxic and was historically used as a poison (Rahidul Hassan, 2023).

b. Arsenic pentoxide (As_2O_5) is a white, odorless solid and also an excellent oxidizing agent used in the manufacture of insecticides, wood preservatives, and dyes.

c. Arsenic acid (H_3AsO_4) is a weak acid and exists as a white, crystalline solid. It is used in the production of certain pharmaceuticals, insecticides, and pigments.

d. Arsenic trichloride ($AsCl_3$) is a colorless, oily liquid with a pungent odor and is used as a precursor in the synthesis of various organic compounds and as a reagent in chemical reactions.

e. Arsenic sulfide (As_2S_3) is a yellow-orange solid compound and is commonly found in nature as the mineral orpiment. It is used as a pigment and in traditional medicine.

These methods allow us to determine the concentration of arsenic in different samples, including water, soil, and biological tissues.

SOURCES OF ARSENIC

Geogenic Sources

Geogenic sources relate to arsenic that occurs naturally in geological formations and that is released into groundwater as a result of natural processes. These processes involve the release of arsenic from the decomposition of rocks and sediments into groundwater, and it is estimated that over 90% of arsenic pollution can be attributed to this natural phenomenon (Shaji et al., 2021). Geological processes can result in the long-term deposition of sediments containing a lot of arsenic in formations that include groundwater. Through leaching and breakdown, these sediments may provide arsenic to the groundwater (Shaji et al., 2021).

Realgar (As_4S_4), arsenopyrite (FeAsS), and pyrite (FeS_2),) are minerals that contain arsenic naturally. Arsenic is released into the nearby groundwater as a result of the weathering and degradation of these minerals over time. Various natural mechanisms lead to the liberation of arsenic, including hydrothermal deposits, volcanic emissions, geothermal events, forest fires, wind-driven dust, and ocean spray. Consequently, this release has the potential to transport arsenic across extensive distances, whether as suspended particles or in gaseous states, through either air or water (Abiye & Bhattacharya, 2019). Due to the arsenic-rich minerals found in volcanic rocks, areas with volcanic activity may have increased levels of arsenic. As they weather, these minerals may emit arsenic into the groundwater. Geothermal fluids that are produced deep under the earth's crust can mobilize arsenic. Arsenic may be carried into groundwater by these fluids when they pass through cracks and porous rocks. According to Maciag et al. (2023),the source of heightened arsenic levels in groundwater can be attributed to minerals such as cordierite and pyrite, which dominate arsenic contamination in the South Mountain Batholith. In addition, human activities like irrigation and groundwater pumping can worsen the release of arsenic into groundwater from geogenic sources. For controlling and reducing arsenic poisoning in groundwater, regular monitoring and knowledge of the regional geological and hydrogeochemical conditions are essential.

Anthropogenic Sources

The contamination of groundwater with arsenic from anthropogenic (human-produced) sources is a serious environmental issue. Although arsenic occurs naturally in the earth's crust, some human activities can mobilize and release arsenic into groundwater, contaminating it. In addition, arsenic can leach into groundwater through industrial processes such as metal smelting and coal combustion. If arsenic-containing insecticides, herbicides, and fertilizers are used in agriculture, arsenic may build up in soil and seep into groundwater. Arsenical pesticides were once widely used in agriculture, but their remnants can still remain in the environment for a very long time (M. U. Khan et al., 2023). Excessive extraction of groundwater for residential and agricultural use (Maity et al., 2020)also contribute As. Arsenic may pollute land and water when industrial waste, such as mine tailings, slag, and other wastes, are improperly disposed of(Abiye & Bhattacharya, 2019). These waste products might not be properly contained and may have significant arsenic contents,

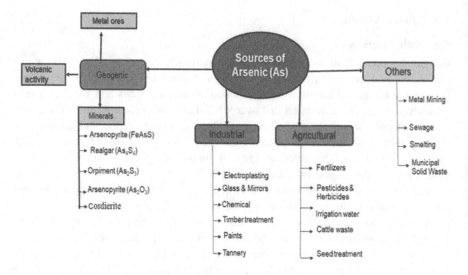

FIGURE 7.1   Sources of arsenic contamination in groundwater.

which might cause groundwater to leach from them. Sources of arsenic contamination in groundwater are given in Figure 7.1.

Untreated or inadequately treated wastewater from homes, businesses, and wastewater treatment plants contains arsenic. These effluents have the potential to contaminate groundwater with arsenic when they are released into the environment or seep into the soil. Sustainable agriculture methods and improved urban planning are needed to address anthropogenic arsenic pollution in groundwater. To locate polluted regions and take the necessary precautions to reduce the hazards to human health and the environment from exposure to arsenic, regular monitoring and testing of groundwater quality are vital.

DISTRIBUTION OF ARSENIC IN INDIA AND THE GLOBE

Groundwater contamination due to arsenic has emerged as a significant global issue, affecting diverse regions. This issue is especially notable in several Asian nations, such as Bangladesh, India, Pakistan, China, Nepal, Vietnam, Myanmar, Thailand, and Cambodia. The impact of this contamination is also observed in regions across Africa, Argentina, North and South America, as well as Pakistan, the USA, and Canada (Nilkarnjanakul et al., 2023; Pramparo et al., 2023; Xing et al., 2023). Notably, Asia, with 32 documented cases, and Europe, with 31, exhibit the highest reported instances of arsenic contamination. This is followed by Africa (20), North America (11), South America (9), and Australia (4) (Shaji et al., 2021). In China, the issue of arsenic contamination is widespread, encompassing over 20 provinces. Significant areas affected by this problem include Xinjiang, Shanxi, Inner Mongolia, Jilin, and Ningxia, as outlined by a comprehensive study led by Zhu and colleagues in 2023. Bangladesh, positioned in the Bengal delta, faces a critical situation with arsenic contamination in its groundwater, impacting 61 out of its 64 districts (Ivy et al., 2023). In Bangladesh alone, nearly 43,000 deaths are attributed to chronic As exposure every year (Preetha et al., 2023).

India, heavily reliant on groundwater for domestic and irrigation purposes, confronts arsenic contamination in 20 states and four Union Territories. The Gangetic Plain, covering approximately 0.25 million square kilometers, serves as a principal source of groundwater, contributing to nearly one-third of the country's approximately 27 million wells. States like West Bengal and Bihar are particularly burdened by extensive arsenic contamination. Other affected regions are Jharkhand, Assam, and Manipur. The problem of increased arsenic concentrations in groundwater within the Gangetic Plain was first identified in the state of West Bengal, specifically in the lower Gangetic Plain region. Subsequent investigations uncovered arsenic contamination in multiple areas across the middle and upper Gangetic Plain, including districts like Patna, Bhojpur, Saran, and Vaishali in Bihar (Pal et al., 2023). Correspondingly, several districts in Uttar Pradesh, encompassing Ballia, Ghazipur, and certain parts of Varanasi, have also reported heightened arsenic levels in their groundwater (Kumar et al., 2010).

HEALTH AND ENVIRONMENTAL CONCERNS ABOUT GROUNDWATER ARSENIC

Arsenic is a highly toxic element that poses significant risks to both humans and animals. Prolonged exposure to this substance can lead to a variety of health problems (Mahamallik & Swain, 2023). Individuals and animals alike face serious consequences when exposed to arsenic, particularly through contaminated groundwater. This exposure can result in the development of abnormal skin conditions, keratinized skin, an elevated risk of cancer, cardiorespiratory diseases, reproductive complications, and cognitive impairments in both adults and children (Bundschuh et al., 2021).

Furthermore, extended contact with arsenic-contaminated water sources heightens the probability of contracting various illnesses, such as conjunctivitis, skin cancer, and kidney cancer (Zhu et al., 2023).

Forms of Arsenic and Their Impact

Arsenic exists in two primary forms: inorganic and organic. Among these, inorganic arsenic compounds are notably more toxic and can trigger a range of health problems over time. Consuming tainted drinking water containing arsenic-laden minerals constitutes the primary pathway of exposure to inorganic arsenic. Inorganic arsenic, particularly in its arsenite form, possesses distinctive characteristics, including solubility, bioaccumulation (K. S. Patel et al., 2023), chemical reactivity, and genotoxicity. These characteristics contribute to its heightened harmful effects in comparison to other arsenic forms.

Long-Term Effects of Exposure

Long-term exposure to elevated levels of inorganic arsenic can result in a condition known as arsenicosis. This condition manifests in a variety of health problems. To better understand the toxicity levels and stages of arsenicosis, refer to Figure 7.2.

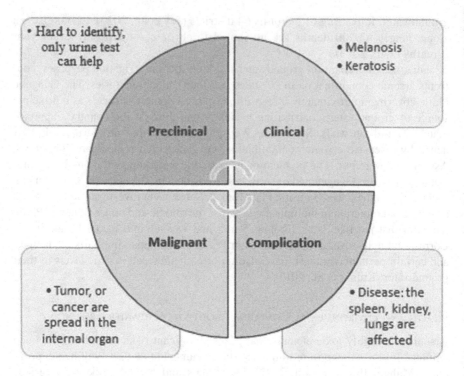

FIGURE 7.2   Stages of arsenicosis.

REMEDIATION METHODS USED FOR ARSENIC

Arsenic has been recognized as an important contaminant in drinking water. Due to the prevalence of high concentrations of arsenic in groundwater, a wide range of technologies have been tried for the removal of arsenic from drinking water (Rahidul Hassan, 2023). At present, a wide array of technologies is in use to address the challenge of polluted groundwater. The primary objective is to discover an economical method for treating water contaminated with arsenic, particularly in developing nations. These diverse technologies rely on a range of physicochemical and biological mechanisms. Arsenic contamination in water has become a major public health problem, so provision of arsenic-free water is urgently needed for the protection of health and well-being of the people, especially for the people residing in the acute arsenic-affected areas (J. Khan et al., 2023). There are many different technologies proposed and used for the removal of arsenic. A variety of technologies are utilized for arsenic remediation from groundwater (Mahamallik & Swain, 2023). These methods encompass physicochemical treatment processes such as oxidation, coagulation-flocculation, adsorption, biological sorption, ion exchange, membrane processes, treatment involving bio-organisms, and electrocoagulation. These approaches have been extensively documented as effective means for addressing arsenic contamination in groundwater (Kumar et al., 2010; Omar et al., 2023). Remediation techniques for arsenic removal are illustrated in Figure 7.3.

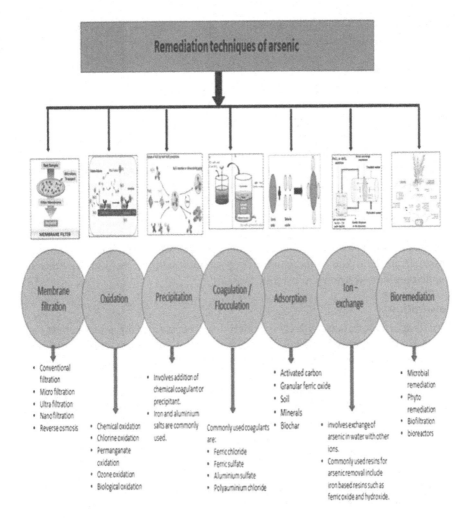

FIGURE 7.3 Remediation techniques for arsenic removal.

Membrane Filtration

Membrane filtration techniques like reverse osmosis, nanofiltration, and electrodialysis are efficient in removing contaminants such as arsenic from water. Membranes are typically made from synthetic material consisting of billions of pores or microscopic holes, which acts as a selective barrier in removing the contaminants. Contaminants, including substances like arsenic and various impurities, are retained within the membrane, enabling only clean and fresh water to permeate. The transportation of molecules through the membrane during the filtration process requires a motivating factor, such as a pressure differential existing between the membrane's two sides. The categorization of membrane filtration methods encompasses four primary types, which are determined by the characteristics of the employed membrane: microfiltration (MF), ultrafiltration (UF), nanofiltration (NF), and reverse osmosis

(RO) (Figure 7.3). Among all membrane filtration techniques, reverse osmosis is the most prevalent and widely used.

Reverse Osmosis

Reverse osmosis is a water purification process which uses a semipermeable membrane to filter out unwanted molecules such as contaminants, sediments, and dirt. This technique is widely employed to eliminate arsenic from potable water and is both cost-effective and efficient. During this procedure, hydraulic pressure is exerted on one side of the semipermeable membrane, compelling water to move across it, while contaminants like arsenic remain trapped within the membrane. Consequently, only purified water is allowed to pass through. It removes approximately 80–95% of the arsenic content, depending upon the device manufacturer, water pH, amount of water filtered, and other factors.

In the reverse osmosis system, there are two solutions of different concentrations separated by a semipermeable membrane. The water containing contaminants passes from the dilute to the concentrated solution with the help of pressure; this applied pressure is known as osmotic pressure.

Adsorption

This is a simple process which is widely used because of its effectiveness, high removal efficiency, and low cost for the removal of substances either from liquid or gaseous solutions. In this process, the contaminated water is passed through the solid platform of solid adsorption media filling an adsorption vessel or column. Iron-based adsorbents, various types of activated carbon, and a number of low-cost materials—including agricultural residues or by-products such as rice husk, coconut shell, almond shell and waste or by-products from industries such as bone char, lignite char, biochar, fly ash, and red mud, also known as mud—have frequently been employed to remediate water contaminated with arsenic (Masood ul Hasan et al., 2023) (Figure 7.4).

As the water passes through the bed of solid adsorption media, the contaminants (such as arsenic) are adsorbed onto the surface of the activated adsorbent. Thus, arsenic-free water passes out from the media. Gradually, the adsorbent column becomes saturated due to arsenic adsorption, so it needs to be regenerated, which is carried out by using 4% caustic soda, NaOH. But on the other side, the sorbents need to be replaced after four to five applications of the regeneration process.

The adsorption process has attracted much attention for the elimination of contaminants such as As in water due to its many advantages. These include the fact that it doesn't need a large volume and additional chemicals, it is a cost-effective process with high removal efficiency, it is easier to set up, and it does not produce harmful by-products (Yeo et al., 2021). Adsorption is a viable arsenic removal technique that has been reported to have more than 95% quantitative efficiency for the removal of As, does not require chemical additions, and is simple in impoverished nations with insufficient skilled labor and unstable electrical supplies. It's important to note that the efficacy of this method is impacted by various factors, including temperature, exposure duration, pH level, the presence of additional chemicals, the quantity of adsorbent and adsorbate used, and the initial arsenic concentration.

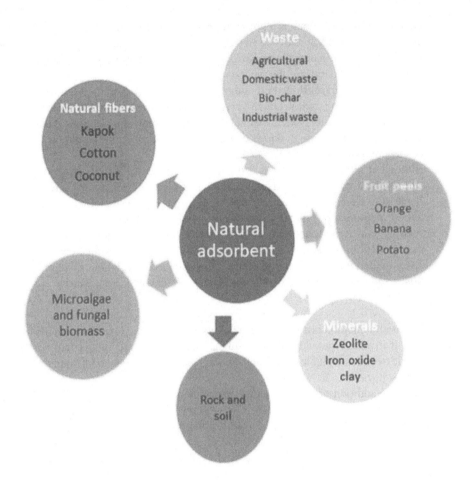

FIGURE 7.4 Natural adsorbents used for As removal.

Oxidation

Oxidation methods are commonly used for arsenic removal from drinking water. These methods involve converting the soluble, toxic forms of arsenic (arsenite) into the less soluble forms (arsenate), which can be easily removed through precipitation or adsorption processes. Some common methods for arsenic removal include:

a. **Chemical oxidation:** Chemical oxidants such as Cl_2, O_3, H_2O_2, or potassium permanganate can be added to the water to convert arsenite to arsenate. This conversion increases the removal efficiency of arsenic during subsequent treatment steps.

b. **Chlorine oxidation:** Chlorine is commonly used as a disinfectant in water treatment. When chlorine is introduced into water, it undergoes a reaction with arsenite, resulting in the formation of arsenate. The reaction rate depends on pH and the chlorine-to-arsenic ratio. After oxidation, the arsenate can be removed using coagulation, adsorption, or other treatment processes.

    c. **Ozone oxidation:** Ozone is a powerful oxidant that can effectively convert arsenite to arsenate. It is dosed into the water, where it reacts with arsenic and other contaminants. Ozone oxidation is often followed by subsequent treatment steps to remove the oxidized arsenic species.

    d. **Permanganate oxidation:** Potassium permanganate ($KMnO_4$) is a strong oxidant that can be used to convert arsenite to arsenate. It is typically added to the water in a controlled manner to ensure complete oxidation. After oxidation, the resulting arsenate can be removed through precipitation or adsorption processes.

    e. **Biological oxidation:** Biological processes, such as microbial oxidation, can also be used to convert arsenite to arsenate. Certain bacteria and microorganisms are capable of catalyzing this conversion. These biological oxidation methods are typically employed in specialized treatment systems, such as constructed wetlands or biofiltration units. It's important to note that oxidation alone may not be sufficient for complete arsenic removal. After oxidation, subsequent treatment steps such as coagulation, precipitation, or adsorption are often employed to remove arsenate from water.

Precipitation

The precipitation method is a widely used technique for arsenic removal from drinking water. It involves the addition of chemical coagulant or precipitant that reacts with arsenic to form insoluble compounds, which then can be removed through sedimentation or filtration.

Prior to the addition of the precipitant, it may be necessary to pretreat the water to adjust its pH or remove any interfering substances. The pH adjustment is crucial because it affects the solubility of arsenic compounds. Chemical coagulants are added to the water to induce the formation of flocs that trap and remove the arsenic. Iron salts, such as $FeCl_3$ or $Fe_2(SO_4)_3$, as well as aluminum salts, such as $Al_2(SO_4)_3$ or poly-aluminum chloride, are often used as coagulants. The chosen coagulant is added to the water, and mixing or gentle agitation is applied to ensure uniform distribution. After the coagulant addition, the water is subjected to gentle mixing or stirring to promote the aggregation of formed precipitates into larger flocs. The flocs formed during flocculation settle under gravity, and the clarified water can be decanted or collected from the top. In some cases, posttreatment may be required to further enhance the quality of the treated water. This can involve additional processes like activated carbon adsorption, pH adjustment, or disinfection to ensure the removal of any residual contaminants and pathogens. It's important to consider that the precipitation method's efficiency can be influenced by factors such as initial arsenic concentration, coagulant dosage, pH, contact time, and water temperature. Furthermore, proper management and disposal of the sludge generated during precipitation are necessary as it contains concentrated arsenic and other contaminants. Appropriate methods should be taken to handle and dispose of the sludge in accordance with local regulations.

Overall, the precipitation method is an effective and widely used technique for arsenic removal from drinking water, offering reliable results when properly implemented and optimized (Figure 7.5).

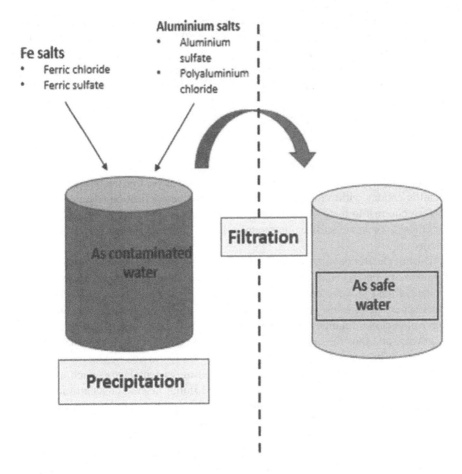

FIGURE 7.5 Precipitation process for As removal.

Coagulation and Flocculation

These are the commonly used chemical processes for the removal of arsenic from drinking water. These processes involve the addition of coagulants and flocculants to destabilize and aggregate the arsenic particles, allowing for their subsequent removal through sedimentation or filtration. The choice of coagulant depends upon various factors such as the water chemistry, pH, and presence of other contaminants (Maurya & Saxena, 2018). Commonly used coagulants for arsenic removal include aluminum-based coagulants and iron-based coagulants. These coagulants can form insoluble hydroxide precipitates, which can absorb arsenic particles. The coagulant is added to the water, typically in a rapid mixing chamber or directly to the water stream. The coagulant destabilizes the negatively charged arsenic particles by neutralizing their charge, allowing them to come together and form larger flocs. In some cases, the addition of a flocculant can aid in the formation of larger, denser flocs. Flocculants are high-molecular-weight polymers that help bind the destabilized particles together, forming larger flocs that settle more readily. The water is gently mixed or stirred to promote the collision and aggregation of the destabilized particles, aiding in the formation of larger flocs. The flocs then undergo sedimentation or settling in a clarifier or sedimentation basin, allowing them to separate from the water. After sedimentation, the water may undergo filtration to remove any remaining suspended particles, including residual arsenic flocs. Different filtration methods such as sand filtration, dual-media filtration, or granular activated carbon (GAC) filtration can be employed.

Electrocoagulation

Electrocoagulation presents an innovative approach compared to traditional chemical precipitation methods. By utilizing iron electrodes, this technique generates metallic cations within the effluent itself through the application of an electric current. In contrast to the chemical precipitation process involving substances like ferric chloride or ferric sulfate, electrocoagulation directly induces the production of anodes.

During electrocoagulation, the Fe(0) anode undergoes oxidation, resulting in the formation of Fe(III). This Fe(III) subsequently reacts with oxyhydroxides and precipitates found in water containing arsenic. This interaction gives rise to binuclear, inner-sphere complexes, which then aggregate to form larger flocs, facilitating the removal of contaminants. After the floc forms it can be easily settled due to the force of gravity, and thus the arsenic can be easily filtered from the contaminated water. It causes pollutants to flocculate and then float. It's worth noting that the electrocoagulation process can be optimized by controlling various factors such as current density, treatment time, electrode spacing, and initial arsenic concentration. Pilot-scale testing is often conducted to determine the optimal operating conditions for a specific water source. Electrocoagulation has been acknowledged as an efficient technique for the removal of arsenic, offering advantages such as simplicity, lower chemical requirements, and the potential for simultaneous removal of other contaminants. However, it is important to monitor and properly handle the metal sludge generated during the process, as it may contain concentrated arsenic and other metals, requiring further disposal and further treatment.

Ion Exchange

Ion exchange is a common method for removing arsenic from drinking water. It involves the exchange of arsenic ions in the water with other ions on a solid exchange medium, known as the ion exchange resin. Here's how ion exchange can be used for arsenic removal:

a. **The selection of ion exchange resin:** Specific ion exchange resins designed to target arsenic removal are used in the process. These resins are typically functionalized with specific chemical groups that have a high affinity for arsenic ions. Commonly used resins for arsenic removal include iron-based resins (ferric oxide/hydroxide) and hybrid anion exchange resins.

b. **Column setup:** The ion exchange resin is packed into a column or vessel, creating a bed of resin through which the water passes. The resin bed is usually contained within a pressure vessel or a packed bed column system.

c. **Pretreatment:** Before coming into contact with the ion exchange resin, it is common to subject the water to pretreatment procedures aimed at enhancing the effectiveness of the ion exchange procedure. This can involve the adjustment of pH or the removal of any interfering substances that might affect the arsenic removal performance.

d. **Arsenic removal:** The water containing arsenic passes through the ion exchange column. During the passage of water through the resin bed, arsenic ions present in the water are drawn toward the functional groups located on the resin material. As a result, these arsenic ions displace other ions that were initially bound to the resin. The equation for the removal of arsenic through ion exchange is as follows:

$$2R - Cl + HAsO_4 \rightarrow R_2HAsO_4 + 2Cl^-$$

e. **Regeneration:** Over time, the ion exchange resin becomes saturated with arsenic ions, reducing its arsenic removal capacity. Regeneration is necessary to restore the resin's capacity for arsenic removal. This involves flushing the resin bed with a regenerant solution that displaces the adsorbed arsenic ions and replaces them with a more easily removable ion, such as chloride or sulfate. The regenerant solution typically contains a strong acid or base.

f. **Disposal of regenerant:** The regenerant solution containing the arsenic and displaced ions needs to be handled and disposed of properly to prevent contamination. It may require appropriate treatment or disposal methods to ensure environmental safety.

Bioremediation

Bioremediation is an alternative approach for removing arsenic from drinking water that utilizes biological processes and organisms to transform or remove the contaminant. Bioremediation offers a sustainable and cost-effective solution to remove arsenic from water sources. It is a technique that uses living organisms, such as bacteria, fungi, or plants, to remove or neutralize pollutants from them into less harmful substances. This approach is used to clean up soil, water, and air contaminated with various pollutants, including oil spills, heavy metals, and organic compounds. This is a sustainable and environmentally friendly method for addressing pollution. The effectiveness of bioremediation depends on various factors, including the types of contaminants, site conditions, the presence of appropriate microbial populations, and the chosen remediation approach. Monitoring and optimization are often required to achieve successful results. The microorganisms that have the ability to remove arsenic from water are identified and isolated. These microorganisms can include bacteria such as *Bacillus*, *Pseudomonas*, and *Clostridium* as well as various types of fungi and algae. The selected microorganisms should have the ability to tolerate and thrive in arsenic-rich environments.

Types of Bioremediation

a. **Microbial bioremediation:** Some types of microorganisms, like bacteria and fungi, can convert various forms of arsenic into less harmful compounds or restrict their mobility. For instance, some bacteria can convert toxic inorganic arsenic (arsenate) into less toxic organic arsenic compounds (arsenobetaine). This process is called microbial methylation. However, it's important to note that the transformed arsenic species may still need to be removed from the water to ensure its safety for drinking.

b. **Biofiltration:** Biofiltration systems utilize microbial communities attached to a filter medium to remove arsenic from water. The microorganisms in the biofilm can interact with the arsenic, either through adsorption or transformation processes, thereby reducing its concentration. The biofiltration systems may employ various filter media, such as activated carbon, sand, or zeolite, to support microbial growth and enhance arsenic removal.

c. **Bioreactors:** Bioreactors are controlled environments where specific microbial processes can be optimized for arsenic removal. Different types of bioreactors, such as packed-bed bioreactors or membrane bioreactors, can be used to create conditions that encourage the growth of arsenic-transforming microorganisms. These microorganisms can convert arsenic species into less toxic forms or immobilize them within the bioreactor.

Bioremediation Pathways for Arsenic Removal

There are various pathways that use specific types of genes and enzymes to carry out the mechanism of arsenic biotransformation. Arsenic is a toxic heavy metal that can contaminate water sources, posing a serious threat to human health. Several bioremediation pathways can be employed to remove arsenic from water. The bioremediation of arsenic involves many pathways.

a. **Uptake:** With the help of microbes, the arsenic is taken up from the groundwater.

b. **Accumulation:** The taken-up arsenic is then accumulated in the microbes. Microbes that have a high accumulating tendency for arsenic are used here for the accumulation pathway of bioremediation.

c. **Reduction:** The reduction of As(V) to As(III) by dissimilatory reduction during anaerobic respiration takes place. Certain bacteria, such as sulfate-reducing bacteria (SRB) and iron-reducing bacteria (IRB), can convert soluble arsenic compounds into less toxic and less mobile forms by reducing arsenic(V) to arsenic(III). This conversion makes arsenic easier to immobilize and precipitate.

d. **Oxidation:** Arsenite oxidase, an enzyme which is responsible for the oxidation of arsenite to arsenate, is synthesized from the bacterial species. Some bacteria are capable of oxidizing arsenic(III) to arsenic(V), which can then be precipitated or adsorbed onto solid surfaces, reducing its mobility and toxicity. This process can be particularly effective in conjunction with other treatment methods. The bacterial species include *Thermus thermophilus, Thermus aquaticus, Crysiogenes arsenates, Geospirillum barnesii* and *Geospirillum arsenophilus.*

e. **Methylation:** The methylation in the arsenic bioremediation pathway is considered the detoxification step, in which the methylated arsenite is converted into methyl arsenate, a less toxic form of arsenic. This process is carried out with the help of a certain specific enzyme, (As III) S-adenosylmethionine methyltransferase. It is an enzymatic methyl donor. The mechanism of conversion of arsenate to trimethylarsine is completed in seven sequential steps. The seven sequential conversions are: arsenate to arsenite, arsenite to methyl arsonate, methyl arsonate to dimethyl arsenate, dimethyl arsenate to dimethyl arsinite, dimethylarsinate to trimethylarsine oxide, trimethylarsine oxide to trimethylarsine. The end product, trimethylarsine, is less toxic and it volatilizes in the environment.

f. **Biosorption:** Biosorption is considered to be one of the effective remedial measures for the removal of the toxic heavy metals. It is an energy-dependent process for the removal of arsenic. Microorganisms can form biofilms on surfaces, creating a matrix that can adsorb and immobilize arsenic from water. Biofilms are beneficial because they concentrate microbial activity and provide a surface for arsenic attachment. Certain biomass, such as algae, fungi, and bacterial biomass, can adsorb arsenic from water through biosorption. These materials have high affinity for arsenic and can effectively remove it from the water. It involves various bacteria, algae, fungi, and yeasts for the binding of the arsenic to different functional groups such as $-COOH$, and $-OH$.

g. **Bioaccumulation:** This is a process in which the toxic heavy metals such as arsenic are accumulated in high amounts in microbes. Some aquatic organisms, such as mussels and shellfish, can accumulate arsenic from water through their feeding processes. Regular harvesting and disposal of these organisms can help in removing arsenic from the water. Bioaccumulation for arsenic removal includes the binding of free As transported by glycerol and phosphate transporters to proteins and peptides.

h. **Volatilization:** This is the final step of the pathways of bioremediation for the removal of arsenic from water. Here the less toxic form of arsenic obtained after the completion of the above steps is released or volatilized into the environment.

The effectiveness of these bioremediation pathways can vary based on factors such as the specific arsenic species present, the concentration of arsenic in the water, the local environmental conditions, and the type of microorganisms or plants used. Additionally, a combination of these pathways or integration with other water treatment methods might be necessary to achieve effective arsenic removal from water sources.

Phytoremediation

Phytoremediation represents an environmentally friendly and economically viable method employed for the extraction of contaminants, including heavy metals like arsenic, from water and soil. Phytoremediation offers a potential solution for mitigating arsenic contamination in groundwater. It involves the use of specially selected plants to uptake, accumulate, and sometimes transform contaminants from the environment. Some aquatic plant and their percentage removal of arsenic given in Figure 7.6.

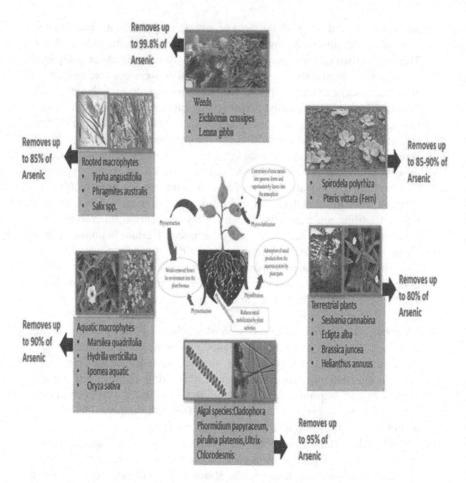

FIGURE 7.6 Phytoremediation process for As removal (Kristanti & Hadibarata, 2023).

Phytoremediation involves using plants to remove or immobilize contaminants. Some plants have the ability to accumulate arsenic in their tissues, effectively removing it from the water. These plants are known as hyperaccumulators. Common hyperaccumulator plants for arsenic include certain ferns, watercress, and some aquatic plants (Ambrožič-Dolinšek et al., 2023; Kristanti & Hadibarata, 2023). After the plants have accumulated arsenic, they can be harvested and disposed of properly to ensure the safe management of the arsenic-containing biomass. Here's a step-by-step process for applying phytoremediation to address arsenic contamination in groundwater.

Process of Phytoremediation

a. **Plant selection:**
   The first step in phytoremediation is to select suitable plant species that have a natural ability to tolerate and accumulate arsenic. Some plant species used for As removal are given in Table 7.1.

TABLE 7.1

Some Plant Species Used for As Removal

Common Name	Scientific Name
Duckweed	Lemna Gibba
Lesser duckweed	Lemna minor
Greater duckweed	Spirodela Polyrhiza
Waterweed	Hydrilla verticillata
–	Marsilea Quadrifolia
–	Phragmites Australis
Butterfly fern	Salvinianatans
Brazilian waterweed	Veronica aquatica
Mosquito fern	Azolla Caroliniana
Water fern	Azolla Filiculoides
Water pepper	Polygonum hydropiper
–	Typha angustifolia
Water lettuce	Pistia stratiotes
–	Chlamydomonas Reinhardtii
–	Dunaliella Salina

Note: Species used for arsenic (As) removal from water.
Sources: Jasrotia et al., 2017; Xi et al., 2023

b. **Site preparation:**
The contaminated water site needs to be prepared for planting. This involves removing any competing vegetation, tilling the soil if applicable, and ensuring that the plants will have access to sufficient sunlight, water, and other suitable conditions required for plant growth so that it can accumulate arsenic or other heavy metals effectively.

c. **Planting:**
The arsenic hyperaccumulator plants mentioned above are then planted in the contaminated water or sediment. The density of planting and arrangement depend on factors such as the plant species and the extent of contamination.

d. **Arsenic uptake:**
Hyperaccumulator plants have the ability to absorb arsenic through their roots from the water. Arsenic exists in water primarily in two forms: arsenite(III) and arsenate(V). These forms can be taken up by the plants and stored in various plant tissues, including roots, stems, and leaves.

e. **Translocation and accumulation:**
After being absorbed by the roots, plants proceed to transport As to different sections of the plant. Some plants accumulate arsenic primarily in their roots, while others can transport it to their aboveground parts. The choice of plant species influences how arsenic is distributed within the plant.

f. **Harvesting and disposal:**
Depending on the phytoremediation strategy, the plants can be harvested once they have accumulated a significant amount of arsenic. The harvested

plant material is then properly disposed of to prevent recontamination. If the goal is to remove arsenic permanently, the harvested plants may be incinerated or disposed of as hazardous waste.

g. **Monitoring and maintenance:**

Regular monitoring of plant health and growth and arsenic levels in both the water and the plants is essential. This helps determine the effectiveness of the phytoremediation process and whether any adjustments or additional treatments are required.

h. **Long-term management:**

Phytoremediation is a relatively slow process, and it might require several growing seasons to achieve significant reduction in arsenic levels. Long-term management involves consistent maintenance and regular monitoring for effective remediation.

There are various pathways of phytoremediation for the removal of contaminants from water (Akhtar et al., 2023) shown in (Figure 7.7).

Pathways of Phytoremediation

a. **Phytoextraction:** This is one of the pathways. Various plants are capable of accumulating and absorbing arsenic in them; these are referred to as hyper-accumulator plants. These are capable of accumulating large amounts of heavy metals such as arsenic from the soil. After the plants have absorbed the pollutants from the soil, they can be harvested or removed from the site, effectively removing the contaminants from the environment.

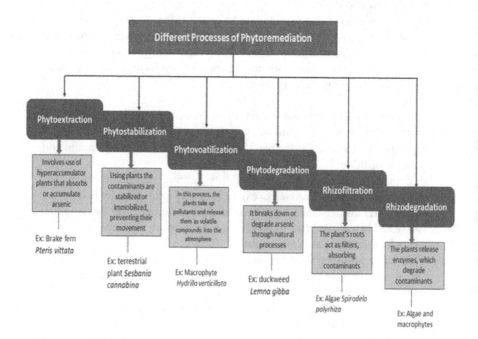

FIGURE 7.7   Phytoremediation process for As removal.

b. **Phytostabilization:** In cases where it is not feasibly and practically possible to remove arsenic from groundwater, then this technique is used. Phytostabilization focuses on using plants to immobilize or stabilize contaminants in the soil, preventing their movement and potential spread. Certain plants can bind arsenic in their root system and prevent it from further leaching and dispersion. The plant's root system binds with the pollutants, reducing their mobility and availability for uptake by other organisms. This approach is often used in combination with other remediation techniques.

c. **Phytovolatilization:** This approach involves using plants to take up pollutants such as arsenic and release them into the atmosphere as less harmful compounds. The pollutants are released into the atmosphere as volatile compounds which are less harmful than the contaminants and are easier to capture and treat in the air. But this approach is not used widely due to the fact that pollutants such as arsenic might be released into the environment.

d. **Phytodegradation:** This is a process where plants help in the degradation or removal of contaminants or pollutants such as arsenic from the environment. Certain plants have the ability to take up and accumulate arsenic in their tissues, effectively removing arsenic from the water. This process can be effective in mitigating arsenic contamination in certain scenarios, but its success depends on factors like plant selection, environmental conditions, and the level of contamination.

e. **Rhizofiltration:** This is a method that uses plant roots to absorb and accumulate contaminants from the groundwater, including arsenic. Certain plants have the ability to take up arsenic and store in their roots. This approach can be effective in remediating arsenic-contaminated groundwater, especially in areas rich in plant species that are known to hyperaccumulate heavy metals. The success of rhizofiltration depends on certain factors such as plant selection, soil contaminants, and the extent of contamination.

f. **Rhizodegradation:** In this process, plants help in remediating contaminants such as arsenic from groundwater by releasing certain compounds or enzymes that enhance the microbial activity in the rhizosphere (root zone). Microbes then break down or immobilize the arsenic, reducing its concentration. This is an eco-friendly approach for environmental cleanup.

DISCUSSION ON THE ADVANTAGES AND DISADVANTAGES OF USING PHYTOREMEDIATION

The choice between bioremediation and phytoremediation should be based on a thorough assessment of the specific site conditions, the type and concentration of arsenic present, regulatory requirements, and available resources. In some cases, a combination of both methods might be used to maximize the effectiveness of arsenic removal. It's important to note that the effectiveness of phytoremediation depends on various factors, including the type of plant used, environmental conditions, the concentration of arsenic, and the specific techniques applied. Phytoremediation is often considered a slower process compared to traditional methods, but it offers a sustainable and cost-effective approach for remediating arsenic-contaminated water sources, especially in areas where other solutions might be impractical.

Phytoremediation presents a promising approach to addressing certain types of environmental contamination, but it's not a one-size-fits-all solution. Its advantages lie in its natural and sustainable nature, cost-effectiveness, aesthetic value, low energy requirements, and long-term benefits. However, its limitations include its slow pace, plant selection, limited applicability, unpredictable results, regulatory approval, chances of contaminant redistribution, and potential uncertainties. Deciding whether to use phytoremediation should be based on a thorough understanding of the contaminants, site conditions, and the specific goals of remediation.

CONCLUSIONS AND FUTURE PROSPECTS

Phytoremediation is a sustainable and environmentally friendly technique for the removal of arsenic from groundwater. It is a promising and environmentally friendly method for removing various pollutants including arsenic from water using plants. The various mechanisms involved, including phytoaccumulation, rhizofiltration, and phytovolatilization, underscore the versatility of plants in addressing arsenic contamination. However, the effectiveness of phytoremediation is influenced by factors such as plant species selection, soil conditions, hydrology, and the presence of co-contaminants. Phytoremediation holds immense promise as a nature-based solution for arsenic removal from groundwater, but there are also challenges to overcome, such as the time it takes for plants to grow and accumulate pollutants, the efficiency of arsenic uptake, and the potential release of accumulated arsenic back into the environment. Ongoing research and innovation will play a crucial role in shaping the future aspects of phytoremediation for arsenic removal from groundwater. Its inherent advantages, coupled with ongoing advancements in plant biology and environmental engineering, position it as a viable component of arsenic-contaminated site remediation strategies, contributing to the restoration of clean and safe groundwater resources.

ACKNOWLEDGEMENT

We are thankful to Uttar Pradesh Council of Science and Technology (UP-CST) for providing the facilities along with finance for this study. The authors are also thankful to Dr. Devendra Sharma, Vice-Chancellor of Shri Ramswaroop Memorial University for the conduction of this study.

REFERENCES

Abiye, T. A., & Bhattacharya, P. (2019). Arsenic concentration in groundwater: Archetypal study from South Africa. *Groundwater for Sustainable Development, 9*, 100246. https://doi.org/10.1016/j.gsd.2019.100246

Akhtar, M. S., Hameed, A., Aslam, S., Ullah, R., & Kashif, A. (2023). Phytoremediation of metal-contaminated soils and water in Pakistan: A review. *Water, Air, & Soil Pollution, 234*(1), 11. https://doi.org/10.1007/s11270-022-06023-8

Ambrožič-Dolinšek, J., Podgrajšek, A., Šabeder, N., Grudnik, Z. M., Urbanek Krajnc, A., Todorović, B., & Ciringer, T. (2023). The potential of *Berula erecta* in vitro for as bioaccumulation and phytoremediation of water environments. *Environmental Pollutants and Bioavailability, 35*(1), 2205010. https://doi.org/10.1080/26395940.2023.2205010

Bundschuh, J., Schneider, J., Alam, M. A., Niazi, N. K., Herath, I., Parvez, F., Tomaszewska, B., Guilherme, L. R. G., Maity, J. P., López, D. L., Cirelli, A. F., Pérez-Carrera, A., Morales-Simfors, N., Alarcón-Herrera, M. T., Baisch, P., Mohan, D., & Mukherjee, A. (2021). Seven potential sources of arsenic pollution in Latin America and their environmental and health impacts. *Science of The Total Environment, 780*, 146274. https://doi.org/10.1016/j.scitotenv.2021.146274

Ivy, N., Mukherjee, T., Bhattacharya, S., Ghosh, A., & Sharma, P. (2023). Arsenic contamination in groundwater and food chain with mitigation options in Bengal delta with special reference to Bangladesh. *Environmental Geochemistry and Health, 45*(5), 1261–1287. https://doi.org/10.1007/s10653-022-01330-9

Jasrotia, S., Kansal, A., & Mehra, A. (2017). Performance of aquatic plant species for phytoremediation of arsenic-contaminated water. *Applied Water Science, 7*(2), 889–896. https://doi.org/10.1007/s13201-015-0300-4

Khan, J., Dwivedi, H., Giri, A., Aggrawal, R., Tiwari, R., & Giri, D. D. (2023). Arsenic contamination in water, health effects and phytoremediation. In *Metals in Water* (pp. 407–429). Elsevier. https://doi.org/10.1016/B978-0-323-95919-3.00021-5

Khan, M. U., Musahib, M., Vishwakarma, R., Rai, N., & Jahan, A. (2023). Hydrochemical characterization, mechanism of mobilization, and natural background level evaluation of arsenic in the aquifers of upper Gangetic plain, India. *Geochemistry, 83*(2), 125952. https://doi.org/10.1016/j.chemer.2023.125952

Kristanti, R. A., & Hadibarata, T. (2023). Phytoremediation of contaminated water using aquatic plants, its mechanism and enhancement. *Current Opinion in Environmental Science & Health, 32*, 100451. https://doi.org/10.1016/j.coesh.2023.100451

Kumar, M., Kumar, P., Ramanathan, A. L., Bhattacharya, P., Thunvik, R., Singh, U. K., Tsujimura, M., & Sracek, O. (2010). Arsenic enrichment in groundwater in the middle Gangetic Plain of Ghazipur District in Uttar Pradesh, India. *Journal of Geochemical Exploration, 105*(3), 83–94. https://doi.org/10.1016/j.gexplo.2010.04.008

Maciag, B. J., Brenan, J. M., Parsons, M. B., & Kennedy, G. W. (2023). Sources of geogenic arsenic in well water associated with granitic bedrock from Nova Scotia, Canada. *Science of The Total Environment, 887*, 163943. https://doi.org/10.1016/j.scitotenv.2023.163943

Mahamallik, P., & Swain, R. (2023). A mini-review on arsenic remediation techniques from water and future trends. *Water Science & Technology, 87*(12), 3108–3123. https://doi.org/10.2166/wst.2023.190

Maity, S., Biswas, R., & Sarkar, A. (2020). Comparative valuation of groundwater quality parameters in Bhojpur, Bihar for arsenic risk assessment. *Chemosphere, 259*, 127398. https://doi.org/10.1016/j.chemosphere.2020.127398

Masood ul Hasan, I., Javed, H., Hussain, M. M., Shakoor, M. B., Bibi, I., Shahid, M., Farwa, Xu, N., Wei, Q., Qiao, J., & Niazi, N. K. (2023). Biochar/nano-zerovalent zinc-based materials for arsenic removal from contaminated water. *International Journal of Phytoremediation, 25*(9), 1155–1164. https://doi.org/10.1080/15226514.2022.2140778

Mohsin, H., Shafique, M., Zaid, M., & Rehman, Y. (2023). Microbial biochemical pathways of arsenic biotransformation and their application for bioremediation. *Folia Microbiologica, 68*(4), 507–535. https://doi.org/10.1007/s12223-023-01068-6

Nilkarnjanakul, W., Watchalayann, P., & Chotpantarat, S. (2023). Urinary arsenic and health risk of the residents association in contaminated-groundwater area of the urbanized coastal aquifer, Thailand. *Chemosphere, 313*, 137313. https://doi.org/10.1016/j.chemosphere.2022.137313

Omar, N. M. A., Othman, M. H. D., Tai, Z. S., Amhamed, A. O. A., Kurniawan, T. A., Puteh, M. H., & Sokri, M. N. M. (2023). Recent progress, bottlenecks, improvement strategies and the way forward of membrane distillation technology for arsenic removal from water: A review. *Journal of Water Process Engineering, 52,* 103504. https://doi.org/10.1016/j.jwpe.2023.103504

Pal, S., Singh, S. K., Singh, P., Pal, S., & Kashiwar, S. R. (2023). Spatial pattern of groundwater arsenic contamination in Patna, Saran, and Vaishali districts of Gangetic plains of Bihar, India. *Environmental Science and Pollution Research.* https://doi.org/10.1007/s11356-022-25105-y

Patel, A. K., Tambat, V. S., Chen, C.-W., Chauhan, A. S., Kumar, P., Vadrale, A. P., Huang, C.-Y., Dong, C.-D., & Singhania, R. R. (2022). Recent advancements in astaxanthin production from microalgae: A review. *Bioresource Technology, 364,* 128030. https://doi.org/10.1016/j.biortech.2022.128030

Patel, K. S., Pandey, P. K., Martín-Ramos, P., Corns, W. T., Varol, S., Bhattacharya, P., & Zhu, Y. (2023). A review on arsenic in the environment: Bio-accumulation, remediation, and disposal. *RSC Advances, 13*(22), 14914–14929. https://doi.org/10.1039/D3RA02018E

Pramparo, S., Blarasin, M., Degiovanni, S., Giacobone, D., Lutri, V., Cabrera, A., & Pascuini, M. (2023). Geochemical processes that explain arsenic in groundwater in a basin developed in the Pampean Mountains and piedmont, Córdoba, Argentina. *Sustainable Water Resources Management, 9*(5), 134. https://doi.org/10.1007/s40899-023-00917-z

Preetha, J. S. Y., Arun, M., Vidya, N., Kowsalya, K., Halka, J., & Ondrasek, G. (2023). Biotechnology advances in bioremediation of arsenic: A review. *Molecules, 28*(3), 1474. https://doi.org/10.3390/molecules28031474

Rahidul Hassan, H. (2023). A review on different arsenic removal techniques used for decontamination of drinking water. *Environmental Pollutants and Bioavailability, 35*(1), 2165964. https://doi.org/10.1080/26395940.2023.2165964

Rajendran, R. M., Garg, S., & Bajpai, S. (2021). Economic feasibility of arsenic removal using nanofiltration membrane: A mini review. *Chemical Papers, 75*(9), 4431–4444. https://doi.org/10.1007/s11696-021-01694-9

Shaji, E., Santosh, M., Sarath, K. V., Prakash, P., Deepchand, V., & Divya, B. V. (2021). Arsenic contamination of groundwater: A global synopsis with focus on the Indian Peninsula. *Geoscience Frontiers, 12*(3), 101079. https://doi.org/10.1016/j.gsf.2020.08.015

Xi, Y., Han, B., Kong, F., You, T., Bi, R., Zeng, X., Wang, S., & Jia, Y. (2023). Enhancement of arsenic uptake and accumulation in green microalga *Chlamydomonas reinhardtii* through heterologous expression of the phosphate transporter *DsPht1. Journal of Hazardous Materials, 459,* 132130. https://doi.org/10.1016/j.jhazmat.2023.132130

Xing, S., Guo, H., & Hu, X. (2023). Sources and enrichment processes of groundwater arsenite and arsenate in fissured bedrock aquifers in the Xunhua-Hualong basin, China. *Applied Geochemistry, 155,* 105708. https://doi.org/10.1016/j.apgeochem.2023.105708

Yeo, K. F. H., Li, C., Zhang, H., Chen, J., Wang, W., & Dong, Y. (2021). Arsenic removal from contaminated water using natural adsorbents: A review. *Coatings, 11*(11), 1407. https://doi.org/10.3390/coatings11111407

Zhu, Y., Yang, Q., Wang, H., Yang, J., Zhang, X., Li, Z., & Martín, J. D. (2023). A hydrochemical and isotopic approach for source identification and health risk assessment of groundwater arsenic pollution in the central Yinchuan basin. *Environmental Research, 231,* 116153. https://doi.org/10.1016/j.envres.2023.116153

8 Exploring the Efficacy of Advanced Phytoremediation Techniques for Sustainable Waste Management

Shikha Gupta and Ayushi Varshney

INTRODUCTION

The advent of industrialization, urbanization, and chemical-based agricultural practices has substantially contributed to the discharge of toxic metal contaminants into the environment. This results in higher levels of metal contaminants beyond the permissible levels in the environment, which causes a serious threat to all life-forms (Sharma et al., 2020).

The presence of heavy metals in the environment is strongly influenced by both natural (geological) and anthropogenic activity. However, global industrial processes have been speculated to be the primary cause of worldwide heavy metal pollution. Infrastructural growth, mining, leather, road maintenance, sewage sludge, smelting, and coal-burning power plants are just a few examples of anthropogenic activities that have accelerated the presence of toxic heavy metal pollutants in the soil (Sabreena et al., 2022; Yan et al., 2020). Furthermore, intensive usage of chemical practices based on pesticides and fertilizers and wastewater as an irrigation source in agriculture has escalated the threshold limits of heavy metals in the soil and groundwaters (Mishra et al., 2019). This has caused widespread concern about the potential implications of heavy metal pollutants on the well-being of the environment (Figure 8.1).

The group of elements known as heavy metals, which have metallic qualities, includes metalloid elements as well as transition metals, actinides, and lanthanides. Typically, metallic elements having specific weights > 5 g/cm^3 and toxicity even at low concentrations are referred to as heavy metals (Mahiya et al., 2014). Heavy metals are a group of metallic elements with relatively high densities, atomic weights, and atomic numbers. Cadmium (Cd), lead (Pb), mercury (Hg), zinc (Zn), copper (Cu), nickel (Ni), and chromium (Cr) are some of the heavy metals (Yan et al., 2020). The continued accumulation in the environment of metal contaminants such as cadmium,

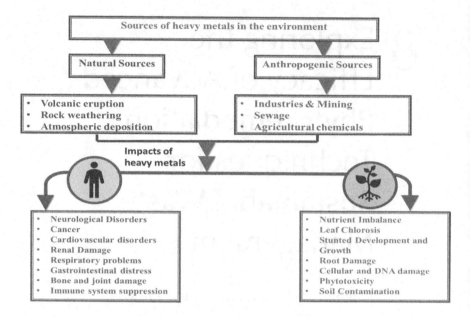

FIGURE 8.1 Sources of heavy metals in the environment and their detrimental impacts on human and plant health.

lead, chromium, arsenic (As), copper, mercury, nickel, and selenium (Se) endangers human health and undermines global environmental sustainability efforts.

The presence of several pollutants in soil, including organic and inorganic pollutants and heavy metals, causes deleterious effects on the environment and the well-being of individuals. Heavy metals pose a long-term hazard to the ecosystem due to their non-biodegradability and persistent nature in soil (Sharma et al., 2020). The development, morphology, and metabolism of microbes are typically negatively impacted by a rise in heavy metal concentration in soil, which results in a decrease in the overall growth and population of beneficial bacteria (Kafle et al., 2022). The escalated concentration of heavy metals in the soil causes declines in soil productivity by negatively influencing the physiochemical properties and redox status of the soil, which in turn affects the population and community of plant-beneficial bacteria (Chu, 2018; Jiang et al., 2019).

Heavy metals accumulate in the soft tissues and are not metabolized by the human body (Varshney et al., 2021). They pose a serious hazard to human health because they can accumulate in the human body through biomagnification and infiltrate the food chain through crops (from producer levels to consumer levels). Heavy metal exposure can have serious consequences for human health and can even be lethal (Singh et al., 2015). Ingestion and inhalation of heavy metals compromise human health by significantly contributing to various disorders involving neurological and renal malfunctioning, growth retardation, immunological and gastrointestinal disorders, and impairment of fertility. Exposure to heavy metals can lead directly to an elevated risk for many cancers, including endometrial and breast cancer (Wani et al., 2023).

In addition, heavy metal contaminants have a deleterious influence on the growth, physiological, and developmental processes in plants. Heavy metal buildup in plants can inevitably cause damage to their physiological systems, including microbial respiration, photosynthetic activity, food uptake, and CO_2 fixation, which in turn cause declines in crop productivity and yield (Varshney et al., 2021).

The deleterious effects caused by these heavy metals on plants and human health are summarized in Table 8.1.

Therefore, remediation measures must be implemented to prevent heavy metals from entering terrestrial, atmospheric, and aquatic habitats, as well as to remediate

TABLE 8.1

Major Toxic and Harmful Effects of Heavy Metal Contaminants on Humans and Plants

S. No.	Heavy Metals	Impact on Plants	Impact on Humans	Reference
1.	Thallium (Th)	Causes growth inhibition, oxidative stress, leaf chlorosis, and the impairment of K homeostasis	Causes alopecia in humans	Chang et al., 2022; Karbowska, 2016
2.	Mercury (Hg)	Causes a decline in plant growth and development and causes oxidative stress	Causes Minamata disease	Ahmad et al., 2018; Sakamoto et al., 2018
3.	Cadmium (Cd)	Promotes oxidative damage, affects plant metabolism, decreases food and water intake, and hinders plant morphology and physiology	Causes itai-itai disease	Haider et al., 2021; Nishijo et al., 2017
4.	Nickel (Ni)	Results in chlorosis and necrosis, and oxidative damage	Causes allergies, nasal and lung cancer, and kidney and cardiovascular diseases	Hassan et al., 2019; Genchi et al., 2020
5.	Copper (Cu)	Interferes with photosynthetic and respiratory processes, promotes cell cytotoxicity	Causes liver and Alzheimer's disease	Mir et al., 2021; Kumar et al., 2021
6.	Zinc (Zn)	Reduced growth, photosynthetic and respiratory rate, imbalanced mineral nutrition, and enhanced generation of ROS	Causes fever, breathing difficulty, nausea, chest pain, and cough	Kaur & Garg, 2021; Hussain et al., 2022
7.	Lead (Pb)	Reduced seed germination, carbon metabolism, induced structural changes in photosynthetic apparatus, decreased antioxidant enzyme activity, and membrane damage	Causes cardiovascular, central nervous system, kidney, and fertility problems	Kumar et al., 2020; Zulfiqar et al., 2019
8.	Chromium (Cr)	Affects plant development, nutrient absorption, increased ROS production, and altered antioxidant potential	Causes cardiovascular, developmental, neurological, and endocrine diseases and immunologic disorders	Ali et al., 2023; Iyer et al., 2023

contaminated soil. Various mechanical and chemical-based methods have been developed to facilitate the reclamation of heavy metal-contaminated soil.

There are many conventional procedures/techniques that can be used to detoxify heavy metal-contaminated areas, including reverse osmosis, chemical precipitation, ion exchange, adsorption, colloidal coagulation, membrane filtration, photocatalysis, and solvent native methods (Yadav et al., 2018). These techniques require extensive maintenance functions and expenses, are laborious, and are often not sustainable (Ali et al., 2013).

In this context, phytoremediation is the direct utilization of living green plants and is an effective, inexpensive, noninvasive, and ecologically benign approach for translocating or stabilizing all hazardous metals and environmental pollutants from polluted soil or groundwater. Phytoremediation is extensively suitable for metal-contaminated areas, with some long-term aesthetic benefits, and it is well-known for its low cost and eco-friendliness; therefore it is applied on a broad scale in locations with high toxic metal concentrations (Alsafran et al., 2022). Numerous research studies have shown the potential of the phytoremediation approach, with the use of different plant species with phytoremediation potential to detoxify various types of heavy metal pollutants (Table 8.2).

TABLE 8.2
List of Plant Species Used in Phytoremediation Employing Various Detoxification Mechanisms for Metal Pollutants

S. No.	Plant Species	Pollutants	Phytoremediation Approach	References
1.	*Rumex dentatus*	Zn, Cd	Phytoextraction	Sajad et al., 2019
2.	*Lolium perenne* L.	Cd, Zn, Pb	Phytoextraction	Zhang et al., 2019
3.	*Typha latifolia, Azolla pinnata, Croton bonplandianum*	Fe, Cu, Ni, Si, Al, Pb, Cr, Cd	Phytoextraction and metal accumulation	Kumari et al., 2016
4.	*Typha domingensis*	Fe, Mn, Zn, Ni, Cd	Phytoextraction	Mojiri, 2012
5.	*Chenopodium album* L.	Pb	ETDA-assisted phytoextraction	Ebrahimi, 2016
6.	*Brassica napus*	Cr	Cr accumulation and adsorption to the root biomass; root enzyme reductase	Perotti et al., 2020
7.	*Argemone mexicana and Solanum nigrum*	Fe	Metal accumulation in aerial parts	Singh et al., 2010
8.	*Trifolium alexandrinum*	Cd, Pb, Cu, Zn	Metal accumulation in roots	Ali et al., 2012
9.	*Ipomoea carnea, Lantana camara,* and *Solanum surattense*	Mn, Fe, Ni, Pb, Cu, Cr, Cd	Metal accumulation in roots and shoots of plants	Pandey et al., 2016
10.	*Saccharum spontaneum* and *Saccharum munja*	Zn, Pb, Cu, Ni, Cd, As	Metal accumulation in root and shoot + Phytostabilization of metals in root system	Banerjee et al., 2020
11.	*Parthenium hysterophorus*	Pb, Ni, Cd	Metal accumulation in root and aerial parts	Ahmad and Al-Othman, 2014

(Continued)

TABLE 8.2 *(Continued)*
List of Plant Species Used in Phytoremediation Employing Various Detoxification Mechanisms for Metal Pollutants

S. No.	Plant Species	Pollutants	Phytoremediation Approach	References
12.	*Sagittaria montevidensis*	P, Mn, Al, V, S, Fe, As, Cu, Mg, Zn, Na, Pb, Cd, Ni, Cr	Phytoextraction + rhizofiltration; bioaccumulating and retaining contaminants in the root tissues	Demarco et al., 2019
13.	*Hydrocotyle ranunculoides* L.	Cu, Zn, Fe, Mn, Na, Cd, Cr, Ni, Pb, Al, As, Co, V	Phytoextraction and translocating the contaminants to shoot system + rhizofiltration	Demarco et al., 2018
14.	*Pistia stratiotes*	Cr, Pb, Ni	Rhizofiltration	Abubakar et al., 2014
15.	*Arundo donax* L.	Cd, Zn	Rhizofiltration by the root system of the plants	Dürešová et al., 2014
16.	*Plectranthus amboinicus*	Pb	Accumulates considerable amount of lead, particularly in the roots, and translocation to the stem and leaf	Ignatius et al., 2014
17.	*Vossia cuspidata*	Cr, Cu, Pb	Phytostabilization	Galal et al., 2017
18.	*Cirsium arvense* and *Salsola soda*	Pb, Fe, Mn, Cu, Zn	Phytoextraction and phytostabilization of heavy metals	Lorestani et al., 2013
19.	*Launaea acanthodes*	As	Phytostabilization	Siyar et al., 2022
20.	*Cousinia congesta*	Mo	Molybdenum remediation by phytostabilization technique	
21.	*Artemisia* sp.	Cu, Mn, Mo, Ni, Pb	Hyperaccumulator plant shows phytoextraction potential of heavy metals	
22.	*Spartina pectinata*	Zn	Phytostabilization	Korzeniowska et al., 2015
23.	*Festuca rubra*	Cd, Co	Phytostabilization	Wyszkowska et al., 2022
24.	*Soybean*	Cu, Zn, Pb, Cd, Mn	Immobilization of heavy metals through amendment-assisted phytostabilization technique	Li et al., 2019
25.	*Atriplex halimus,* and *Medicago lupulina*	Pb, Zn, Ni	Phytostabilization with low metal translocation to aerial parts	Amer et al., 2013
26.	*Zea mays, Sorghum sudanense, Vetiveria zizanoides*	Polycyclic aromatic hydrocarbons (PAHs)	Rhizodegradation of PAHs with the help of bacteria of genera *Pseudomonas, Sphingomonas*	Sivaram et al., 2020
27.	*Cajanus cajan*	Petroleum oily sludge-spiked soil	Rhizodegradation of petroleum oily sludge-contaminated soil	Allamin et al., 2020

In this chapter, we describe the concept of the phytoremediation process in plants and the different mechanisms involved in phytoremediation, with major emphasis on the recent advancements in phytoremediation strategies.

PHYTOREMEDIATION: SUSTAINABLE ALTERNATIVES TO HEAVY METAL POLLUTION

The term *phytoremediation* is an amalgamation of two terms: *phyto* means plant, while *remedium* is a Latin suffix that means to restore. Traditionally, the term refers to the utilization of plants and associated microorganisms to minimize the potent negative impacts of heavy metal pollutants manifested on the environment (Sarwar et al., 2017). The phytoremediation technique has received considerable attention for being an economically feasible, aesthetically friendly, environmentally benign, and sustainable approach to remediate heavy metals and metalloids from soil and water sources.

Heavy metals can be classified as either essential or nonessential elements based on their function in biological systems. In contrast to nonessential heavy metals like Pb, Cd, and Cr, which are highly toxic and have no recognized role in biological systems, Cu, Fe, Mn, Ni, and Zn are needed for physiological and metabolic processes, but they become toxic when present in excess (Fasani et al., 2018). In general, plants use avoidance and tolerance so that the heavy metal levels can be maintained without exceeding the optimum limits (Khan et al., 2023) (Figure 8.2). Plants can inhibit and restrict the uptake and movement of HMs into their tissues through avoidance. The avoidance strategy is sometimes referred to as an innate defense mechanism since it prevents migration to the plant's higher parts by limiting the absorption of heavy metals through the root system. Various defense mechanisms, such as root sorption, metal precipitation, and exclusion are involved in the avoidance process. Root sorption functions through root exudates that form complexes with heavy metals to lessen their bioavailability and toxicity. In addition, heavy metal transport from the

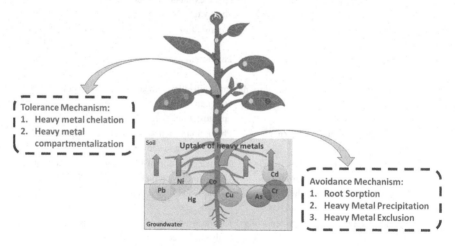

FIGURE 8.2 Detoxification strategies employed by plants to withstand abiotic stress induced by heavy metal contaminants.

soil to the roots and shoots is prevented by metal precipitation through adsorption, absorption, and chelation processes. The metal exclusion mechanism functions as a protective mechanism against heavy metals to lessen toxicity between the root and shoot systems (Khan et al., 2023; Yan et al., 2020). The tolerance strategy was deployed in order to withstand heavy metal toxicity once the heavy metal has penetrated the plant cell. The mechanisms involved in this are (i) heavy metal chelation through various organic and inorganic ligands present inside the cytosol, and (ii) heavy metal compartmentalization in metabolically inert organelles such as vacuoles, leaf petioles, leaf sheaths, and trichomes for storage without inducing toxicity in the plant (Sabreena et al., 2022; Yan et al., 2020).

PHYTOREMEDIATION TECHNIQUES

Based on their mechanism, distinct phytoremediation methods can be categorized as (1) phytoextraction, (2) phytostabilization, (3) phytodegradation (phytotransformation), (4) phytovolatilization, and (5) rhizofiltration, as shown in Figure 8.3. The involved mechanisms are succinctly described and expanded upon below.

PHYTOEXTRACTION

The phytoextraction process encompasses the intake and accrual of heavy metal from the contaminated site such as soil, sediments, and water resources and the transport of heavy metals to aerial parts such as shoots and leaves of the plants and subsequently harvesting of the plant biomass for safe disposal (Dhanwal et al., 2017). The phytoextraction process is categorized as natural or continuous phytoextraction and induced or phytochelatin-mediated phytoextraction (Kanwar et al., 2020). Heavy metal phytoextraction involves the following steps: Intake of heavy metals by plant roots, mobilization of heavy metals in the rhizosphere, translocation of heavy metal ions from roots to aerial sections of plants, and sequestration and compartmentation

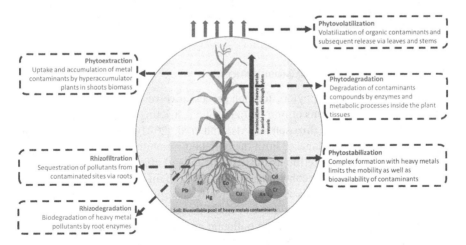

FIGURE 8.3   Traditional approaches to phytoremediation and fate of heavy metal contaminants.

of heavy metal ions in plant tissues such as vacuoles, cell walls, cell membranes, and other metabolically inert areas of plant tissues (Yan et al., 2020). The general mechanisms involved in the accumulation of hazardous heavy metals include metal cation absorption followed by metal-phytochelatin complex (M-PC) or metal–ligand complex formation inside the plant cell (Asgari Lajayer et al., 2019; Khan et al., 2023). Following the formation of the M-PC, these intricate molecules are translocated to the plant's vacuole and stored or transformed into a less harmful state. The effectiveness of phytoextraction is influenced by soil physiochemical properties such as pH, cation exchange capacity (CEC), oxi-reduction potential, texture, and heavy metal bioavailability as well as plant species (Bortoloti and Baron, 2022). Plant hyperaccumulators have the potential to accumulate heavy metals in their roots and shoot tissues in concentrations between 100 and 500 times more than other plants without compromising their development and functioning. The phytoextraction potential of plant species is generally regulated by two parameters, plant biomass and concentration of heavy metals in aboveground parts of plants. Therefore, plant species employed for phytoextraction should not only withstand and successfully accumulate heavy metal from contaminated sites but also have the potential to enhance the biomass of aboveground parts and be profitable (Yadav et al., 2018). There are already more than 450 plant species known to be potential metal hyperaccumulators with many belonging to families such as Asteraceae, Brasicaceae, Euphorbiaceae, Fabaceae, Flacourticeae, and Violaceae, which have been shown to accumulate higher quantities of heavy metals (Wani et al., 2023). Table 8.2 has shown the plant species with phytoextraction potential during phytoremediation of metal contaminants.

However, there are other issues to take into account, such as using edible crops for phytoextraction. Such an activity ought to be avoided because heavy metals bioaccumulate in the plant's edible portion and enter the food chain as a result, potentially harming human health. Therefore, choosing nonedible hyperaccumulators is essential for the effective and secure phytoremediation of heavy metals. Furthermore, the contaminated biomass that was harvested following the phytoextraction process contained higher levels of heavy metals that may pose serious risks to the environment and human health. There are other methods that help to store the contaminated biomass material in landfills, like neutralization and pyrolysis processes.

PHYTOSTABILIZATION/PHYTOIMMOBILIZATION

The phytostabilization or phytorestoration process involves the conversion of pollutants into less harmful forms or immobilizing and restricting the passage of contaminants within the roots or rhizosphere (Alsafran et al., 2022). The procedure mainly relies on the actions of plants to limit the pollutant's mobility and bioavailability in polluted areas to minimize the associated detrimental effects—for example, minimizing the risk of metals contaminating the food chain (Patra et al., 2020). The stabilization of heavy metals extensively entails adsorption, binding, or coprecipitation of pollutants with soil additives (biosolids, manures, and composts), which in turn constrain the mobility of the pollutant and thereby prevent the contamination of the environment such as soil and water resources (Wani et al., 2023). It not only improves the hydraulic capability for upward transport of pollutants, but it also reduces pollutant mobility through physical and chemical root absorption (Yadav et al., 2018). Contaminants may be immobilized in the rhizosphere by phytochemical exudates (conjugation of heavy metals with sugars, proteins, and amino acid derivatives), retained on the root surface by transport proteins, or biological mechanisms may sequester them inside the vacuoles of root cells. A trivalent complex of As-tris-thiolate was formed in the rhizosphere by the plant root epidermis, for example, immobilizing arsenic (As) by binding it to ferric sulfate inside vacuoles (Hammond et al., 2018). The plant must be able to adapt to the soil conditions and develop quickly with an extended lifespan for phytostabilization to be effective. Table 8.2 provides a list of plants known to employ phytostabilization methods.

PHYTODEGRADATION AND RHIZODEGRADATION

In phytodegradation, organic pollutants such as herbicides or chlorinated hydrocarbons, after being sequestered by plants, are subjected to breakdown or transformation either through metabolic processes or through degrading enzymes such as dehalogenase, peroxidase, nitroreductase, nitrilase, and phosphatase (Mahjoub, 2014). Rhizodegradation is the term for the biodegradation or mineralization of organic pollutants in the rhizosphere. It entails the passive accrual of pollutants into the plant tissue via the root system into small and less toxic by-products. The phytodegradation process is regulated by several factors, including absorption effectiveness, transpiration rate, and other physical and chemical characteristics of the soil (Rai et al., 2021).

PHYTOVOLATILIZATION

Phytovolatilization is the process by which mainly organic pollutants are converted into various volatile chemicals and released into the atmosphere during the process called transpiration through stomata (Ali et al., 2013). There are several phases involved in phytovolatilization. Plants first absorb toxins—for example, organic pollutants such as acetone, phenol, and chlorinated benzene from the soil—convert them into volatile forms, and then release the contaminants to the atmosphere through a process known as volatilization. It can be carried out either directly, comprising volatilization of organic compounds by the stem and leaves through wounds, the epidermis, suberin, and other dermal layers, or indirectly through plant root–soil interactions. Several plants could release neutral forms of harmful organic pollutants into the environment through the volatilization mechanism (Limmer and Burken, 2016).

RHIZOFILTRATION

Rhizofiltration is the process of remediating aqueous pollutants from wastewater, groundwater, or surface water by adsorbing, concentrating, and precipitating onto the roots or other submerged organs of metal-tolerant aquatic plants in a saturated zone (Rezania et al., 2016). In rhizofiltration, the root tissues of plants act as a sink for metal contaminants, thus limiting their passage to other habitats. The groundwater or surface water resources are cleaned with this method via the adsorption or absorption of impurities by plant roots. This technique works best for cleaning up soil and water that have been heavily contaminated with fertilizers like nitrogen and phosphorus. The process of rhizofiltration is regulated by various physical factors such as the pH of the rhizosphere, root exudates, and root turnover (Mahajan and Kaushal, 2018). This process entails the symbiotic association between roots and microbes to decontaminate the metal pollutants from the site. Bacterial species belonging to genera *Pseudomonas, Mycobacterium* spp., and *Rhodococcus* spp. are used in the rhizofiltration process. Rhizofiltration is typically used by aquatic plants belonging to the species hyacinth, Azolla, duckweed, cattail, and poplar as well as terrestrial plants in phytoremediation (Meitei and Prasad, 2021).

MODERN APPROACHES FOR ADVANCEMENT OF PHYTOREMEDIATION

Traditional phytoremediation approaches lack large-scale applications due to a number of downsides. Species of naturally occurring hyperaccumulator plants either grow slowly and produce little aboveground biomass or are poorly adapted to a range of environmental conditions. In addition, conventional phytoremediation has certain limitations, which include long-term processes to reclaim metal-contaminated soil and variability in metal bioconcentrations due to the influence of physiochemical properties of soil.

These inescapable restrictions force researchers to alter classic phytoremediation methods in order to reduce these restrictions and ensure that phytoremediation is applied widely. There are several contemporary strategies to address the aforementioned restrictions and improve the phytoremediation processes. Numerous biotechnological advancements and approaches, such as genomic approaches, proteomic approaches, transcriptomics, metabolomics, transgenic plant-based approaches, nanoparticle-based approaches, and plant-microbe based approaches, have been shown to improve plant phytoremediation capabilities against heavy metals.

GENETIC ENGINEERING: ROLE OF TRANSGENIC PLANTS IN PHYTOREMEDIATION

Genetic engineering is the use of biotechnological methods to change an organism's genetic makeup. It is a vital method that can be used to significantly increase the phytoremediation capacity of many plant species. Genetic engineering has significantly contributed to enhancing the potential of plants in eliminating or detoxifying harmful inorganic and organic contaminants from the environment (Bortoloti and Baron, 2022).

Direct DNA methods of gene transfer or *Agrobacterium tumefaciens*-mediated transformation can be used to introduce specific genes from microorganisms, plants, and animals to create transgenic plants (Sarwar et al., 2017). Transgenic plants are typically created to either boost immobilization or improve plant tolerance to heavy metals, allowing for greater translocation and accumulation in aerial components of plants. The use of genetic engineering in phytoremediation aids in the development of plants with a large biomass content, a strong root system, and high levels of tolerance that can be cultivated under various stressful conditions. The plant genetic engineering technology is used to identify and harness potential genes for biosynthetic pathways for phytohormone synthesis; gene encoding phytohormones is a useful strategy for developing transgenic plants with increased biomass grown under heavy metal stress conditions (Subašić et al., 2022). Additionally, this method relies on transgenic plants that have certain genes overexpressed that are involved in metal uptake, absorption, translocation, vacuolar sequestration, and plant tolerance of heavy metal stress (Kumar et al., 2023). Transgenic plants with improved transport mechanisms and increased metal tolerance have the ability to sequester more and more metals in cellular organelles with inert metabolic activity, such as vacuoles (Jan et al., 2016). Effective phytoremediation is facilitated by the overexpression of genes for metal transporters (*ZIP, HMAs, MATE, OPT, CDF, ABC, CAX, NRAMP, COPT* families, etc.) and metal chelators (phytochelatins, metallothioneins, glutathione, amino acids, and organic acids) (Yan et al., 2020). The phytovolatilization ability of genetically modified plants can also be improved by overexpressing genes involved in the biotransformation of toxic metalloids into less toxic volatile forms (Kaur et al., 2019). Heavy metals can cause excessive ROS generation and oxidative stress, hence heavy metal tolerance is usually demonstrated by the strength of the oxidative stress defense mechanism. Therefore, the genetic engineering approach is used to develop transgenic plants that overexpress genes involved in antioxidant machinery to enhance the antioxidant activity of plants against metal stress-induced oxidative damage (Yan et al., 2020).

Such transgenic plants can be employed in phytoremediation to revitalize the metal-contaminated soil. A comprehensive understanding of the mechanism for metal tolerance and detoxification in plants is crucial to develop transgenic metal hyperaccumulator plants. Table 8.3 provides a list of genetically altered plants utilized in phytoremediation investigations.

PLANT–MICROBES INTERACTION

Plant-associated soil microbes may affect the availability and uptake of heavy metals by plants in the rhizosphere. Plant growth-promoting microbes increase plants' efficiency in the phytoremediation of metal contaminants and enhance their tolerance to metal stress. Numerous microorganisms have been identified as crucial inoculants for enhancing plant development and the phytoremediation process, including

TABLE 8.3
Genetically Engineered Plants Used in Phytoremediation Studies for Metal Detoxification

Transgenic Plant	Target Gene	Results	Reference
Orzya sativa	*HsSL* and *AbSMT*	Enhanced selenate and selenite stress tolerance	Li et al., 2022
Arabidopsis thaliana	*AtPCS1* and *AtABCC1*	Enhanced arsenic tolerance and accumulation	Song et al., 2010
Nicotiana tabacum L.	Metallothionein gene *PaMT3-1*	Enhanced the translocation of cadmium to leaves	Zhi et al., 2020
Nicotiana tabacum L.	ZNT1 zinc transporter	Enhanced zinc and cadmium tolerance and accumulation	Lin et al., 2016
Nicotiana tabacum L.	Metal tolerance/transport protein *OsMTP1*	Enhanced biomass growth and hyperaccumulation of cadmium	Ye et al., 2020

rhizobacteria, mycorrhizae, and yeast. Species from a wide range of genera make up plant growth-promoting bacteria (PGPB), which also includes free-living bacteria, bacteria that form particular symbiotic interactions with plants, bacterial endophytes that can colonize a plant's internal tissues, and cyanobacteria (blue-green algae) (Glick, 2012). Plant growth-promoting bacteria play a pivotal role in improving phytoremediation efficiency by increasing plant growth and biomass, increasing plant tolerance to metals, and improving nutrient acquisition and heavy metal translocation through the production of organic acids, enzymes, antibiotics, phytohormones, and siderophores (Alves et al., 2022). Danyal et al. (2023) have reported that apart from augmenting growth, plant growth-promoting bacteria aid in the biodegradation, sequestration, mobilization, volatilization, and solubilization of industrial effluents, which enhances the phytoremediation process. Ali et al. (2021) have reported that coinoculation of biochar and *Bacillus* sp. MN54 has significantly helped in degrading petroleum hydrocarbons and improved the growth of maize plants in petroleum-contaminated soil. An additional significant microbial population that can help plants with phytoremediation is arbuscular mycorrhizal fungi (AMF). Through the broad hyphal network, the presence of AMF in rhizospheres enhances the absorptive surface area of plant roots, improving water and nutrient uptake as well as heavy metal bioavailability. A physical barrier created by mycorrhizae prevents heavy metals from attaching to fungal mycelia, limiting the metals' bioavailability, translocation, and bioaccumulation in the plants. This is another way that mycorrhizae demonstrate their function in phytoremediation (Bhantana et al., 2021). A recent study by Rasouli et al. (2023) has demonstrated that AMF inoculation is an effective way to increase the phytoremediation of lead and nickel by lavender plants.

NANOTECHNOLOGY-BASED APPROACHES: NANO-PHYTOREMEDIATION

Nanotechnology is a vast branch of study that focuses on materials and applications that occur on a very small scale. Nano-phytoremediation is a method for cleaning up environmental toxins that combines nanotechnology and phytotechnology. Due to their enormous surface area and high reactivity compared to their bulk form, nanomaterials are helpful for the remediation process and may rapidly penetrate contaminated zones (Srivastav et al., 2018). Various nanoparticles and nanomaterials have been shown in numerous studies to significantly detoxify or remediate the organic, inorganic, and heavy metal contaminants in soils. Application of plant growth-promoting rhizobacteria (PGPR) and titanium dioxide nanoparticles (TiO_2 NPs) has been reported to improve the growth of *Trifolium repens* plants grown in Cd-contaminated soil by enhancing cadmium uptake and accumulation by the plant (Zand et al., 2020). Likewise, Ma et al. (2022) observed improved growth and biomass production in sunflower plants in chromium-contaminated soil when treated with a combined application of cerium dioxide (CeO_2) nanoparticles (NPs) and *Staphylococcus aureus*. Mohammadi et al. (2020) demonstrated that the application of nano-zerovalent iron (Fe^0 nanoparticles) to *Helianthus annuus* ameliorated chromium toxicity by reducing chromium uptake and enhancing the antioxidant enzyme activity of plants. Similarly, zinc nanoparticles are used to attenuate the toxic impact of lead (Pb) accumulation in *Persicaria hydropiper*. Zinc nanoparticles improved the chlorophyll and carotenoid content as well as enhanced Pb accumulation in roots, stems, and leaves of plants (Hussain et al., 2021).

MULTI-OMICS-BASED APPROACHES

Numerous omics approaches comprising genomics (study of complete gene makeup of plants), proteomics (study of proteins), and metabolomics (metabolite profile) have been implicated to get an in-depth and overall understanding of mechanistic action and pathways involved in plants to cope with heavy metal/metalloid toxicity (Yadav et al., 2023). Recent advancements in omics-based studies have provided a comprehensive understanding of the symbiotic or bipartite relationship and mechanisms involved between plant metabolism and the soil microbial community (Bell et al., 2014). The application of a multi-omics approach can help to unravel phenotypic changes in plants in response to metals/metalloids toxicity, reveal stress-responsive mechanisms at the genomic level, comprehend what is happening at the transcript and proteome levels, and provide information about the interaction of metabolites with the phenotype (Raza et al., 2022). Omics-based approaches that entail genomics, proteomics, and metabolomics have been taken into consideration in order to improve potential remediation technologies for heavy metal contamination (Yadav et al., 2023). Genomics is the field of study which entails the entire genome of organisms. It is a multidisciplinary field that concentrates on genome function, DNA sequencing techniques, recombinant DNA, genome editing tools, and high-throughput analysis for sequencing, assembling, and analyzing the genome's structure and function (Saxena et al., 2020). The identification of genes, enzymes, or other molecular components involved in triggering toxic metal/metalloid stress can be aided by genomics (Bell et al., 2014).

Genome editing generally enables researchers to change the regulation of gene expression at particular regions and promotes a new understanding of the functional genomics of plants. The most recent advancement, clustered regularly interspersed palindromic repeats (CRISPR)-assisted phytoremediation, has surely made advances with a wide range of applications in comparison to other techniques such as trait mapping techniques (QTL mapping) and genome-wide association analyses (Naz et al., 2022). The CRISPR-Cas9 system has modified target genes by various gene editing techniques such as knock-in/out, deletion, insertion, and substitution mutations, which in turn aids in identifying and exploring specific genes and their functions that underlie the metal-scavenging potential of many crops and phytoremediation of heavy metals (Soda et al., 2018; Mansoor et al., 2022). Transgenic plants engineered with CRISPR-Cas9 were employed to reduce, immobilize, and sequester the HMs. For instance, the metallochaperones gene *OsHIPP16* was edited by CRISPR-Cas9 in *Oryza sativa* to reduce Cd accumulation in mutant lines and overexpression lines to reduce the harmful impact (Cao et al., 2022). Belykh et al. (2019) highlights the use of CRISPR-Cas9-driven control of gene expression to manage metal homeostasis in plants.

Overall, CRISPR-Cas9-mediated genome editing has demonstrated significant potential to modify the genome of many species for increased agronomic traits (plant biomass, growth rate, yield, productivity) and heavy metal stress resilience (Sarma et al., 2021).

Proteomics provides information about the stress-inducible proteins and their role in reducing the toxicity of toxic metals and metalloids. It also gives information about protein profiles of the organism from the cellular to the organ level, pertaining to its physiological state under pollutant-induced stress conditions (Singh et al., 2016). Metabolomics offers an understanding of the variably regulated metabolites and complex metabolic activities that occur within the plant under heavy metal stress conditions (Bell et al., 2014). The high-throughput omics studies provide a meticulous data target gene (genetic determinants), proteins, and metabolites related to heavy metal stress which aid in developing transgenic plants with multifarious traits.

CONCLUSION AND FUTURE PERSPECTIVE

Over the past few decades, there has been an increase in anthropogenic activity, industrialization, and contemporary agricultural methods, which has elevated the risk of heavy metal toxicity to living organisms. Heavy metal contamination is a critical concern for agricultural production and food safety due to its harmful effects, rapid buildup, and persistent nature in the environment.

Phytoremediation offers a number of advantages and has been shown to be a potential method for reclaiming heavy metal-contaminated soil. It is a practical, affordable, aesthetically pleasing, and environmentally sustainable approach. Traditional phytoremediation utilizes hyperaccumulator plants that are capable of extracting and harvesting high amounts of harmful metals as well as other inorganic and organic contaminants from soil. Additionally, a number of studies have suggested several phytoremediation strategies, such as phytoextraction, phytofiltration, phytodegradation, and phytovolatilization, to lessen the concerns caused by heavy metals in water and soil. However, a single strategy is neither feasible nor sufficient for effective heavy metal-polluted soil cleanup. Due to the fact that naturally occurring hyperaccumulators are typically slow growing and produce relatively less harvestable aboveground biomass output, conventional phytoremediation technologies are less cost-effective when used on large scales. Therefore, it is advised to combine several existing phytoremediation strategies with recent biotechnological approaches for speedy and effective decontamination of heavy metal from the environment. The improvements in genetic engineering and the advancement of agronomic strategies would tremendously aid in making this technology more effective and manageable. Genetic engineering develops transgenic plants with traits such as high biomass production, higher metal accumulation, tolerance against metal toxicity, and good climatic adaptation. Future phytoremediation will be highly successful and comprehensive only if many technologies, including microbe-assisted, nanotechnology, and chelate-assisted phytoextraction are combined in order to make these remedial measures more practical, efficient, and affordable. In-depth research is required to comprehend the dynamics of plant–pollutant interactions, plant–microbe interactions, and ways to dispose of waste properly with the least amount of environmental harm. Moreover, for successful implementation of a phytoremediation program, several issues must be addressed, including risk assessment, lay persons' comprehension, acceptance, and awareness of this environmentally friendly technology, as well as networking among scientists, stakeholders, industrialists, governments, and nongovernmental organizations.

REFERENCES

Abubakar MM, Ahmad MM, Getso BU. Rhizofiltration of heavy metals from eutrophic water using *Pistia stratiotes* in a controlled environment. IOSR Journal of Environmental Science, Toxicology and Food Technology. 2014;8(6):27–34.

Ahmad A, Al-Othman AA. Remediation rates and translocation of heavy metals from contaminated soil through *Parthenium hysterophorus*. Chemistry and Ecology. 2014 May 19;30(4):317–27.

Ahmad P, Ahanger MA, Egamberdieva D, Alam P, Alyemeni MN, Ashraf M. Modification of osmolytes and antioxidant enzymes by 24-epibrassinolide in chickpea seedlings under mercury (Hg) toxicity. Journal of Plant Growth Regulation. 2018 Mar;37:309–22.

Ali H, Khan E, Sajad MA. Phytoremediation of heavy metals—concepts and applications. Chemosphere. 2013 May 1;91(7):869–81.

Ali H, Naseer M, Sajad MA. Phytoremediation of heavy metals by *Trifolium alexandrinum*. International Journal of Environmental Sciences. 2012;2(3):1459–69.

Ali MH, Khan MI, Bashir S, Azam M, Naveed M, Qadri R, Bashir S, Mehmood F, Shoukat MA, Li Y, Alkahtani J. Biochar and *Bacillus* sp. MN54 assisted phytoremediation of diesel and plant growth promotion of maize in hydrocarbons contaminated soil. Agronomy. 2021 Sep 8;11(9):1795.

Ali S, Mir RA, Tyagi A, Manzar N, Kashyap AS, Mushtaq M, Raina A, Park S, Sharma S, Mir ZA, Lone SA. Chromium toxicity in plants: Signaling, mitigation, and future perspectives. Plants. 2023 Mar 29;12(7):1502.

Allamin IA, Halmi MI, Yasid NA, Ahmad SA, Abdullah SR, Shukor Y. Rhizodegradation of petroleum oily sludge-contaminated soil using *Cajanus cajan* increases the diversity of soil microbial community. Scientific Reports. 2020 Mar 5;10(1):4094.

Alsafran M, Usman K, Ahmed B, Rizwan M, Saleem MH, Al Jabri H. Understanding the phytoremediation mechanisms of potentially toxic elements: A proteomic overview of recent advances. Frontiers in Plant Science. 2022 May 6;13:881242.

Alves AR, Yin Q, Oliveira RS, Silva EF, Novo LA. Plant growth-promoting bacteria in phytoremediation of metal-polluted soils: Current knowledge and future directions. Science of The Total Environment. 2022 Sep 10;838:156435.

Amer N, Chami ZA, Bitar LA, Mondelli D, Dumontet S. Evaluation of *Atriplex halimus*, *Medicago lupulina* and *Portulaca oleracea* for phytoremediation of Ni, Pb, and Zn. International Journal of Phytoremediation. 2013 May 1;15(5):498–512.

Asgari Lajayer B, Khadem Moghadam N, Maghsoodi MR, Ghorbanpour M, Kariman K. Phytoextraction of heavy metals from contaminated soil, water and atmosphere using ornamental plants: Mechanisms and efficiency improvement strategies. Environmental Science and Pollution Research. 2019 Mar 1;26:8468–84.

Banerjee R, Jana A, De A, Mukherjee A. Phytoextraction of heavy metals from coal fly ash for restoration of fly ash dumpsites. Bioremediation Journal. 2020 Jan 2;24(1):41–9.

Bell TH, Joly S, Pitre FE, Yergeau E. Increasing phytoremediation efficiency and reliability using novel omics approaches. Trends in Biotechnology. 2014 May 1;32(5):271–80.

Belykh ES, Maystrenko TA, Velegzhaninov IO. Recent trends in enhancing the resistance of cultivated plants to heavy metal stress by transgenesis and transcriptional programming. Molecular Biotechnology. 2019 Oct;61(10):725–41.

Bhantana P, Rana MS, Sun XC, Moussa MG, Saleem MH, Syaifudin M, Shah A, Poudel A, Pun AB, Bhat MA, Mandal DL. Arbuscular mycorrhizal fungi and its major role in plant growth, zinc nutrition, phosphorous regulation and phytoremediation. Symbiosis. 2021 May;84:19–37.

Bortoloti G, Baron D. Phytoremediation of toxic heavy metals by Brassica plants: A biochemical and physiological approach. Environmental Advances. 2022; 8, 100204.

Cao HW, Zhao YN, Liu XS, Rono JK, Yang ZM. A metal chaperone OsHIPP16 detoxifies cadmium by repressing its accumulation in rice crops. Environmental Pollution. 2022 Oct 15;311:120058.

Chang HF, Tseng SC, Tang MT, Hsiao SS, Lee DC, Wang SL, Yeh KC. Physiology and Molecular Basis of Thallium Toxicity and Accumulation in Arabidopsis Thaliana. Available at SSRN 4196841. 2022.

Chu D. Effects of heavy metals on soil microbial community. In IOP Conference Series: Earth and Environmental Science 2018 Feb 1 (Vol. 113, p. 012009). IOP Publishing.

Danyal Y, Mahmood K, Ullah S, Rahim A, Raheem G, Khan AH, Ullah A. Phytoremediation of industrial effluents assisted by plant growth promoting bacteria. Environmental Science and Pollution Research. 2023 Jan;30(3):5296–311.

Demarco CF, Afonso TF, Pieniz S, Quadro MS, Camargo FA, Andreazza R. Phytoremediation of heavy metals and nutrients by the *Sagittaria montevidensis* into an anthropogenic contaminated site at Southern of Brazil. International Journal of Phytoremediation. 2019 Sep 19;21(11):1145–52.

Demarco CF, Afonso TF, Pieniz S, Quadro MS, Camargo FAO, Andreazza R. *In situ* phytoremediation characterization of heavy metals promoted by Hydrocotyleranunculoides at Santa Bárbara stream, an anthropogenic polluted site in southern of Brazil. Environmental Science and Pollution Research. 2018;25(28):28312–21.

Dhanwal P, Kumar A, Dudeja S, Chhokar V, Beniwal V. Recent Advances in Phytoremediation Technology. In: Kumar R, Sharma AK, Ahluwalia SS, editors. Advances in Environmental Biotechnology. Springer Singapore: Singapore; 2017. pp. 227–41.

Dürešová Z, Šuňovská A, Horník M, Pipíška M, Gubišová M, Gubiš J, Hostin S. Rhizofiltration potential of *Arundo donax* for cadmium and zinc removal from contaminated wastewater. Chemical Papers. 2014 Nov; 68:1452–62.

Ebrahimi M. Enhanced phytoremediation capacity of *Chenopodium album* L. grown on Pb-contaminated soils using EDTA and reduction of leaching risk. Soil and Sediment Contamination: An International Journal. 2016 Aug 17;25(6):652–67.

Fasani E, Manara A, Martini F, Furini A, DalCorso G. The potential of genetic engineering of plants for the remediation of soils contaminated with heavy metals. Plant, Cell & Environment. 2018 May;41(5):1201–32.

Galal TM, Gharib FA, Ghazi SM, Mansour KH. Phytostabilization of heavy metals by the emergent macrophyte *Vossia cuspidata* (Roxb.) Griff.: A phytoremediation approach. International Journal of Phytoremediation. 2017 Nov 2;19(11):992–9.

Genchi G, Carocci A, Lauria G, Sinicropi MS, Catalano A. Nickel: Human health and environmental toxicology. International Journal of Environmental Research and Public Health. 2020 Feb;17(3):679.

Glick BR. Plant growth-promoting bacteria: Mechanisms and applications. Scientifica (Cairo). 2012 Sep;2012:963401.

Haider FU, Liqun C, Coulter JA, Cheema SA, Wu J, Zhang R, Wenjun M, Farooq M. Cadmium toxicity in plants: Impacts and remediation strategies. Ecotoxicology and Environmental Safety. 2021 Mar 15;211:111887.

Hammond CM, Root RA, Maier RM, Chorover J. Mechanisms of arsenic sequestration by *Prosopis juliflora* during the phytostabilization of metalliferous mine tailings. Environmental Science & Technology. 2018 Feb 6;52(3):1156–64.

Hassan MU, Chattha MU, Khan I, Chattha MB, Aamer M, Nawaz M, Ali A, Khan MA, Khan TA. Nickel toxicity in plants: Reasons, toxic effects, tolerance mechanisms, and remediation possibilities—A review. Environmental Science and Pollution Research. 2019 May 1;26:12673–88.

Hussain F, Hadi F, Rongliang Q. Effects of zinc oxide nanoparticles on antioxidants, chlorophyll contents, and proline in *Persicaria hydropiper* L. and its potential for Pb phytoremediation. Environmental Science and Pollution Research. 2021 Jul;28:34697–713.

Hussain S, Khan M, Sheikh TM, Mumtaz MZ, Chohan TA, Shamim S, Liu Y. Zinc essentiality, toxicity, and its bacterial bioremediation: A comprehensive insight. Frontiers in Microbiology. 2022 May 31;13:900740.

Ignatius A, Arunbabu V, Neethu J, Ramasamy EV. Rhizofiltration of lead using an aromatic medicinal plant *Plectranthus amboinicus* cultured in a hydroponic nutrient film technique (NFT) system. Environmental Science and Pollution Research. 2014 Nov; 21:13007–16.

Iyer M, Anand U, Thiruvenkataswamy S, Babu HW, Narayanasamy A, Prajapati VK, Tiwari CK, Gopalakrishnan AV, Bontempi E, Sonne C, Barceló D. A review of chromium (Cr) epigenetic toxicity and health hazards. Science of The Total Environment. 2023 Jul 15;882:163483.

Jan S, Rashid B, Azooz MM, Hossain MA, Ahmad P. Genetic Strategies for Advancing Phytoremediation Potential in Plants: A Recent Update. In: Ahmad P, editor. Plant Metal Interaction, Elsevier; 2016 Jan 1. pp. 431–54.

Jiang B, Adebayo A, Jia J, Xing Y, Deng S, Guo L, Liang Y, Zhang D. Impacts of heavy metals and soil properties at a Nigerian e-waste site on soil microbial community. Journal of Hazardous Materials. 2019 Jan 15;362:187–95.

Kafle A, Timilsina A, Gautam A, Adhikari K, Bhattarai A, Aryal N. Phytoremediation: Mechanisms, plant selection and enhancement by natural and synthetic agents. Environmental Advances. 2022 Jul 1;8:100203.

Kanwar VS, Sharma A, Srivastav AL, Rani L. Phytoremediation of toxic metals present in soil and water environment: A critical review. Environmental Science and Pollution Research. 2020 Dec;27:44835–60.

Karbowska B. Presence of thallium in the environment: Sources of contaminations, distribution and monitoring methods. Environmental Monitoring and Assessment. 2016 Nov;188:1–9.

Kaur H, Garg N. Zinc toxicity in plants: A review. Planta. 2021 Jun;253(6):129.

Kaur R, Yadav P, Kohli SK, Kumar V, Bakshi P, Mir BA, Thukral AK, Bhardwaj R. Emerging Trends and Tools in Transgenic Plant Technology for Phytoremediation of Toxic Metals and Metalloids. In: Prasad MNV, editor. Transgenic plant technology for remediation of toxic metals and metalloids. Academic Press; 2019 Jan 1. pp. 63–88.

Khan IU, Qi SS, Gul F, Manan S, Rono JK, Naz M, Shi XN, Zhang H, Dai ZC, Du DL. A green approach used for heavy metals 'phytoremediation' via invasive plant species to mitigate environmental pollution: A review. Plants. 2023 Feb 6;12(4):725.

Korzeniowska J, Stanislawska-Glubiak E. Phytoremediation potential of Miscanthus × giganteus and *Spartina pectinata* in soil contaminated with heavy metals. Environmental Science and Pollution Research. 2015 Aug;22(15):11648–57.

Kumar A, Kumar A, MMS CP, Chaturvedi AK, Shabnam AA, Subrahmanyam G, Mondal R, Gupta DK, Malyan SK, Kumar SS, A. Khan S. Lead toxicity: Health hazards, influence on food chain, and sustainable remediation approaches. International Journal of Environmental Research and Public Health. 2020 Apr;17(7):2179.

Kumar K, Shinde A, Aeron V, Verma A, Arif NS. Genetic engineering of plants for phytoremediation: Advances and challenges. Journal of Plant Biochemistry and Biotechnology. 2023 Mar;32(1):12–30.

Kumar V, Pandita S, Sidhu GP, Sharma A, Khanna K, Kaur P, Bali AS, Setia R. Copper bioavailability, uptake, toxicity and tolerance in plants: A comprehensive review. Chemosphere. 2021 Jan 1;262:127810.

Kumari A, Lal B, Rai UN. Assessment of native plant species for phytoremediation of heavy metals growing in the vicinity of NTPC sites, Kahalgaon, India. International Journal of Phytoremediation. 2016 Jun 2;18(6):592–7.

Li X, Wang X, Chen Y, Yang X, Cui Z. Optimization of combined phytoremediation for heavy metal contaminated mine tailings by a field-scale orthogonal experiment. Ecotoxicology and Environmental Safety. 2019 Jan 30;168:1–8.

Li Z, Tian Y, Wang B, Peng R, Xu J, Fu X, Han H, Wang L, Zhang W, Deng Y, Wang Y. Enhanced phytoremediation of selenium using genetically engineered rice plants. Journal of Plant Physiology. 2022 Apr 1;271:153665.

Limmer M, Burken J. Phytovolatilization of organic contaminants. Environmental Science & Technology. 2016 Jul 5;50(13):6632–43.

Lin Y-F, Hassan Z, Talukdar S, Schat H, Aarts MGM. Expression of the ZNT1 zinc transporter from the metal hyperaccumulator *Noccaea caerulescens* confers enhanced zinc and cadmium tolerance and accumulation to *Arabidopsis thaliana*. PLoS One. 2016;11(3):e0149750.

Lorestani B, Yousefi N, Cheraghi M, Farmany A. Phytoextraction and phytostabilization potential of plants grown in the vicinity of heavy metal-contaminated soils: A case study at an industrial town site. Environmental Monitoring and Assessment. 2013 Dec;185:10217–23.

Ma J, Alshaya H, Okla MK, Alwasel YA, Chen F, Adrees M, Hussain A, Hameed S, Shahid MJ. Application of cerium dioxide nanoparticles and chromium-resistant bacteria reduced chromium toxicity in sunflower plants. Frontiers in Plant Science. 2022 May;13:876119.

Mahajan P, Kaushal J. Role of phytoremediation in reducing cadmium toxicity in soil and water. Journal of Toxicology. 2018 Oct 23;2018.

Mahiya S, Lofrano G, Sharma SK. Heavy metals in water, their adverse health effects and biosorptive removal: A review. International Journal of Chemistry. 2014;3(1):132–49.

Mahjoub B. Plants for soil remediation. In Gaspard S, Ncibi MC, editors. Biomass for Sustainable Applications: Pollution Remediation and Energy. The Royal Society of Chemistry; 2014. pp. 106–43.

Mansoor S, Khan NF, Farooq I, Kaur N, Manhas S, Raina S, Khan IF. Phytoremediation at molecular level. In: Phytoremediation. Academic Press; 2022. pp. 65–90.

Meitei MD, Prasad MN. Potential of *Typha latifolia* L. for phytofiltration of iron-contaminated waters in laboratory-scale constructed microcosm conditions. Applied Water Science. 2021 Feb;11:47.

Mir AR, Pichtel J, Hayat S. Copper: Uptake, toxicity and tolerance in plants and management of Cu-contaminated soil. Biometals. 2021 Aug;34(4):737–59.

Mishra S, Bharagava RN, More N, Yadav A, Zainith S, Mani S, Chowdhary P. Heavy Metal Contamination: An Alarming Threat to Environment and Human Health. In: Sobti R, Arora N, Kothari R, editors. Environmental Biotechnology: For Sustainable Future. Singapore: Springer; 2019. pp. 103–25.

Mohammadi H, Amani-Ghadim AR, Matin AA, Ghorbanpour M. Fe 0 nanoparticles improve physiological and antioxidative attributes of sunflower (Helianthus annuus) plants grown in soil spiked with hexavalent chromium. 3 Biotech. 2020 Jan;10:1–1.

Mojiri A. Phytoremediation of heavy metals from municipal wastewater by *Typha domingensis*. African Journal of Microbiology Research. 2012 Jan 23;6(3):643–7.

Naz M, Benavides-Mendoza A, Tariq M, Zhou J, Wang J, Qi S, Dai Z, Du D. CRISPR/Cas9 technology as an innovative approach to enhancing the phytoremediation: Concepts and implications. Journal of Environmental Management. 2022 Dec 1;323:116296.

Nishijo M, Nakagawa H, Suwazono Y, Nogawa K, Kido T. Causes of death in patients with Itai-itai disease suffering from severe chronic cadmium poisoning: A nested case–control analysis of a follow-up study in Japan. BMJ open. 2017 Jul 1;7(7):e015694.

Pandey SK, Bhattacharya T, Chakraborty S. Metal phytoremediation potential of naturally growing plants on fly ash dumpsite of Patratu thermal power station, Jharkhand, India. International Journal of Phytoremediation. 2016 Jan 2;18(1):87–93.

Patra DK, Pradhan C, Patra HK. Toxic metal decontamination by phytoremediation approach: Concept, challenges, opportunities and future perspectives. Environmental Technology & Innovation. 2020 May 1;18:100672.

Perotti R, Paisio CE, Agostini E, Fernandez MI, González PS. CR (VI) phytoremediation by hairy roots of *Brassica napus*: Assessing efficiency, mechanisms involved, and post-removal toxicity. Environmental Science and Pollution Research. 2020 Mar; 27:9465–74.

Rai GK, Bhat BA, Mushtaq M, Tariq L, Rai PK, Basu U, Dar AA, Islam ST, Dar TU, Bhat JA. Insights into decontamination of soils by phytoremediation: A detailed account on heavy metal toxicity and mitigation strategies. Physiologia Plantarum. 2021 Sep;173(1):287–304.

Rasouli F, Hassanpouraghdam MB, Pirsarandib Y, Aazami MA, Asadi M, Ercisli S, Mehrabani LV, Puglisi I, Baglieri A. Improvements in the biochemical responses and Pb and Ni phytoremediation of lavender (*Lavandula angustifolia* L.) plants through *Funneliformis mosseae* inoculation. BMC Plant Biology. 2023 May 13;23(1):252.

Raza A, Tabassum J, Zahid Z, Charagh S, Bashir S, Barmukh R, Khan RS, Barbosa Jr F, Zhang C, Chen H, Zhuang W. Advances in "omics" approaches for improving toxic metals/metalloids tolerance in plants. Frontiers in Plant Science. 2022 Jan 4;12:794373.

Rezania S, Taib SM, Din MF, Dahalan FA, Kamyab H. Comprehensive review on phyto-technology: Heavy metals removal by diverse aquatic plants species from wastewater. Journal of Hazardous Materials. 2016 Nov 15;318:587–99.

Sabreena, Hassan S, Bhat SA, Kumar V, Ganai BA, Ameen F. Phytoremediation of heavy metals: An indispensable contrivance in green remediation technology. Plants. 2022 May 6;11(9):1255.

Sajad MA, Khan MS, Khan MA. 48. Evaluation of lead phytoremediation potential of Rumex dentatus: A greenhouse experiment. Pure and Applied Biology (PAB). 2019 Jun 1;8(2):1499–504.

Sakamoto M, Nakamura M, Murata K. Mercury as a global pollutant and mercury exposure assessment and health effects. Nihon Eiseigaku Zasshi/Japanese Journal of Hygiene. 2018 Jan 1;73(3):258–64.

Sarma H, Islam NF, Prasad R, Prasad MN, Ma LQ, Rinklebe J. Enhancing phytoremedia-tion of hazardous metal (loid) s using genome engineering CRISPR–Cas9 technology. Journal of Hazardous Materials. 2021 Jul 15;414:125493.

Sarwar N, Imran M, Shaheen MR, Ishaque W, Kamran MA, Matloob A, Rehim A, Hussain S. Phytoremediation strategies for soils contaminated with heavy metals: Modifications and future perspectives. Chemosphere. 2017 Mar 1;171:710–21.

Saxena P, Singh NK, Singh AK, Pandey S, Thanki A, Yadav TC. Recent advances in phytore-mediation using genome engineering CRISPR–Cas9 technology. In: Pandey VC, Singh V, editors. Bioremediation of Pollutants. Elsevier; 2020 Jan 1:125–41.

Sharma P, Tripathi S, Chandra R. Phytoremediation potential of heavy metal accumulator plants for waste management in the pulp and paper industry. Heliyon. 2020 Jul 1;6(7).

Singh R, Singh DP, Kumar N, Bhargava SK, Barman SC. Accumulation and transloca-tion of heavy metals in soil and plants from fly ash contaminated area. Journal of Environmental Biology. 2010 Jul 1;31(4):421–30.

Singh R, Singh S, Parihar P, Singh VP, Prasad SM. Arsenic contamination, consequences and remediation techniques: A review. Ecotoxicology and Environmental Safety. 2015 Feb 1;112:247–70.

Singh S, Parihar P, Singh R, Singh VP, Prasad SM. Heavy metal tolerance in plants: Role of transcriptomics, proteomics, metabolomics, and ionomics. Frontiers in Plant Science. 2016 Feb 8;6:1143.

Sivaram AK, Subashchandrabose SR, Logeshwaran P, Lockington R, Naidu R, Megharaj M. Rhizodegradation of PAHs differentially altered by C3 and C4 plants. Scientific Reports. 2020 Sep 30;10(1):16109.

Siyar R, DoulatiArdejani F, Norouzi P, Maghsoudy S, Yavarzadeh M, Taherdangkoo R, Butscher C. Phytoremediation potential of native hyperaccumulator plants growing on heavy metal-contaminated soil of Khatunabad copper smelter and refinery, Iran. Water. 2022 Nov 8;14(22):3597.

Soda N, Verma L, Giri J. CRISPR-Cas9 based plant genome editing: Significance, opportunities and recent advances. Plant Physiology and Biochemistry. 2018 Oct 1;131:2–11.

Song W-Y, Park J, Mendoza-Cózatl DG, Suter-Grotemeyer M, Shim D, Hörtensteiner S, et al. Arsenic tolerance in Arabidopsis is mediated by two ABCC-type phytochelatin transporters. Proceedings of the National Academy of Sciences. 2010;107(49):21187–92.

Srivastav A, Yadav KK, Yadav S, Gupta N, Singh JK, Katiyar R, et al. Nano-phytoremediation of Pollutants from Contaminated Soil Environment: Current Scenario and Future Prospects. In: Ansari AA, Gill SS, Gill R, R. Lanza G, Newman L, editors. Phytoremediation: Management of Environmental Contaminants, Volume 6. Cham: Springer International Publishing; 2018. pp. 383–401.

Subašić M, Šamec D, Selović A, Karalija E. Phytoremediation of cadmium polluted soils: Current status and approaches for enhancing. Soil Systems. 2022 Jan 4;6(1):3.

Varshney A, Mohan S, Dahiya P. Growth and antioxidant responses in plants induced by heavy metals present in fly ash. Energy, Ecology and Environment. 2021 Apr;6:92–110.

Wani ZA, Ahmad Z, Asgher M, Bhat JA, Sharma M, Kumar A, Sharma V, Kumar A, Pant S, Lukatkin AS, Anjum NA. Phytoremediation of Potentially Toxic Elements: Role, Status and Concerns. Plants. 2023 Jan 17;12(3):429.

Wyszkowska J, Boros-Lajszner E, Kucharski J. Calorific value of *Festuca rubra* biomass in the phytostabilization of soil contaminated with nickel, cobalt and cadmium which disrupt the microbiological and biochemical properties of soil. Energies. 2022 May 9;15(9):3445.

Yadav KK, Gupta N, Kumar A, Reece LM, Singh N, Rezania S, Khan SA. Mechanistic understanding and holistic approach of phytoremediation: A review on application and future prospects. Ecological Engineering. 2018 Sep 1;120:274–98.

Yadav R, Singh G, Santal AR, Singh NP. Omics approaches in effective selection and generation of potential plants for phytoremediation of heavy metal from contaminated resources. Journal of Environmental Management. 2023 Jun 15;336:117730.

Yan A, Wang Y, Tan SN, Mohd Yusof ML, Ghosh S, Chen Z. Phytoremediation: A promising approach for revegetation of heavy metal-polluted land. Frontiers in Plant Science. 2020 Apr 30;11:359.

Ye P, Wang M, Zhang T, Liu X, Jiang H, Sun Y, Cheng X, Yan Q. Enhanced cadmium accumulation and tolerance in transgenic hairy roots of *Solanum nigrum* L. expressing iron-regulated transporter gene IRT1. Life. 2020 Dec 3;10(12):324.

Zand AD, Mikaeili Tabrizi A, Vaezi Heir A. Application of titanium dioxide nanoparticles to promote phytoremediation of Cd-polluted soil: Contribution of PGPR inoculation. Bioremediation Journal. 2020 Jul 2;24(2-3):171–89.

Zhang J, Yang N, Geng Y, Zhou J, Lei J. Effects of the combined pollution of cadmium, lead and zinc on the phytoextraction efficiency of ryegrass (*Lolium perenne* L.). RSC advances. 2019;9(36):20603–11.

Zhi J, Liu X, Yin P, Yang R, Liu J, Xu J. Overexpression of the metallothionein gene PaMT3-1 from *Phytolacca americana* enhances plant tolerance to cadmium. Plant Cell, Tissue and Organ Culture (PCTOC). 2020 Oct;143:211–8.

Zulfiqar U, Farooq M, Hussain S, Maqsood M, Hussain M, Ishfaq M, Ahmad M, Anjum MZ. Lead toxicity in plants: Impacts and remediation. Journal of Environmental Management. 2019 Nov 15;250:109557.

9 Phytoremediation
A Biotechnological Strategy to Control Soil Pollution by Heavy Metals

Priyanka Mohapatra and Sujata Mohanty

INTRODUCTION

The rise in the population globally has led to an increase in demand for the necessities of life such as food, water, shelter, timber, and energy. This demand has created the need for rapid industrialization, urbanization, mining activities, and increased agriculture (Kafle et al., 2022; Sarwar et al., 2017). In fact, at present it is impossible to escape from the harmful chemicals and heavy metals which are the by-products of these anthropogenic activities (Sarwar et al., 2017). These by-products include both organic and inorganic contaminants. Although organic contaminants are biodegradable, inorganic contaminants include heavy metals that cannot be degraded, and they pose a grave environmental concern as they accumulate and lead to soil contamination (Ghosh and Singh, 2005).

Metals and metalloids with an atomic density more than 6 g/cm^3 are categorized as heavy metals (Pinto et al., 2015). Metals such as Cu, Fe, Mn, Ni, Zn, Pb, Cr, Hg, As, and Cd are listed as heavy metals (Fasani et al., 2018). Heavy metals occur naturally in the rocks and are dispersed in nature via soil erosion and volcanic eruptions. The main sources of heavy metals in cities are mostly from industrial emissions and vehicular traffic, whereas in rural areas the occurrence of heavy metals is due to tanneries, mining, drilling, sewage sludge, warfare, batteries, pesticides, and fertilizers (Muthusaravanan et al., 2018). Heavy metals found in soil cannot be degraded. For example, lead (Pb) maintains very high levels in the soil for more than 150 years and is retained in the soil for more than 1000 years (Nandakumar et al., 1995). Cadmium (Cd) has a soil retention lifespan of 18 years (Forstner, 1995). Soil contamination by heavy metals has negative effects on the activity of microbes and their biodiversity (Chen et al., 2014). Because the soil microorganisms transform organic carbon, which is essential for soil fertility, a decrease in the population of soil microbes results in a decline of soil fertility (Xu et al., 2018). Moreover the consumption of plants that are cultivated in lands contaminated by heavy metals (Cd, Cr, and Zn) poses a threat to human health as a result of the accumulation of these metals in the cultivated food crops (Wang et al., 2003).

The entry of these heavy metals into the food chain is dangerous for human beings because they form free radicals, thus causing oxidative stress, and they can take the place of essential elements in enzymes, disturbing their functions (Henry, 2000; Sarwar

DOI: 10.1201/9781003442295-9

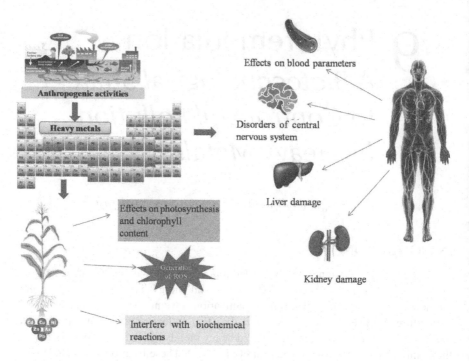

FIGURE 9.1   Effects of heavy metals on plants and humans. (Photos by Priyanka Mohapatra.)

et al., 2010). Another vital issue associated with heavy metals is their ability to biomagnify in the food chain (Majed et al., 2016). We conclude that heavy metals ruin soil properties, reduce agricultural productivity, upset the balance of the ecosystem, and damage human health if people are exposed directly or through the food web (Ahmad et al., 2016; Mao et al., 2015). The effects of heavy metals on plants and animals are shown in Figure 9.1. Therefore, in view of the serious outcomes of soil contamination with heavy metals, soil remediation strategies like soil washing/flushing, surface capping, vitrification, landfilling, phytoremediation, and bioremediation have evolved and are used for cleanup and revival of the contaminated areas (Gomes, 2012; Liu et al., 2018).

Most of the cleanup methods are expensive, and they may release secondary pollutants into the environment or be accompanied by detrimental effects on the soil structure, biological activities, and soil fertility (Alaboudi et al., 2018; Pulford and Watson, 2003). Therefore, it is vital to establish economical and green strategies (Purakayastha and Chhonkar, 2010). Thus in this chapter we give a detailed account of the technique popularly known as phytoremediation, which uses green plants to remediate heavy metal–contaminated soils.

MECHANISM FOR ACCUMULATION OF HEAVY METALS IN PLANTS

UPTAKE OF HEAVY METALS BY ROOT

The soil contains heavy metals in the form of ions. Several factors like pH, organic content, and microorganisms present in soil affect the availability and mobility of these heavy metal ions to the roots of the plant (Luo et al., 2014). The first barrier for the heavy metal ions is the root cell walls. These cell walls have cellulose and pectin that bind with the heavy metal ions and help in their entry into the root cells of the plant. The heavy metals get into the roots via the symplastic and apoplastic pathways. The apoplast is the outer layer of the plasma membrane containing an aqueous layer where the metal ions can pass freely. The symplast is formed by the channels that connect the cytoplasm of two cells. These channels are called plasmodesmata. They help in the movement of sugar, amino acids, and heavy metal ions (Chen et al., 2013).

After passing through the root cell walls, the heavy metals cross into the plasma membrane and reach the cytosol via transmembrane carriers that help in the uptake of other essential ions, such as, Ca^{2+}, Fe^{2+}, Mg^{2+}, Cu^{2+}, Mn^{2+}, and Zn^{2+} (Clemens, 2006; Gallego et al., 2012). There are no specific transporters for nonessential metal ions and heavy metal ions, so they enter the plant through the transporters used by nutritional or essential metal ions (Gallego et al., 2012). The uptake of toxic heavy metals in plant roots is caused when there is low availability of nutritional metal ions in the soil (He et al., 2015). In case of *Typha latifolia*, the uptake of Pb^{2+} and Cd^{2+} are enhanced when the soil levels of Ca^{2+} are low (Rodriguez-Hernandez et al., 2015). Furthermore, it has been suggested that in plants, the Ca^{2+} channels are the main players that help in the uptake of toxic heavy metals (Rodriguez-Hernandez et al., 2015).

Transport of Heavy Metal Ions

The heavy metals from the cytosols of the root cells move to the cellular organelles like mitochondria, vacuoles, and chloroplasts inside the cells. The translocation of heavy metals to neighboring cells is facilitated by transporters or plasmodesmata present on the plasma membrane. The heavy metals on reaching the xylem vessels get translocated to the aerial parts by the transpiration stream (Choppala et al., 2014).

The loading of xylem in plant roots with heavy metals is regarded as the checkpoint for the transport of metals to the shoot system (Yamaguchi et al., 2011). Some plants often show a high capacity for xylem loading of heavy metals and transport to the shoot system; these plants are referred to as hyperaccumulators. On the other hand, non-hyperaccumulators exhibit a low capacity for heavy metal xylem loading. Hyperaccumulators transport the absorbed heavy metal ions to vacuoles located in the root cells, thus resulting in high accumulation of heavy metals in roots (Yamaguchi et al., 2011).

Sequestration and Detoxification of Heavy Metals

After transport, heavy metals accumulate in plant cells. This can cause harmful effects on the plants, such as chlorosis, stunted growth, a reduction in photosynthesis, increases in proteolysis lipid peroxidation, and disturbances in the balance between antioxidants and reactive oxygen species (ROS), leading to cell death (Gallego et al., 2012; Iannone et al., 2015). The plants during their evolution have developed various strategies to reduce heavy metal toxicity by sequestration and detoxification (Luo et al., 2016). Sequestration takes place at the cell wall and inside cell organelles such as the Golgi apparatus and vacuoles (Ovečka and Takáč, 2014). The main sequestration locations are the cell walls of the root cells because they are the first site where the heavy metals enter (Huguet et al., 2012; Vázquez et al., 2006). The main components of the cell wall are cellulose, lignin, pectin, hemicellulose, and suberin. Heavy metal ions entering the root cells have strong affinity for suberin and pectin (Baxter et al., 2009; Chen et al., 2013).

The metal ions sequestered in the root cell walls may reduce the entry of more metal ions into the cytoplasm of the cell, thus decreasing the migration of these ions in the direction of the xylem and their distribution to the shoot system. This can lower the toxicity caused by heavy metals (Gallego et al., 2012). Additionally, heavy metal sequestration in cell organelles, specifically in vacuoles, plays a major role in the storage and detoxification of metals (Saraswat and Rai, 2011). As vacuoles in plant cells fill a sizeable volume and have much less of a role in biochemical reactions than other cellular compartments, they are a perfect location for the storage of toxic components (Ovečka and Takáč, 2014). Therefore, when heavy metals enter into the cytosol, they are directly sequestered in vacuoles.

On entering the vacuole, the heavy metals are chelated by glutathione (GSH), metallothioneins (MTs), and phytochelatins (PCs) and are converted to high- or low-molecular-weight (HMW or LMW) compounds or metal chelates (Huang et al., 2012; Song et al., 2014). Thus, they are stored in vacuoles as high- or low-molecular-weight compounds or metal chelates (Gallego et al., 2012; Huguet et al., 2012; Song et al., 2014). These complexes are further broken down by the acids in the vacuole (Chiang et al., 2011; Choppala et al., 2014). Moreover, the Golgi apparatus is associated with the exocytosis of heavy metals (Peiter et al., 2007). The detoxification step includes chelation of heavy metals and scavenging of increased ROS by activating the antioxidant system (GSH-ascorbate cycle, proline and phenolics) (Gratão et al., 2005; Lin and Aarts, 2012).

METHODS OF HEAVY METAL REMEDIATION

The level of heavy metals in our surroundings is increasing every year (Govindasamy et al., 2011). Hence, treatment of the soil containing high concentrations of heavy metals is necessary to lower their negative effect on the environment. It is a tough job because it is costly and needs skilled labor with technical knowledge (Barcelo and Poschenrieder, 2003). Until now, several approaches have been employed for this purpose. The conventional approaches are listed below and presented in Figure 9.2.

Physical methods can be used to treat a broad range of heavy metals. The main disadvantage of this method is that the contaminants need to be processed, and it is therefore expensive. Separation by this method depends on the particle size. Some very commonly used physical methods are heat treatment, electroremediation, vitrification, and soil replacement (Sharma et al., 2018). In chemical methods the chemical characteristics of the heavy metal contaminants are changed to make them harmless (Leštan et al., 2008). Though these methods are highly efficient, they produce harmful by-products that need further processing. These methods include precipitation, chemical extraction, ion exchange, soil amendment, chemical leaching, and nanoremediation (Sharma et al., 2018).

Biological methods use plants, animals, and microorganisms for elimination of heavy metals from the surroundings. They are cheaper and do not generate secondary

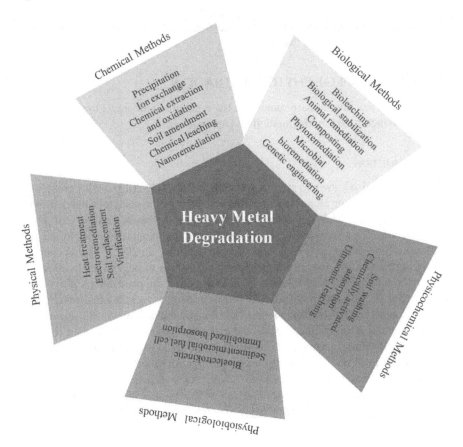

FIGURE 9.2 Different methods for the remediation of heavy metal–contaminated soil. (Photos by Priyanka Mohapatra.)

contaminants (Abbas et al., 2014). Bioleaching, composting, biological stabilization, animal remediation, microbial remediation, biofilm-based remediation, integrated biosystem remediation, phytoremediation, and genetic engineering are the various forms of biological methods for treating heavy metal–contaminated soil (Alvarez et al., 2017; Chen et al., 2015; Dixit et al., 2015; Guarino and Sciarrillo, 2017; Ibuot et al., 2017;Li et al., 2017; Rozas et al., 2017; Suthar et al., 2014).

Physicochemical methods combine physical and chemical methods of extraction of heavy metals from the soil (Dermont et al., 2008). Soil washing, ultrasonic leaching, and chemically activated adsorption are few physicochemical methods (Nayak et al., 2017; Wuana and Okieimen, 2011). Physiobiological methods area fusion of physical and biological methods of treating soil pollution; some examples are the bioelectrokinetic approach, sediment microbial fuel cells, and immobilized biosorption (Abbas et al., 2017; Aryal and Liakopoulou-Kyriakides, 2015; Huang et al., 2017). Traditional methods are costly and can be ineffective, often leading to the production of secondary wastes which need further processing. These disadvantages have paved the way for methods like phytoremediation. Phytoremediation is an environmentally friendly and economical method that employs green plants to eliminate the pollutants from soil (Ashraf et al., 2019).

CLASSIFICATION OF PHYTOREMEDIATION

Plants use various methods to remediate the contaminated sites. Phytoremediation can be classified based on the method and application.

PHYTOEXTRACTION

This method is also known as phytoaccumulation (Ali et al., 2013). In this process, the heavy metal is taken up along with nutrients and water by the plant root and distributed to the shoot system. The heavy metal accumulates in the plant, thus the term phytoaccumulation. The absorbed metals are translocated to the inactive sites in the plant such as cell membrane, cell wall, and vacuole (Kafle et al., 2022). The mechanisms of phytoextraction involve accumulation and absorption of heavy metal cations and the formation of metal–phytochelatin complexes within the plant cell (Asgari Lajayer et al., 2019). The complex molecules are transferred to the vacuole of the plant cell for storage (Yadav, 2010). The plant biomass and the amount of heavy metal accumulated in the plant above the soil determines the extraction potential of the plant (Li et al., 2010). After the plant reaches a desirable height, the aerial parts that have accumulated heavy metals are harvested and removed, thus assuring the removal of heavy metal contaminants permanently (Jabeen et al., 2009). Plants such as *Calendula officinalis, Helianthus annuus, Brassica juncea, Cynodon dactylon*, and *Arabidopsis thaliana* can accumulate Pb, Cr, Ni, Cu, and Cd (Shah and Daverey, 2020).

PHYTOSTABILIZATION

This process lowers the mobility and bioavailability of soil heavy metals by immobilizing their off-site movement by using plants (Pulford and Watson, 2003). The roots of plants that can survive in contaminated soil grow deep into the contaminated zone and change the physical, chemical, and biological properties of the soil, thus lowering the toxicity of the heavy metals present (Jabeen et al., 2009). Countries like China and Europe use phytostabilization for remediation of Zn, Cu, and Pb contaminated soils (Dickinson et al., 2009). Hyperaccumulators are the most efficient plants for phytostabilization. They collect the heavy metals in the roots, thus restricting their movement into the food web and their leaching into groundwater (Ali et al., 2013; Mahar et al., 2016). Some plants detoxify the organic chemicals and heavy metals by secreting glutathione, which combines with the contaminants to render them less toxic. Apart from this, the presence of pollutants stimulates the release of certain enzymes from the roots that detoxify the contaminants (Kafle et al., 2022). The most common species used for phytostabilization of metals like Pb, Cu, and Zn are *Agrostis* spp. and *Festuca* spp. (Mahar et al., 2016). *Nerium oleander* are used for phytostabilizing heavy metals like Cr and Ni (Elloumi et al., 2017). Currently, the use of soil additives like dolomite, chalcedonite, and halloysite are known to aid in the process of phytostabilization of heavy metals (Sharma et al., 2018).

PHYTOVOLATILIZATION

The pollutants are up taken by the plants from the soil and converted to a volatile form, followed by their release into the environment. The main drawback of this method is that it doesn't eliminate the contaminants completely, and it releases poisonous heavy metals back to the soil (Sharma et al., 2018). This method is mostly used for heavy metals like Hg and Se (Ali et al., 2013). *Pteris vittata* was used to treat As at laboratory scale, but it released Se into the environment (Sakakibara et al., 2010).

PHYTODEGRADATION

In this process, the pollutants are absorbed by the plants and broken down into simpler and less toxic compounds. This is followed by distribution within the plant tissue. But this method can be used for organic pollutants only and cannot be used for heavy metal degradation as they are non-biodegradable (Doty et al., 2000). When the degradation of organic contaminants takes place in the vicinity of the root (rhizosphere), it is called rhizodegradation (Ali et al., 2013).

PLANTS USED IN PHYTOREMEDIATION

The ability of plants to accumulate heavy metals varies among species and genotypes (Laghlimi et al., 2015; Nouri ct al., 2009). Some features that influence the accumulation of heavy metals in the plants are: (1) the architecture of the root, (2) the ability to use water efficiently, (3) the chemistry of the rhizosphere, (4) the affinity of transporter proteins present in roots for metals, (5) the translocation of metals within the plant post xylem loading, (6) age, and (7) growth stage (Hamon and McLaughlin, 2003; Nouri et al., 2009).

The depth to which the root grows determines the extent to which soil can be cleaned. For grasses, the cleaning depths are not more than three feet; for shrubs, the depths that can be cleaned are less than 10 feet, and for bigger trees the roots can reach as deep as 20 feet (Laghlimi et al., 2015). Therefore, plants should be selected for phytoremediation based on the site of contaminants (Sharma and Reddy, 2004). Many works are available that have reported separate patterns of accumulation and mobility of different heavy metals within the same plant (Pulford and Watson, 2003; Yang et al., 2015). For example, Pb and Cd have a tendency to collect in the root, while Cu and Zn are translocated to the leaves within the same plant (Pulford and Watson, 2003). Plants like *Robinia pseudoacacia* and *Populus purdomii* Rehd. are ideal for phytoremediation applications of heavy metals. The maximum concentrations of Cu and Zn, 140.85 mg/kg and 432.08 mg/kg, respectively, were accumulated in the leaves of *Populus purdomii* Rehd.(Yang et al., 2015). Substantial amounts of Cd (3.86 mg/kg) and Pb (712.37 mg/kg) were accumulated in the roots of *Populus simonii* Carrand *Robinia pseudoacacia* Linn., respectively (Yang et al., 2015). The highest concentrations of Cu and Cd collected in *Robinia pseudoacacia* were 6418.2 mg/kg and 4699.8 mg/kg, respectively (Serbula et al., 2012). *Robinia pseudoacacia* is an ideal plant for phytoremediation applications because of its fast growth rate, high biomass production, and high transpiration rates (Dadea et al., 2017).

Centella asiatica and *Eichhornia crassipes* are aquatic macrophytic species mostly found in swampy areas contaminated with heavy metals. They grow very rapidly and can accumulate substantial concentrations of heavy metals from the soil (Muthusaravanan et al., 2018). Mokhtar et al. (2011) reported that *Centella asiatica* and *Eichhornia crassipes* removed 99.6% and 97.3% of copper, respectively. Plants like Indian mustard (*Brassica juncea*) and musk-grass (*Chara canescens*) can absorb heavy metals and metalloid contaminants such as Hg and Se and change them into gaseous form, which is followed by their release into the environment (Ghosh and Singh, 2005). A list of the plants used in phytoremediation studies is given in Table 9.1.

TABLE 9.1
List of Some Plants Used for Phytoremediation Studies

Scientific Name	Common Name	Mechanism	Remarks	Reference
Sesbania drummondii	Rattlebush	Phytoextraction	The uptake and accumulation of Pb was enhanced by EDTA.	Barlow et al. (2000)
Lolium perenne	Ryegrass	Phytoextraction	Increased survival time in soils containing Ni, Cu, and Fe.	Hernández et al. (2019)
Helianthus annus	Sunflower	Phytostabilization	It is for low concentrations of metals like Cu, Zn, Pb, Hg, As, and Cd; vermicompost is used as a supplement.	Jadia and Fulekar (2008)
Polypogon monspeliensis	Rabbifoot grass	Phytovolatilization	Dimethylchloroarsine ($AsCl(CH_3)_2$) and pentamethylarsine ($As(CH_3)_5$) are volatilized. More toxic organic forms of As are not volatilized.	Ruppert et al. (2013)
Vigna unguiculata	Cowpea	Phytostabilization	For treating Pb and zinc present in mine wastes.	Kshirsagar and Aery (2007)
Nicotiana tabacum	Tobacco	Phytoextraction	Cd accumulated in stems and leaves.	Yang et al. (2019b)
Brassica juncea	Indian mustard	Phytovolatization	Hg was taken up and volatilized from the contaminated sites.	Moreno et al. (2005)
Salix alba		Phytostabilization	Phytostabilized Cd and Cu.	Mataruga et al. (2020)
Pistacia lentiscus	Mastic tree	Phytostabilization	Lowers metal mobility of Zn, Hg, Pb.	Concas et al. (2015)
Cyperus rotundus	Nut grass	Phytoaccumulation	Removal of Cr and Cd from contaminated sites.	Muthsaravanan et al. (2018)
Zea mays	Maize	Phytoaccumulation	Good accumulators of Pb.	Shahandeh and Hossner (2000)

CURRENT TRENDS IN PHYTOREMEDIATION

PHYTOREMEDIATION THROUGH GENETICALLY ENGINEERED PLANTS

In recent times, genetic engineering has proved to be a cutting edge technology that has gained attention to improve the methods of phytoremediation technology (da Conceição Gomes et al., 2016).

According to previous reports, green plants can detoxify heavy metals. Various attempts have been made to develop genetically modified crops (GM crops) that can tolerate heavy metals present in the soil (Gerhardt et al., 2017). Genetic engineering involves modification of the genetic materials of plants to enhance their heavy metal tolerance potential and metal accumulation capacity (da Conceição Gomes et al., 2016; Kumar and Prasad, 2019).

Overexpression of various genes, like genes that encode antioxidant enzymes, phytochelatins, vascular heavy metal transporters, and heavy metals exporters can help in manipulating the toxicity of heavy metals in plants (Raza et al., 2021). The overexpression of yeast cadmium factor 1 (YCF1) in *Arabidopsis thaliana* improved the tolerance of the plant to Cd(II) and Pb(II) (Song et al., 2003). *YCF1* gene transported Cd into the vacuole by combining with glutathione (GSH). Moreover, tolerance to Pb was also achieved when homologous proteins were encoded to NtCBP4 by the cyclic nucleotide-gated ion channel 1 (CNGC1 gene) present in the T-DNA mutants of *Arabidopsis* (Zeng et al., 2015). Two bacterial genes, merA and merB, encoding mercuric ion reductase and organo-mercurial lyase, respectively, have been used for phytoremediation of mercury. They are known to reduce the toxicity of mercury (Dhankher et al., 2011). *A. thaliana* and tobacco that expressed modified merA were found to reduce Hg levels significantly. They were resistant to $HgCl_2$ (25–250 μM) and lowered Hg toxicity completely by volatilization compared to the control plants. Similarly, the expression of modified merB in *Arabidopsis* displayed resistance to phenyl-mercuric acetate and monomethylmercuric chloride as compared to the control (Bizily et al., 2000).

Yellow stripe-like transporters (YSLs) play a role in the uptake and transport of metals in plants. *SnYSL3*, a member from this YSL gene family, has been acquired and characterized by *Solanum nigrum*, which is a Cd hyperaccumu-lator. When *SnYSL3* expresses itself, it encodes a protein (plasma membrane-localized) that forms various metal–nicotianamine complexes. An upregulation in the expression of *SnYSL3* was observed in the presence of high Cd levels in soil, proving its role in tolerating Cd metal stress (Feng et al., 2017). The over-expression of wheat *TaPCS1* gene in tobacco shrub encoded the enzyme PC synthase that improved their tolerance to Cd and Pb (Raza et al., 2021). Though these reports prove the efficiency of genetic engineering for phytoremediation, the research conducted to date is limited to the hydroponic and *in situ* environ-ments. In order to understand the potential of transgenic plants for phytoreme-diation in real environmental conditions, more studies have to be undertaken (Kafle et al., 2022). Some plants that have been genetically engineered for phy-toremediation are listed in Table 9.2.

TABLE 9.2

List of Some Transgenic Plants Used for Phytoremediation Studies

Scientific Name	Gene	Source	Degradation	Reference
Arabidopsis thaliana	MerC	Bacteria	Hg	Ozyigit et al. (2021)
Arabidopsis thaliana	ZmUBP15-16-19	*Zea mays*	Cd	Kong et al. (2019)
Nicotiana tabacum	PtoEXPA12	*Populus tomentosa*	Cd	Zhang et al. (2018)
Brassica juncea	AtACBP1 AtACBP4	*Arabidopsis thaliana*	Pb	Du et al. (2015)
Arabidopsis thaliana; *Populus tomentosa*	PtABCC1	*Populus trichocarpa*	Hg	Sun et al. (2018)
Arabidopsis thaliana	YSL	*Miscanthus sacchariflorus*	Cd	Chen et al. (2018)
Nicotiana tabacum	PjHMT	*Prosopis juliflora*	Cd	Keeran et al. (2017)
Oryza sativa	ricMT	*Oryza sativa*	Cd, Cu	Zhang et al. (2017)
Nicotiana tabacum	PtPC	*Populus tomentosa*	Cd	Chen et al. (2015)
Arabidopsis thaliana	PCs1	*Brassica napus*	Cd	Bai et al. (2019)

Application of Modeling for Phytoremediation of Heavy Metal–Contaminated Soil

Although phytoremediation is an economical, sustainable method that requires minimal resources for the elimination of soil pollutants, its traditional practice is inhibited by its lengthy treatment time and changing efficiencies (Tang, 2023). Therefore, it has already made an entrance in the field of mathematical modeling of the various chemical and biological methods (Urbaniak et al., 2017). The employment of modeling in the field of environmental science can help us to avoid human biases and errors (Chirakkara et al., 2016). There are different types of models that give us different outcomes. For example, we can build a model for phytoremediation to determine the accumulation of heavy metals in a particular plant part or their collective distribution in the entire plant (Hasanuzzaman et al., 2017).

Nowadays, various scientific studies have employed many prediction and optimization methods, based on mathematical, statistical, artificial, and computational models, such as RSM, ANN, and GA. These models are cheaper, less complicated, and accurate (Manzoor et al., 2023).

In an investigation by Titah et al. (2018), response surface methodology (RSM) and an artificial neural network (ANN) were optimized to predict the removal of As by *Ludwigia octovalvis* using arsenic-contaminated soil in a pilot reed bed system. The concentration of As in the soil was predicted to be 39 mg/kg by both the models. The removal values of As by ANN and RSM were 71.4% and 72.6%, respectively. The experimental value of As removal by *L. octovalvis* was 70.6%. The differences between the predicted and experimental values for ANN and RSM were 1.87% and 3.49%. Thus, it was concluded that the ANN model showed better prediction and fitting ability in comparison to RSM. In another work, the efficiency of two models (ANN and RSM) were compared to optimize the extraction of Pb from Pb-polluted soil by using garden geranium (*Pelargonium hortorum*). ANN and RSM were employed to model the extraction of Pb from the contaminated soil. The predicted values of ANN and RSM were nearly 86.05% and 36%, respectively, with respect to the experimental values. For the optimization, ANN was found to be more accurate, reliable, and efficient in comparison to RSM as the correlation coefficient (R^2) of ANN and RSM were 0.99 and 0.90 respectively (Manzoor et al., 2023). A multi-layer perceptron neural network (MLP) model was used to predict the bioconcentration factor (BCF) of *Salix dasyclados* L. for the bioaccumulation of heavy metals. The structure of MLP obtained was 18-17-9. The RMSE (root-mean-square errors) were negligible for prediction, validation, and testing phases, proving the efficiency of the ANN (MLP model) (Złoch et al., 2017). Three mathematical models—RSM-BBD, ANN, and random forest (RF)—were built to predict the phytoaccumulation efficiency of *Ophiopogon japonicas* to accumulate Zn from the soil enriched by EDTA. Of these, the RF model was found to be the most efficient and accurate with an accuracy of 88.23% (Janani et al., 2019). A list of successful applications of a few selected models for evaluation of phytoremediation efficiency is given in Table 9.1.

APPLICATION OF NANOTECHNOLOGY IN PHYTOREMEDIATION

Phytoremediation of soil polluted with Pb, Cr, Cd, Ni, and Zn could be improved by implementing nanomaterials. The most studied heavy metals are Pb and Cd as these two metals are most commonly found in contaminated sites (Song et al., 2019). *Lolium perenne* L. (ryegrass) is commonly used for remediation of Pb from polluted soil owing to its low cost, rapid growth, and high tolerance to Pb. The use of nanomaterials has been found to be efficient in enhancing Pb phytoextraction by ryegrass. A study was conducted by Liang et al. (2017) to study the effects of nano-hydroxyapatite on the phytoextraction of Pb by ryegrass. They conducted the experiment over durations of 1, 1.5, 2, 3, and 12 months. The results of their experiment showed that there was a significant increase in the accumulation of Pb by using 0.2% (w/w) nano-hydroxyapatite after 1.5 months. The removal rates of Pb present in the soil by the control group (comprising ryegrass soil) varied from 16.74% to 31.76%. But with the help of nano-hydroxyapatite, Pb removal by ryegrass was found to be over 30% one month after application. In fact, after three months the removal rate increased to 44.39%.

In another study, different concentrations (0, 100, 200, 500, 1000, and 2000 mg/kg) of nZVI particles were used to help in the Pb phytoextraction by ryegrass. After 45 days of treatment, it was observed that low concentrations of nZVI (100, 200, and 500 mg/kg) improve the Pb accumulation in ryegrass. The highest accumulation of Pb (1175.40 µg per pot) was obtained with 100 mg/kg of nZVI particles. But when the concentrations of the nZVI particles increased (1000 and 2000 mg/kg), it caused severe oxidative stress in the plant, resulting in a decrease in the accumulation of the Pb (Huang et al., 2018).

An investigation was carried out by Singh and Lee (2016) to study the effect of TiO_2 nanoparticles on the accumulation of Cd in soybean plants. Several concentrations (100, 200, and 300 mg/kg) of TiO_2 nanoparticles were added to the soil, and the distribution and accumulation of Cd in plants were noted on the 60th day after sowing. The results showed that there was an increase in the accumulation of Cd in the

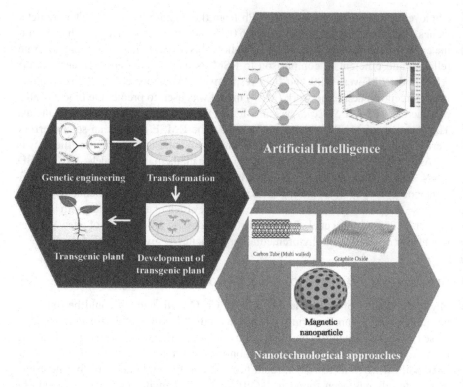

FIGURE 9.3   Current trends in phytoremediation. (Photos by Priyanka Mohapatra.)

shoots (1.9, 2.1, 2.6 times) and roots (2.5, 2.6, and 3.3 times), respectively. When the concentration of nanoparticles was 300 mg/kg, the Cd accumulation was recorded to be the maximum (1534.7 mg/g). An experiment was conducted to study the effect of ramie in combination with different concentrations of nZVI particles (stabilized by starch) on Cd-contaminated soil. It was observed that the augmentation of nZVI particles enhanced accumulation of Cd in the leaves, stems, and roots by 31–73%, 16–50%, and 29–52%, respectively.

The latest trends in the field of phytoremediation has been presented in Figure 9.3.

ECONOMIC IMPORTANCE OF PHYTOREMEDIATION

The cost of phytoremediation is fluctuating. The types of pollutants, their concentration, circumstances of the site, and priority of the contaminant causing a harmful effect in the food chain influence the cost of treatment by phytoremediation (Farraji et al., 2016). According to several reports, the cost of phytoremediation is less when compared to other traditional techniques for treating heavy metals. Phytoremediation was estimated to be 10 times cheaper than chemical treatment. For removing waste, the price range for phytoremediation was between $20 and $40 per ton, whereas in cases of chemical treatment the price was $100 to $500 per ton (Raskin and Ensley, 2000). The cost for treating petroleum hydrocarbons by phytoremediation was

$2500 to $15,000/hectare, which was much less than the cost of bioremediation by microbes ($20,000 to $60,000/hectare) (Cunningham et al., 1996). Furthermore, the removal of Pb by excavation from the contaminated sites was $141,500 to $182,100. This was extremely costly in comparison to phytoremediation. In another report by Mulbry et al. (2008), the treatment of dairy wastewater by phytoremediation was found to be more economical than the traditional methods.

Several studies have explored the phytoremediation capacity of energy-efficient crops, and consequences faced by the pollutants present in these plants (Saxena et al., 2020). Maize cultivation for phytoremediation in the Netherlands (Campine region) and Belgium could produce 2 90,000–38,000 KV of renewable energy (thermal and electrical) per hectare every year; this could be used to replace coal as a fuel (Meers et al., 2010). Additionally, the quality of polluted lands can be improved by phytoremediation, and this helps in the production of food (Gomes, 2012). Phytoremediation that targets the soil phosphorus and nitrogen will generate compost and fertilizers rich in nutrients that may be employed to improve the fertility of the soils (Riaz et al., 2022).

A phytoremediation project was carried out for treating soil contaminated with heavy metals such as Pb, Cd, and As. After two years the levels of Pb, Cd, and As in the soil decreased to concentrations lower than the national standards. The total expenditure of the project was $75,375.20 per hm^2; this was cheaper in comparison to other technologies (Wan et al., 2016).

The cost for cleaning one acre of land (50 cm depth) in the United States by phytoremediation was $60,000 to $1,000,000, which was lower than the amount spent on soil excavation ($4,000,000) (Jabeen et al., 2009; Salt et al., 1995)

The phytoremediation potential of the US and European markets was estimated to beapproximately$36 to $54 billion (Henry, 2000). There are several companies around the world that offer services for phytoremediation. These include Over Verde Ltd (Flintshire, UK); Clean Biotec (Spain); Ecolotree (United States); Symbio Greentech (India); BioRemed AB (Sweden); and Bioclear earth (Netherlands), among others (Riaz et al., 2022).

The cost of phytoremediation is determined by the evaluation procedure, plants to be cultivated on the treated soil, and speed of cleaning (Lewandowski et al., 2006). The economic factors of the crops to be used for phytoremediation should be assessed by policymakers and researchers on the basis of their returns and benefits to farmers and society (Sarwar et al., 2017).

FUTURE AND CONCLUSIONS

Heavy metals are considered to be hazardous, and their presence in the environment is considered to be a threat to humankind because of their toxicity, bioaccumulation, and persistence. Metals such as Pb, As, Cd, and Hg are the most toxic metals and are nonessential elements for living organisms. Phytoremediation is a sustainable, cost-effective, socially acceptable, and environmentally friendly technology for eliminating pollution in air, water, and soil. This method employs plants for treating environmental contaminants through various mechanisms such as phytostabilization, phytovolatilization, phytoextraction, and phytodegradation. The remediation capacity of plants varies according to their target contaminants. Hence, the selection of plants according to the soil contaminants and location is very important. This chapter has listed various plants that have been reported to be efficient for remediation of a wide range of pollutants by using some specific mechanism of phytoremediation. To improve the phytoremediation potential of the plants, various biotechnological approaches like genetic engineering, nanotechnology, and modeling have been widely used. Though a lot of work has been done in the field of phytoremediation, still more substantial research must be undertaken in field conditions for selecting plants, discovering new genes, and developing transgenic plants that will help us to understand the metabolic activities responsible for heavy metal tolerance in hyperaccumulator plants and to decipher new dimensions for phytoremediation.

REFERENCES

Abbas, S.H., Ismail, I.M., Mostafa, T.M. and Sulaymon, A.H., 2014. Biosorption of heavy metals: A review. *J Chem Sci Technol, 3*(4), pp. 74–102.

Abbas, S.Z., Rafatullah, M., Ismail, N. and Syakir, M.I., 2017. A review on sediment microbial fuel cells as a new source of sustainable energy and heavy metal remediation: Mechanisms and future prospective. *International Journal of Energy Research, 41*(9), pp. 1242–1264.

Ahmad, R., Tehsin, Z., Malik, S.T., Asad, S.A., Shahzad, M., Bilal, M., Shah, M.M. and Khan, S.A., 2016. Phytoremediation potential of hemp (*Cannabis sativa* L.): Identification and characterization of heavy metals responsive genes. *Clean - Soil, Air, Water, 44*, pp. 195–201.

Alaboudi, K.A., Ahmed, B. and Brodie, G., 2018. Phytoremediation of Pb and Cd contaminated soils by using sunflower (*Helianthus annuus*) plant. *Annals of Agricultural Sciences, 63*(1), pp. 123–127.

Ali, H., Khan, E. and Sajad, M.A., 2013. Phytoremediation of heavy metals—Concepts and applications. *Chemosphere, 91*(7), pp. 869–881.

Alvarez, A., Saez, J.M., Costa, J.S.D., Colin, V.L., Fuentes, M.S., Cuozzo, S.A., Benimeli, C.S., Polti, M.A. and Amoroso, M.J., 2017. Actinobacteria: Current research and perspectives for bioremediation of pesticides and heavy metals. *Chemosphere, 166*, pp. 41–62.

Aryal, M. and Liakopoulou-Kyriakides, M., 2015. Bioremoval of heavy metals by bacterial biomass. *Environmental Monitoring and Assessment, 187*, pp. 1–26.

Asgari Lajayer, B., Khadem Moghadam, N., Maghsoodi, M.R., Ghorbanpour, M. and Kariman, K., 2019. Phytoextraction of heavy metals from contaminated soil, water and atmosphere using ornamental plants: Mechanisms and efficiency improvement strategies. *Environmental Science and Pollution Research, 26*, pp. 8468–8484.

Ashraf, S., Ali, Q., Zahir, Z.A., Ashraf, S. and Asghar, H.N., 2019. Phytoremediation: Environmentally sustainable way for reclamation of heavy metal polluted soils. *Ecotoxicology and Environmental Safety, 174*, pp. 714–727.

Bai, J., Wang, X., Wang, R., Wang, J., Le, S. and Zhao, Y., 2019. Overexpression of three duplicated BnPCS genes enhanced Cd accumulation and translocation in *Arabidopsis thaliana* mutant *cad1-3*. *Bulletin of Environmental Contamination and Toxicology, 102*, pp. 146–152.

Barcelo, J. and Poschenrieder, C., 2003. Phytoremediation: Principles and perspectives. *Contribution to Science, 2*, pp. 333–344.

Barlow, R., Bryant, N., Andersland, J. and Sahi, S. (2000). Lead hyperaccumulation by *Sesbania drummondii*. Paper presented at the Proceedings of the 2000 Conference on Hazardous Waste Research.

Baxter, I., Hosmani, P.S., Rus, A., Lahner, B., Borevitz, J.O., Muthukumar, B., Mickelbart, M.V., Schreiber, L., Franke, R.B. and Salt, D.E., 2009. Root suberin forms an extracellular barrier that affects water relations and mineral nutrition in Arabidopsis. *PLoS Genetics*, *5*(5), p. e1000492.

Bizily, S.P., Rugh, C.L. and Meagher, R.B., 2000. Phytodetoxification of hazardous organomercurials by genetically engineered plants. *Nature Biotechnology*, *18*(2), pp. 213–217.

Chen, G., Liu, Y., Wang, R., Zhang, J. and Owens, G., 2013. Cadmium adsorption by willow root: The role of cell walls and their subfractions. *Environmental Science and Pollution Research*, *20*, pp. 5665–5672.

Chen, L., Luo, S., Li, X., Wan, Y., Chen, J. and Liu, C., 2014. Interaction of Cd hyperaccumulator *Solanum nigrum* L. and functional endophyte Pseudomonas sp. Lk9 on soil heavy metals uptake. *Soil Biology and Biochemistry*, *68*, pp. 300–308.

Chen, M., Xu, P., Zeng, G., Yang, C., Huang, D. and Zhang, J., 2015. Bioremediation of soils contaminated with polycyclic aromatic hydrocarbons, petroleum, pesticides, chlorophenols and heavy metals by composting: Applications, microbes and future research needs. *Biotechnology Advances*, *33*(6), pp. 745–755.

Chen, H., Zhang, C., Guo, H., Hu, Y., He, Y. and Jiang, D., 2018. Overexpression of a *Miscanthus sacchariflorus* yellow stripe-like transporter MsYSL1 enhances resistance of Arabidopsis to cadmium by mediating metal ion reallocation. *Plant Growth Regulation*, *85*, pp. 101–111.

Chiang, P.N., Chiu, C.Y., Wang, M.K. and Chen, B.T., 2011. Low-molecular-weight organic acids exuded by Millet (*Setaria italica* (L.) Beauv.) roots and their effect on the remediation of cadmium-contaminated soil. *Soil Science*, *176*(1), pp. 33–38.

Chirakkara, R.A., Cameselle, C. and Reddy, K.R., 2016. Assessing the applicability of phytoremediation of soils with mixed organic and heavy metal contaminants. *Reviews in Environmental Science and Bio/Technology*, *15*, pp. 299–326.

Choppala, G., Saifullah, Bolan, N., Bibi, S., Iqbal, M., Rengel, Z., Kunhikrishnan, A., Ashwath, N. and Ok, Y.S., 2014. Cellular mechanisms in higher plants governing tolerance to cadmium toxicity. *Critical Reviews in Plant Sciences*, *33*(5), pp. 374–391.

Clemens, S., 2006. Toxic metal accumulation, responses to exposure and mechanisms of tolerance in plants. *Biochimie*, *88*(11), pp. 1707–1719.

Concas, S., Lattanzi, P., Bacchetta, G., Barbafieri, M. and Vacca, A., 2015. Zn, Pb and Hg contents of *Pistacia lentiscus* L. grown on heavy metal-rich soils: Implications for phytostabilization. *Water, Air, & Soil Pollution*, *226*, pp. 1–15.

Cunningham, S.D., Anderson, T.A., Schwab, A.P. and Hsu, F.C., 1996. Phytoremediation of soils contaminated with organic pollutants. *Advances in Agronomy*, *56*(1), pp. 55–114.

da Conceição Gomes, M.A., Hauser-Davis, R.A., de Souza, A.N. and Vitória, A.P., 2016. Metal phytoremediation: General strategies, genetically modified plants and applications in metal nanoparticle contamination. *Ecotoxicology and Environmental Safety*, *134*, pp. 133–147.

Dadea, C., Russo, A., Tagliavini, M., Mimmo, T. and Zerbe, S., 2017. Tree species as tools for biomonitoring and phytoremediation in urban environments: A review with special regard to heavy metals. *Arboriculture & Urban Forestry*, *43*(4), pp. 155–167.

Dermont, G., Bergeron, M., Mercier, G. and Richer-Laflèche, M., 2008. Soil washing for metal removal: A review of physical/chemical technologies and field applications. *Journal of Hazardous Materials*, *152*(1), pp. 1–31.

Dhankher, O.P., Pilon-Smits, E.A.H., Meagher, R.B. and Doty, S., 2011. Biotechnological approaches for phytoremediation. In: Altman, A. and Hasegawa, P.M. (Eds.), *Plant Biotechnology and Agriculture* (pp. 309–328). Academic Press, Oxford.

Dickinson, N.M., Baker, A.J., Doronila, A., Laidlaw, S. and Reeves, R.D., 2009. Phytoremediation of inorganics: Realism and synergies. *International Journal of Phytoremediation*, *11*(2), pp. 97–114.

Dixit, R., Wasiullah, X., Malaviya, D., Pandiyan, K., Singh, U.B., Sahu, A., Shukla, R., Singh, B.P., Rai, J.P., Sharma, P.K. and Lade, H., 2015. Bioremediation of heavy metals from soil and aquatic environment: An overview of principles and criteria of fundamental processes. *Sustainability*, *7*(2), pp. 2189–2212.

Doty, S.L., Shang, T.Q., Wilson, A.M., Tangen, J., Westergreen, A.D., Newman, L.A., Strand, S.E. and Gordon, M.P., 2000. Enhanced metabolism of halogenated hydrocarbons in transgenic plants containing mammalian cytochrome P450 2E1. *Proceedings of the National Academy of Sciences*, *97*(12), pp. 6287–6291.

Du, Z.Y., Chen, M.X., Chen, Q.F., Gu, J.D. and Chye, M.L., 2015. Expression of Arabidopsis acyl-CoA-binding proteins AtACBP$_1$ and AtACBP$_4$ confers Pb(II) accumulation in *Brassica juncea* roots. *Plant, Cell & Environment*, *38*(1), pp. 101–117.

Elloumi, N., Belhaj, D., Mseddi, S., Zouari, M., Abdallah, F.B., Woodward, S. and Kallel, M., 2017. Response of *Nerium oleander* to phosphogypsum amendment and its potential use for phytoremediation. *Ecological Engineering*, *99*, pp. 164–171.

Farraji, H., Zaman, N.Q., Tajuddin, R. and Faraji, H., 2016. Advantages and disadvantages of phytoremediation: A concise review. *International Journal of Environmental & Technological Science*, *2*, pp. 69–75.

Fasani, E., Manara, A., Martini, F., Furini, A. and DalCorso, G., 2018. The potential of genetic engineering of plants for the remediation of soils contaminated with heavy metals. *Plant, Cell & Environment*, 41, 1201–1232. https://doi.org/10.1111/pce.12963

Feng, S., Tan, J., Zhang, Y., Liang, S., Xiang, S., Wang, H. and Chai, T., 2017. Isolation and characterization of a novel cadmium-regulated yellow stripe-like transporter (SnYSL3) in *Solanum nigrum*. *Plant Cell Reports*, *36*, pp. 281–296.

Forstner, U., 1995. Land contamination by metals: Global scope and magnitude of the problem. In: Allen, H.E., Huang, C.P., Bailey, G.W. and Bowers, A.R. (Eds.), *Metal Speciation and Contamination of Soil* (pp. 1–33). CRC, Boca Raton.

Gallego, S.M., Pena, L.B., Barcia, R.A., Azpilicueta, C.E., Iannone, M.F., Rosales, E.P., Zawoznik, M.S., Groppa, M.D. and Benavides, M.P., 2012. Unravelling cadmium toxicity and tolerance in plants: Insight into regulatory mechanisms. *Environmental and Experimental Botany*, *83*, pp. 33–46.

Gerhardt, K.E., Gerwing, P.D. and Greenberg, B.M., 2017. Opinion: Taking phytoremediation from proven technology to accepted practice. *Plant Science*, *256*, pp. 170–185.

Ghosh, M. and Singh, S.P., 2005. A review on phytoremediation of heavy metals and utilization of it's by products. *Asian Journal of Energy &Environment*,*6*(4), p. 18.

Gomes, H.I., 2012. Phytoremediation for bioenergy: Challenges and opportunities. *Environmental Technology Reviews*, *1*(1), pp. 59–66.

Govindasamy, C., Arulpriya, M., Ruban, P., Francisca, L.J. and Ilayaraja, A., 2011. Concentration of heavy metals in seagrasses tissue of the Palk Strait, Bay of Bengal. *International Journal of Environmental Sciences*, *2*, pp. 145–153.

Gratão, P.L., Polle, A., Lea, P.J. and Azevedo, R.A., 2005. Making the life of heavy metal-stressed plants a little easier. *Functional Plant Biology*, *32*(6), pp. 481–494.

Guarino, C. and Sciarrillo, R., 2017. Effectiveness of in situ application of an Integrated Phytoremediation System (IPS) by adding a selected blend of rhizosphere microbes to heavily multi-contaminated soils. *Ecological Engineering*, *99*, pp. 70–82.

Hamon, R. and McLaughlin, M., 2003. Food crop edibility on the ok Tedi/Fly river flood plain. *Report for OK*, *6*(6).

Hasanuzzaman, M., Nahar, K., Gill, S.S., Alharby, H.F., Razafindrabe, B.H. and Fujita, M., 2017. Hydrogen peroxide pretreatment mitigates cadmium-induced oxidative stress in *Brassica napus* L.: An intrinsic study on antioxidant defense and glyoxalase systems. *Frontiers in Plant Science*, *8*, p. 115.

He, J., Li, H., Ma, C., Zhang, Y., Polle, A., Rennenberg, H., Cheng, X. and Luo, Z.B., 2015. Overexpression of bacterial γ-glutamylcysteine synthetase mediates changes in cadmium influx, allocation and detoxification in poplar. *New Phytologist*, *205*(1), pp. 240–254.

Henry, J.R., 2000. An overview of the phytoremediation of lead and mercury (pp. 1–31). Washington, DC: US Environmental Protection Agency, Office of Solid Waste and Emergency Response, Technology Innovation Office.

Hernández, A., Loera, N., Contreras, M., Fischer, L. and Sánchez, D., 2019. Comparison between *Lactuca sativa* L. and *Lolium perenne*: Phytoextraction capacity of Ni, Fe, and Co from galvanoplastic industry. In: Wang, T., et al. (Eds.), *Energy Technology 2019*(pp. 137–147). Springer.

Huang, D., Qin, X., Peng, Z., Liu, Y., Gong, X., Zeng, G., Huang, C., Cheng, M., Xue, W., Wang, X. and Hu, Z., 2018. Nanoscale zero-valent iron assisted phytoremediation of Pb in sediment: Impacts on metal accumulation and antioxidative system of *Lolium perenne*. *Ecotoxicology and Environmental Safety*, *153*, pp. 229–237.

Huang, J., Zhang, Y., Peng, J.S., Zhong, C., Yi, H.Y., Ow, D.W. and Gong, J.M., 2012. Fission yeast HMT1 lowers seed cadmium through phytochelatin-dependent vacuolar sequestrationin Arabidopsis. *Plant Physiology*,*158*, pp. 1779–1788.

Huang, T., Peng, Q., Yu, L. and Li, D., 2017. The detoxification of heavy metals in the phosphate tailing-contaminated soil through sequential microbial pretreatment and electrokinetic remediation. *Soil and Sediment Contamination: An International Journal*, *26*(3), pp. 308–322.

Huguet, S., Bert, V., Laboudigue, A., Barthès, V., Isaure, M.P., Llorens, I., Schat, H. and Sarret, G., 2012. Cd speciation and localization in the hyperaccumulator *Arabidopsis halleri*. *Environmental and Experimental Botany*, *82*, pp. 54–65.

Iannone, M.F., Groppa, M.D. and Benavides, M.P., 2015. Cadmium induces different biochemical responses in wild type and catalase-deficient tobacco plants. *Environmental and Experimental Botany*, *109*, pp. 201–211.

Ibuot, A., Dean, A.P., McIntosh, O.A. and Pittman, J.K., 2017. Metal bioremediation by CrMTP4 over-expressing *Chlamydomonas reinhardtii* in comparison to natural wastewater-tolerant microalgae strains. *Algal Research*, *24*, pp. 89–96.

Jabeen, R., Ahmad, A. and Iqbal, M., 2009. Phytoremediation of heavy metals: Physiological and molecular mechanisms. *The Botanical Review*, *75*, pp. 339–364.

Jadia, C.D. and Fulekar, M.H., 2008. Phytoremediation: The application of vermicompost to remove zinc, cadmium, copper, nickel and lead by sunflower plant. *Environmental Engineering & Management Journal (EEMJ)*, *7*(5), pp. 547–558.

Janani, K., Sivarajasekar, N., Muthusaravanan, S., Ram, K., Prakashmaran, J., Sivamani, S., Dhakal, N., Shahnaz, T. and Selvaraju, N., 2019. Optimization of EDTA enriched phytoaccumulation of zinc by *Ophiopogon japonicus*: Comparison of response surface, artificial neural network and random forest models. *Bioresource Technology Reports*, *7*, p. 100265.

Kafle, A., Timilsina, A., Gautam, A., Adhikari, K., Bhattarai, A. and Aryal, N., 2022. Phytoremediation: Mechanisms, plant selection and enhancement by natural and synthetic agents. *Environmental Advances*, *8*, p. 100203.

Keeran, N.S., Ganesan, G. and Parida, A.K., 2017. A novel heavy metal ATPase peptide from *Prosopis juliflora* is involved in metal uptake in yeast and tobacco. *Transgenic Research*, *26*, pp. 247–261.

Kong, J., Jin, J., Dong, Q., Qiu, J., Li, Y., Yang, Y., Shi, Y., Si, W., Gu, L., Yang, F. and Cheng, B., 2019. Maize factors ZmUBP15, ZmUBP16 and ZmUBP19 play important roles for plants to tolerance the cadmium stress and salt stress. *Plant Science*, *280*, pp. 77–89.

Kshirsagar, S. and Aery, N.C., 2007. Phytostabilization of mine waste: Growth and physiological responses of *Vigna unguiculata* (L.) Walp. *Journal of Environmental Biology*, *28*(3), p. 651.

Kumar, A. and Prasad, M.N.V., 2019. Plant genetic engineering approach for the Pb and Zn remediation: Defense reactions and detoxification mechanisms. In *Transgenic Plant*

Technology for Remediation of Toxic Metals and Metalloids (pp. 359–380). Academic Press.

Laghlimi, M., Baghdad, B., El Hadi, H. and Bouabdli, A., 2015. Phytoremediation mechanisms of heavy metal contaminated soils: A review. *Open Journal of Ecology, 5*(08), p. 375.

Leštan, D., Luo, C.L. and Li, X.D., 2008. The use of chelating agents in the remediation of metal-contaminated soils: A review. *Environmental Pollution, 153*(1), pp. 3–13.

Lewandowski, I., Schmidt, U., Londo, M. and Faaij, A., 2006. The economic value of the phytoremediation function–assessed by the example of cadmium remediation by willow (*Salix* ssp). *Agricultural Systems, 89*(1), pp. 68–89.

Li, J., Liao, B., Lan, C., Ye, Z., Baker, A. and Shu, W., 2010. Cadmium tolerance and accumulation in cultivars of a high-biomass tropical tree (*Averrhoa carambola*) and its potential for phytoextraction. *Journal of Environmental Quality, 39*, pp. 1262–1268.

Li, X., Dai, L., Zhang, C., Zeng, G., Liu, Y., Zhou, C., Xu, W., Wu, Y., Tang, X., Liu, W. and Lan, S., 2017. Enhanced biological stabilization of heavy metals in sediment using immobilized sulfate reducing bacteria beads with inner cohesive nutrient. *Journal of Hazardous Materials, 324*, pp. 340–347.

Liang, J., Yang, Z., Tang, L., Zeng, G., Yu, M., Li, X., Wu, H., Qian, Y., Li, X. and Luo, Y., 2017. Changes in heavy metal mobility and availability from contaminated wetland soil remediated with combined biochar-compost. *Chemosphere, 181*, pp. 281–288.

Lin, Y.F. and Aarts, M.G., 2012. The molecular mechanism of zinc and cadmium stress response in plants. *Cellular and Molecular Life Sciences, 69*, pp. 3187–3206.

Liu, L., Li, W., Song, W. and Guo, M., 2018. Remediation techniques for heavy metal contaminated soils: Principles and applicability. *Science of the Total Environment, 633*, pp. 206–219.

Luo, Z.B., He, J., Polle, A. and Rennenberg, H., 2016. Heavy metal accumulation and signal transduction in herbaceous and woody plants: Paving the way for enhancing phytoremediation efficiency. *Biotechnology Advances, 34*(6), pp. 1131–1148.

Luo, Z.B., Wu, C., Zhang, C., Li, H., Lipka, U. and Polle, A., 2014. The role of ectomycorrhizas in heavy metal stress tolerance of host plants. *Environmental and Experimental Botany, 108*, pp. 47–62.

Mahar, A., Wang, P., Ali, A., Awasthi, M.K., Lahori, A.H., Wang, Q., Li, R. and Zhang, Z., 2016. Challenges and opportunities in the phytoremediation of heavy metals contaminated soils: A review. *Ecotoxicology and Environmental Safety, 126*, pp. 111–121.

Majed, N., Real, M., Isreq, H., Akter, M. and Azam, H.M., 2016. Food adulteration and biomagnification of environmental contaminants: A comprehensive risk framework for Bangladesh. *Frontiers in Environmental Science, 4*, p. 34.

Manzoor, M., Kamboh, U.R., Gulshan, S., Tomforde, S., Gul, I., Siddiqui, A. and Arshad, M., 2023. Optimizing sustainable phytoextraction of lead from contaminated soil using response surface methodology (RSM) and artificial neural network (ANN). *Sustainability, 15*(14), p. 11049.

Mao, X., Jiang, R., Xiao, W. and Yu, J., 2015. Use of surfactants for the remediation of contaminated soils: A review. *Journal of Hazardous Materials, 285*, pp. 419–435.

Mataruga, Z., Jarić, S., Marković, M., Pavlović, M., Pavlović, D., Jakovljević, K., Mitrović, M. and Pavlović, P., 2020. Evaluation of *Salix alba, Juglans regia* and *Populus nigra* as biomonitors of PTEs in the riparian soils of the Sava River. *Environmental Monitoring and Assessment, 192*, pp. 1–20.

Meers, E., Slycken, S.V., Adriaensen, K., Ruttens, A., Vangronsveld, J., Laing, G.D., Witters, N., Thewys, T. and Tack, F.M.G., 2010.Theuseofbio-energycrops (*Zea mays*) for 'phytoremediation' of heavy metals on moderately contaminated soils: A field experiment. *Chemosphere, 78*, pp. 35–41.

Mokhtar, H., Morad, N. and Fizri, F.F.A., 2011. Phytoaccumulation of copper from aqueous solutionsusing *Eichhornia crassipes* and *Centella asiatica*. *International Journal of Environmental Science and Development, 2*(3), p. 205.

Moreno, F.N., Anderson, C.W., Stewart, R.B. and Robinson, B.H., 2005. Mercury volatilisation and phytoextraction from base-metal mine tailings. *Environmental Pollution*, 136(2), pp. 341–352.

Mulbry, W., Kondrad, S., Pizarro, C. and Kebede-Westhead, E., 2008. Treatment of dairy manure effluent using freshwater algae: Algal productivity and recovery of manure nutrients using pilot-scale algal turf scrubbers. *Bioresource Technology*, 99(17), pp. 8137–8142.

Muthusaravanan, S., Sivarajasekar, N., Vivek, J.S., Paramasivan, T., Naushad, M., Prakashmaran, J., Gayathri, V. and Al-Duaij, O.K., 2018. Phytoremediation of heavy metals: Mechanisms, methods and enhancements. *Environmental Chemistry Letters*, 16, pp. 1339–1359.

Nandakumar, P.B.A., Dushenkov, V., Motto, H. and Raskin, I., 1995. Phytoextraction: The use of plants to remove heavy metals from soils. *Environmental Science & Technology*, 29, pp. 1232–1238.

Nayak, A., Bhushan, B., Gupta, V. and Sharma, P., 2017. Chemically activated carbon from lignocellulosic wastes for heavy metal wastewater remediation: Effect of activation conditions. *Journal of Colloid and Interface Science*, 493, pp. 228–240.

Nouri, J., Khorasani, N., Lorestani, B., Karami, M., Hassani, A.H. and Yousefi, N., 2009. Accumulation of heavy metals in soil and uptake by plant species with phytoremediation potential. *Environmental Earth Sciences*, 59, pp. 315–323.

Ovečka, M. and Takáč, T., 2014. Managing heavy metal toxicity stress in plants: Biological and biotechnological tools. *Biotechnology Advances*, 32(1), pp. 73–86.

Ozyigit, I.I., Can, H. and Dogan, I., 2021. Phytoremediation using genetically engineered plants to remove metals: A review. *Environmental Chemistry Letters*, 19(1), pp. 669–698.

Peiter, E., Montanini, B., Gobert, A., Pedas, P., Husted, S., Maathuis, F.J., Blaudez, D., Chalot, M. and Sanders, D., 2007. A secretory pathway-localized cation diffusion facilitator confers plant manganese tolerance. *Proceedings of the National Academy of Sciences*, 104(20), pp. 8532–8537.

Pinto, A.P., De Varennes, A., Fonseca, R. and Teixeira, D.M., 2015. Phytoremediation of soils contaminated with heavy metals: Techniques and strategies. In Ansari, A., Gill, S., Gill, R., Lanza, G., Newman, L. (Eds.), Phytoremediation: Management of Environmental Contaminants (Volume 1, pp. 133–155). Springer.

Pulford, I.D. and Watson, C., 2003. Phytoremediation of heavy metal-contaminated land by trees—A review. *Environment International*, 29(4), pp. 529–540.

Purakayastha, T.J. and Chhonkar, P.K., 2010. Phytoremediation of heavy metal contaminated soils. In *Soil Heavy Metals* (pp. 389–429). Springer.

Raskin, I. and Ensley, B.D., 2000. *Phytoremediation of Toxic Metals*. John Wiley and Sons.

Raza, A., Habib, M., Charagh, S. and Kakavand, S.N., 2021. Genetic engineering of plants to tolerate toxic metals and metalloids. In *Handbook of Bioremediation* (pp. 411–436). Academic Press.

Riaz, U., Athar, T., Mustafa, U. and Iqbal, R., 2022. Economic feasibility of phytoremediation. In *Phytoremediation* (pp. 481–502). Academic Press.

Rodriguez-Hernandez, M.C., Bonifas, I., Alfaro-De la Torre, M.C., Flores-Flores, J.L., Bañuelos-Hernández, B. and Patiño-Rodríguez, O., 2015. Increased accumulation of cadmium and lead under Ca and Fe deficiency in *Typha latifolia*: A study of two pore channel (TPC1) gene responses. *Environmental and Experimental Botany*, 115, pp. 38–48.

Rozas, E.E., Mendes, M.A., Nascimento, C.A., Espinosa, D.C., Oliveira, R., Oliveira, G. and Custodio, M.R., 2017. Bioleaching of electronic waste using bacteria isolated from the marine sponge *Hymeniacidon heliophila* (Porifera). *Journal of Hazardous Materials*, 329, pp. 120–130.

Ruppert, L., Lin, Z.Q., Dixon, R.P. and Johnson, K.A., 2013. Assessment of solid phase microfiber extraction fibers for the monitoring of volatile organoarsinicals emitted from a plant–soil system. *Journal of Hazardous Materials*, 262, pp. 1230–1236.

Sakakibara, M., Watanabe, A., Inoue, M., Sano, S. and Kaise, T., 2010. January. Phytoextraction and phytovolatilization of arsenic from As-contaminated soils by *Pteris vittata*. In *Proceedings of the annual international conference on soils, sediments, water and energy* (Vol. 12, No. 1, p. 26)

Salt, DE., Blaylock, M., Kumar, N.P.B.A., Dushenkov, V., Ensley, B.D., Chet, I. and Raskin, I., 1995. Phytoremediation: a novel strategy for the removal of toxic metals from the environment using plants. *Biotechnology, 13*, pp. 468–475.

Saraswat, S. and Rai, J.P.N., 2011. Complexation and detoxification of Zn and Cd in metal accumulating plants. *Reviews in Environmental Science and Biotechnology, 10*, pp. 327–339.

Sarwar, N., Imran, M., Shaheen, M.R., Ishaque, W., Kamran, M.A., Matloob, A., Rehim, A. and Hussain, S., 2017. Phytoremediation strategies for soils contaminated with heavy metals: Modifications and future perspectives. *Chemosphere, 171*, pp. 710–721.

Sarwar, N., Saifullah, Malhi, S.S., Zia, M.H., Naeem, A., Bibi, S. and Farid, G., 2010. Role of plant nutrients in minimizing cadmium accumulation by plant. *Journal of Science of Food and Agriculture, 90*, pp. 925–937.

Saxena, G., Purchase, D., Mulla, S.I., Saratale, G.D. and Bharagava, R.N., 2020. Phytoremediation of heavy metal-contaminated sites: Eco-environmental concerns, field studies, sustainability issues, and future prospects. *Reviews of Environmental Contamination and Toxicology, 249*, pp. 71–131.

Serbula, S.M., Miljkovic, D.D., Kovacevic, R.M. and Ilic, A.A., 2012. Assessment of airborne heavy metal pollution using plant parts and topsoil. *Ecotoxicology and Environmental Safety, 76*, pp. 209–214.

Shah, V. and Daverey, A., 2020. Phytoremediation: A multidisciplinary approach to clean up heavy metal contaminated soil. *Environmental Technology & Innovation, 18*, p. 100774.

Shahandeh, H. and Hossner, L.R., 2000. Plant screening for chromium phytoremediation. *International Journal of Phytoremediation, 2*(1), pp. 31–51.

Sharma, H.D. and Reddy, K.R., 2004. *Geoenvironmental Engineering: Site Remediation, Waste Containment, and Emerging Waste Management Technologies*. John Wiley & Sons.

Sharma, S., Tiwari, S., Hasan, A., Saxena, V. and Pandey, L.M., 2018. Recent advances in conventional and contemporary methods for remediation of heavy metal-contaminated soils. *3 Biotech, 8*, pp. 1–18.

Singh, J. and Lee, B.K., 2016. Influence of nano-TiO$_2$ particles on the bioaccumulation of Cd in soybean plants (Glycine max): A possible mechanism for the removal of Cd from the contaminated soil. *Journal of Environmental Management, 170*, pp. 88–96.

Song, B., Xu, P., Chen, M., Tang, W., Zeng, G., Gong, J., Zhang, P. and Ye, S., 2019. Using nanomaterials to facilitate the phytoremediation of contaminated soil. *Critical Reviews in Environmental Science and Technology, 49*(9), pp. 791–824.

Song, W.Y., Ju Sohn, E., Martinoia, E., Jik Lee, Y., Yang, Y.Y., Jasinski, M., Forestier, C., Hwang, I. and Lee, Y., 2003. Engineering tolerance and accumulation of lead and cadmium in transgenic plants. *Nature Biotechnology, 21*(8), pp. 914–919.

Song, W.Y., Mendoza-CÓZATL, D.G., Lee, Y., Schroeder, J.I., Ah, S.N., Lee, H.S., Wicker, T. and Martinoia, E., 2014. Phytochelatin–metal (loid) transport into vacuoles shows different substrate preferences in barley and Arabidopsis. *Plant, Cell & Environment, 37*(5), pp. 1192–1201.

Sun, L., Ma, Y., Wang, H., Huang, W., Wang, X., Han, L., Sun, W., Han, E. and Wang, B., 2018. Overexpression of PtABCC1 contributes to mercury tolerance and accumulation in Arabidopsis and poplar. *Biochemical and Biophysical Research Communications, 497*(4), pp. 997–1002.

Suthar, S., Sajwan, P. and Kumar, K., 2014. Vermiremediation of heavy metals in wastewater sludge from paper and pulp industry using earthworm *Eisenia fetida*. *Ecotoxicology and Environmental Safety, 109*, pp. 177–184.

Tang, K.H.D., 2023. Phytoremediation: Where do we go from here? *Biocatalysis and Agricultural Biotechnology, 50*, p. 102721.

Titah, H.S., Halmi, M.I.E.B., Abdullah, S.R.S., Hasan, H.A., Idris, M. and Anuar, N., 2018. Statistical optimization of the phytoremediation of arsenic by *Ludwigia octovalvis*-in a pilot reed bed using response surface methodology (RSM) versus an artificial neural network (ANN). *International Journal of Phytoremediation, 20*(7), pp. 721–729.

Urbaniak, M., Zieliński, M. and Wyrwicka, A., 2017. The influence of the Cucurbitaceae on mitigating the phytotoxicity and PCDD/PCDF content of soil amended with sewage sludge. *International Journal of Phytoremediation, 19*(3), pp. 207–213.

Vázquez, S., Goldsbrough, P. and Carpena, R.O., 2006. Assessing the relative contributions of phytochelatins and the cell wall to cadmium resistance in white lupin. *Physiologia Plantarum, 128*(3), pp. 487–495.

Wan, X., Lei, M. and Chen, T., 2016. Cost–benefit calculation of phytoremediation technology for heavy-metal-contaminated soil. *Science of the Total Environment, 563*, pp. 796–802.

Wang, Q.R., Cui, Y.S., Liu, X.M., Dong, Y.T. and Christie, P., 2003. Soil contamination and plant uptake of heavy metals at polluted sites in China. *Journal of Environmental Science Health - Part A, 38*, pp. 823–838.

Wuana, R.A. and Okieimen, F.E., 2011. Heavy metals in contaminated soils: A review of sources, chemistry, risks and best available strategies for remediation. *International Scholarly Research Notices, 2011*.

Xu, Y., Seshadri, B., Sarkar, B., Wang, H., Rumpel, C., Sparks, D., Farrell, M., Hall, T., Yang, X. and Bolan, N., 2018. Biochar modulates heavy metal toxicity and improves microbial carbon use efficiency in soil. *Science of the Total Environment, 621*, pp. 148–159.

Yadav, S.K., 2010. Heavy metals toxicity in plants: An overview on the role of glutathione and phytochelatins in heavy metal stress tolerance of plants. *South African Journal of Botany, 76*(2), pp. 167–179.

Yamaguchi, N., Mori, S., Baba, K., Kaburagi-Yada, S., Arao, T., Kitajima, N., Hokura, A. and Terada, Y., 2011. Cadmium distribution in the root tissues of solanaceous plants with contrasting root-to-shoot Cd translocation efficiencies. *Environmental and Experimental Botany, 71*(2), pp. 198–206.

Yang, Y., Ge, Y., Tu, P., Zeng, H., Zhou, X., Zou, D., Wang, K. and Zeng, Q., 2019. Phytoextraction of Cd from a contaminated soil by tobacco and safe use of its metal-enriched biomass. *Journal of Hazardous Materials, 363*, pp. 385–393.

Yang, Y., Liang, Y., Ghosh, A., Song, Y., Chen, H. and Tang, M., 2015. Assessment of arbuscular mycorrhizal fungi status and heavy metal accumulation characteristics of tree species in a lead–zinc mine area: Potential applications for phytoremediation. *Environmental Science and Pollution Research, 22*, pp. 13179–13193.

Zeng, H., Xu, L., Singh, A., Wang, H., Du, L. and Poovaiah, B.W., 2015. Involvement of calmodulin and calmodulin-like proteins in plant responses to abiotic stresses. *Frontiers in Plant Science, 6*, p. 600.

Zhang, H., Ding, Y., Zhi, J., Li, X., Liu, H. and Xu, J., 2018. Over-expression of the poplar expansin gene *PtoEXPA12* in tobacco plants enhanced cadmium accumulation. *International Journal of Biological Macromolecules, 116*, pp. 676–682.

Zhang, H., Lv, S., Xu, H., Hou, D., Li, Y. and Wang, F., 2017. H_2O_2 is involved in the metallothionein-mediated rice tolerance to copper and cadmium toxicity. *International Journal of Molecular Sciences, 18*(10), p. 2083.

Złoch, M., Kowalkowski, T., Tyburski, J. and Hrynkiewicz, K., 2017. Modeling of phytoextraction efficiency of microbially stimulated *Salix dasyclados* L. in the soils with different speciation of heavy metals. *International Journal of Phytoremediation, 19*(12), pp. 1150–1164.

10 Rhizofiltration
A Sustainable Green Technology for Remediation of Heavy Metals from Aquatic Systems

Trinath Biswal

INTRODUCTION

The contamination of groundwater and other water bodies by heavy metals (HMs) is now a major global issue. Although many techniques have been developed for the remediation of HMs from water bodies, such as chemical processes, ion exchange, and microbiological precipitation methods, none of these have been satisfactory. All these methodologies exhibit differing effectiveness for different kinds of metals, and these methods are usually expensive if the standards of cleanup are high [1, 2]. There are some higher plants in aquatic ecosystems whose root systems can be utilized for the removal of HMs and organic contaminants from water bodies. These plants include water hyacinth (*Eichhornia crassipes*), duckweed (*Lemna minor*), water velvet (*Azolla pinnata*), and pennywort (*Hydrocotyle umbellata*). But the process of removing HMs with these plant species is quite slow due to the small size of the plants and the slower rate of growth of the plant roots. Indian mustard (*Brassica juncea*) is an important terrestrial plant whose roots grow rapidly and can remove HMs efficiently. Rhizofiltration is also analogous to phytoextraction; however, it is mainly associated with the remediation of contaminated groundwater but not soil. In this process, the remediation of contaminants can be done by absorption, precipitation, and concentration of contaminants through the roots or root systems of the plant species. The presence of heavy metals is highly dangerous to living organisms, including animals and humans, because of their properties of biomagnification, bioaccumulation, and persistence within the environment. Continuing exposure to heavy metals has a very serious impact on our respiratory system and causes cancer, cardiovascular disorders, and other ailments. Even in low concentrations, heavy metals and their ions result in teratogens, toxins, carcinogens, and mutagens for the human body [3, 4]. The technique of rhizofiltration is a sustainable pathway for the removal of organic contaminants and heavy metals (HMs) from groundwater, which therefore protects human health and the environment. But the traditional techniques of remediation of HMs from water are highly expensive, require toxic materials, generate hazardous by-products in the form of effluents, and

DOI: 10.1201/9781003442295-10

are inadequate to remove most of the HMs from the groundwater. Rhizofiltration is a cost-effective, environmentally friendly green technology that remediates HMs from aquatic bodies. *B. juncea* was found to be an ideal plant species for rhizofiltration due to its ability to accumulate a high concentration of Pb and other HMs in its plant roots. Furthermore, a number of plant species in the Brassicaceae family possess the capability of removing HMs such as Pb, Cd, and Cu. Metals are natural components of the environment, and their presence in water bodies is highly hazardous to the health of human beings, animals, and aquatic biotic communities. The excessive addition of HMs to the water system is mainly due to anthropogenic activities like the discharge of industrial wastewater effluents, the combustion of fossil fuels, the discharge of urban effluents, and sewage wastewater. The permissible values of HMs in natural water resources are <1.0 µg/L. If the HMs present in water bodies exceed the permissible limit, then they are highly toxic to human health. The degradation of biological ecosystems is primarily due to organic contaminants and, secondarily, to the persistence of HMs in the ecosystem. The existence of HMs in the ecosystem has numerous negative impacts on both the terrestrial and aquatic environments [5, 6]. The HMs enter the food chain and food web of different ecosystems and cause hazards to human health and livestock, either indirectly or directly, and result in a great threat to the living communities around the globe. There are some natural phenomena, like volcanic eruptions and weathering of rocks, that release HMs into the groundwater via geochemical recycling. Phytoremediation is a sustainable technology that causes the removal of HMs from aquatic systems. There are various kinds of phytoremediation, including phytoextraction, phytodegradation, phytovolatization, phytostabilization, and rhizofiltration. But the process of rhizofiltration is most important because it retards the mobility of the pollutants, restricts their movement in the groundwater, and hence decreases their bioavailability for entering the food chain and food web [7].

BASIC CONCEPT REGARDING RHIZOFILTRATION

Phytoremediation is an energy-saving, environmentally friendly, cost-effective, and promising technology for the removal of HMs from aquatic systems and wastewater effluents. In the process of phytoremediation, one of the mechanisms is rhizofiltration, which uses the root zone (rhizosphere) of the plant species for the remediation of contamination through absorption, precipitation, and concentration. Rhizofiltration (hydraulic control) is a technique based on the capability of the plant root system to accumulate and absorb HM contaminants from water bodies. Root exudates can precipitate organic contaminants and HMs and be adsorbed by them, causing an improvement in the quality of water. There are many metals that promote the growth of plants, and key components of different kinds of proteins and enzymes are known as essential metals. However, the concentration of nonessential HMs causes HM-promoted phytotoxicity, which replaces essential elements, inactivates enzymes, and blocks functional groups. The mechanism of rhizofiltration cleans out HMs such as Cd, Ni, Cr, V, Cu, and Pb and some radionuclides such as Sr, U, and Cs. Trees with long roots can absorb huge amounts of water and HMs, along with other toxic organic contaminants. The increase in the number of oxygen radicals interferes with the activities of electron transport, which is one of the important HMs (Cr, Ni, Cd, Pb) associated with phytotoxicity. The HMs can be detoxified through the thiol groups present in the protein of the plant roots. *In situ* rhizofiltration is effective for polluted surface water, but *ex situ* rhizofiltration is effective for polluted supply water and groundwater resources. Root exudates promote the heavy metal precipitation on the surface of the roots and are effective for *ex situ* rhizofiltration. *Typha angustifolia* is an important aquatic plant species that results in potential rhizofiltration for the removal of HMs. It absorbs Zn, Cd, and Pb effectively via its root system and reduces the metal concentration in water bodies. In addition to *Typha angustifolia*, there are many aquatic plant species such as water fern (*Azolla*), water hyacinth (*Eichhornia*), and water lettuce (*Pistia*) that can also remediate HMs from water. *Pistia* has an excellent capacity for phytoextraction, rhizofiltration, and phytostabilization of Pb and As, whereas *Azolla* and *Eichhornia* can absorb Cu and Ni efficiently from polluted water sources. Underground polluted plumes can be recovered and filtered by using rhizofiltration. It was found that the hydroponically cultivated roots of terrestrial plants possess a higher efficiency for remediating HMs than the root systems of other plants. The method of rhizofiltration begins with the recognition and extraction of groundwater with a pump-and-treat (P&T) system. For effective rhizofiltration, specific plant species having a high adsorption capability for HMs from contaminated water sources and wastewater effluents are hydroponically cultivated with the help of greenhouses. Therefore, plant species with high efficiency for HM remediation have been planted at the sources or sites of contaminated water resources. Subsequently, the plant species are removed from the contaminated areas when the roots of the plants are saturated with pollutants, and after that, these are properly disposed of. Table 10.1 illustrates the various plant species used in rhizofiltration for the removal of toxic HM contaminants (Hg, As, Cd, Fe, Cr, and Pb) and organic pollutants like polyaromatic hydrocarbons (PAHc) and chlorinated biphenyl [8–10].

Data Needed for Rhizofiltration

- Depth of the source of contaminated water.
- Kinds of HMs present in the contaminated water resources.
- Level or degree of contamination and its monitoring.
- The plant species must be emergent, aquatic, and subemergent.
- The hydraulic retention time and sorption by the root system of plants must be successfully designed and developed.
- Rhizofiltration is specifically effective in cases of low concentrations of HMs and huge volumes of water resources [25].

MECHANISM OF RHIZOFILTRATION

Rhizofiltration is a kind of phytoremediation that involves the filtering of groundwater, surface water, and wastewater effluents through the root systems of plants to remediate toxic organic substances, HMs, and excess nutrients. The continuous buildup of potentially toxic elements (PTEs) such as cadmium (Cd), selenium (Se), mercury (Hg), lead (Pb), copper (Cu), chromium (Cr), nickel (Ni), and

TABLE 10.1
Plants Used for Remediation of HMs and Toxic Organic Materials through Rhizofiltration

HMs to Be Remediated	Specific Plant Species	Reference
Cadmium (Cd)	*Brassica juncea, Azolla filiculoides* Lam, *Setaria italica* (L.) Beauv., *Salvinia auriculata* Aubl., *Pistia stratiotes* L., *Salvinia minima* Baker, and *domingensis*	[11]
Arsenic (As)	*Cynara cardunculus*, Chinese brake fern (*Pteris vittata*), *Stratiotes, Spirodela polyrhiza*, and *Eichhornia crassipes*	[12]
Lead (Pb)	*Pistia stratiotes* L., *Brassica juncea, Azolla filiculoides* Lam, *Oxycaryum cubense* (Poepp. & Kunth) Palla, *Salvinia minima* Baker, *Azolla pinnata, Carex pendula, Salvinia auriculata* Aubl., and *Typha domingensis*	[13]
Zinc (Zn)	*Typha domingensis, Eichhornia crassipes* (Mart.) Solms, *Helianthus annuus, Jatropha curcas, Tagetes erecta*, and *Pongamia pinnata*	[14]
Aluminum (Al)	*Pistia stratiotes* L., *Typha domingensis, Centella asiatica, Scirpus grossus, Thypa angustifolia*, and *Ipomoea aquatica*	[15]
Nickel (Ni)	*Eichhornia crassipes* (Mart.) Solms, spinach, sunflower, *Jatropha curcas, Alocasia puber*, and *Pongamia pinnata*	[16]
Iron (Fe)	*Pistia stratiotes* L., *Ipomoea aquatica, Typha domingensis*, and *Centella asiatica*	[17]
Copper (Cu)	*Eichhornia crassipes* (Mart.) Solms, *Limnocharis flava, Brassica napus* L., and *Elodea*	[18]
Manganese (Mn)	*Pistia stratiotes* L., hyacinth, *Cnidoscolus multilobus*, duckweed, *Platanus mexicana, Azolla, Solanum diversifolium, Asclepias curassavica* L., cattail, *Pluchea symphytifolia*, and poplar	[19]
Antimony (Sb)	*Brassica rapa, Cynodon dactylon*, and *Amaranthus mangostanus*	[20]
Mercury (Hg)	*Salvinia natans, Lemna minor*, Indian mustard, *Sansevieria trifasciata*, sunflower, *Celosia plumosa*, tobacco, spinach, and rye	[21]
Chromium (Cr)	Sunflower, rye, tobacco, Indian mustard, *Typha angustifolia, Canna, Juncus acutus* L., *Leersia hexandra*, and spinach	[22]
Synthetic dyes	*Bryophyllum fedtschenkoi, Tagetes erecta, Arundo donax* L., *Catharanthus roseus, Trachyspermum ammi, Hibiscus rosa-sinensis*, and *Chrysanthemum indicum*	[23]
Radionuclides	Sunflower, tobacco, *Lactuca sativa*, spinach, *Raphanus sativus* L., rye and Indian mustard, *Brassica campestris* L., and *Oenanthe javanica*	[24]

arsenic (As) is the cause of severe impacts on human health and challenges to global environmental sustainability. These heavy metals persist for a long time in water bodies and cause adverse effects on human health and the surrounding environment [26].

- The mechanism of rhizofiltration for remediation of HMs from water is based on the impact of biochemical and physical factors associated with the root systems of plant species.

- The efficiency or effectiveness of the mechanism of rhizofiltration is determined by the effectiveness of root system creation, which causes the intake of HMs into the plant body through plant roots.
- The various factors, such as root exudates and pH changes in the rhizosphere, are the causes of the absorption and precipitation of HMs onto the surfaces of root systems.
- The root exudates or environment of the root system

The root systems or root exudates may create a biogeochemical environment that causes the precipitation of toxic pollutants into the water resources or roots of the plant species. When the roots or root systems are completely saturated by HMs, then the plants are picked for disposal. The exudates, like simple organic acids and phenolic compounds, are released from the entire cell at the time of the decay of the roots. These root exudates can also alter the speciation of the metals and the consumption of metal ions of various kinds with the instantaneous release of protons that cause acidification of the medium and facilitate the bioavailability and transport of metals. Mostly, some specific genes present in plant species result in the effective accumulation of metals. Some organic acids and glutathione metabolism play a vital role in the tolerance of metal retention in plant roots. The native environmental conditions, including temperature, intensity of light, rainfall, pH, and wind velocity, also play a vital role in the acceptance of HMs by the root systems of the plants. This method involves the growth of plant species hydroponically and transplanting the HMs into the contaminated water, where the plant species used can concentrate, precipitate, and absorb the HMs in their roots, root systems, and shoots. The root system delivers a huge surface area, which facilitates the accumulation and absorption of nutrients and required quantities of water, along with other nonessential pollutants and HMs. To adapt the plant species, an adequate root system has to be created where plant species are planted in polluted water. The plant species are planted in contaminated zones of water bodies where the root systems of the plants consume water along with HMs, organic contaminants, and nutrients. Plants used for rhizofiltration in human-made wetlands and lagoons are termed *in situ* rhizofiltration, whereas in *ex situ* rhizofiltration, the polluted water containing HMs is diverted into the system containing an array of plants that can purify the polluted water. The development of industrial sectors or the emission of toxic pollutants is not the actual problem, but the key issue is the addition of toxic pollutants to a healthy and clean environment. Restraint in the transportation of pollutants is a suitable model strategy for limiting pollutants in a clean environment. Rhizofiltration and phytostabilization were found to be the most efficient sustainable green methodology for decreasing the movement of toxic pollutants, particularly HMs. Basically, HMs are discharged due to extensive industrial activity through industrial wastewater effluents and solid wastes. Therefore, it is vital to keep these industrial wastewater effluents and solid wastes out of clean human-use water resources. The planting of some specific plant species not only prevents the transport of pollutants but also helps improve the air quality and helps maintain an adequate temperature near the plant area. Likewise, rhizofiltration is the most efficient and cost-effective technique for limiting the toxic pollutants moving along the medium of groundwater or surface water. The locations of most industries can be phytostabilized through the proper planting of specific trees or plants according to the kinds of pollutants, topography, and soil conditions there. The basic principle behind the processes of phytostabilization and rhizofiltration is the accumulation of pollutants in the plant tissues, which reduces the bioavailability and mobility of metals by accumulating them within the root system or capturing them inside the substrate molecules. The possible impact of the accumulation of excessive HMs is a reduction in their ability to be transported through water. Both approaches have a comparable primary mechanism in the method of complexation with root exudates or root systems and chelation due to the molecules that bind the molecules of HMs. Further, these mechanisms involve cell wall bonding and microbes that happen in phytostabilization; otherwise, the formation of antioxidant materials and the compartmentation of toxic pollutants occur in rhizofiltration. Again, the mobility of trace elements, organic wastes, texture, temperature, and redox potential are various key factors that influence the mechanism of rhizofiltration [27–29].

Limitations of Rhizofiltration

- The pH of the contaminated water was properly adjusted to constantly allow for the accumulation of optimum HMs in water bodies.
- The speciation of chemicals and the interaction of the HMs and other contaminants present in the water or wastewater must be properly understood for effective application.
- A properly engineered system is necessary to control and regulate the flow rate and concentration of wastewater effluent or contaminated water.
- Basically, plants in a terrestrial ecosystem are suitable for rhizofiltration and grow properly in a nursery or greenhouse system.
- Saturated plant species must be periodically removed, replaced, properly disposed of.
- Rhizofiltration studies are limited to greenhouse systems and laboratories, and its practical application in the field is a challenge.
- The absorption of toxic contaminants generally occurs on or into the root systems of the surrounding water bodies in the rhizosphere (root zone).
- Specified plant species are planted or cultivated in polluted groundwater where the root system of the plant accumulates HMs via absorption. Rhizofiltration is effective if the concentration of HMs is low and the volume of the water is large. The plant species translocate HMs to the shoots that are never utilized in situations where the HMs and organic material contamination is comparatively higher.
- The products associated with bioavailability and toxicity after the biodegradation of the pollutants are not always recognized. The by-products obtained after degradation may be bioaccumulated in aquatic organisms and animals or can be mobilized in groundwater.
- It is necessary to know about the various materials produced at the time of degradation in the metabolic cycle of plant species that do not enter the food chain or food web.

- If the root system of the plants is not able to penetrate to that depth of the water where the pollutants are persisting, then the water must be removed from the ground, and trees must be planted for remediation of the ground-water [30, 31].

Advantages of Rhizofiltration

- The method of rhizofiltration is used for remediation of toxic HMs and organic pollutants, which may be carried out either *in situ,* where specific plants or trees are grown directly in the polluted water body, or *ex situ,* where plants or trees are allowed to grow off-site and later on, after proper growth, are introduced to the polluted water sources.
- This method of water remediation is comparatively cost-effective and requires lower operational and capital costs, depending on the kinds of con-taminants present in the water bodies.
- In some cases, specific pollutants present in the water bodies were reduced significantly within a very short interval of time, and in other cases, they required a longer time. For example, the root of a sunflower plant decreases the uranium level within approximately 24 hours.
- The method of treatment of water is basically pleasing and reduces leaching and water infiltration of pollutants.
- After harvesting the saturated plant, the crop can be converted to biofuel briquettes, which are a proper substitute for fossil fuel [32, 33].

GENETIC ENGINEERING IN RHIZOFILTRATION

HM contamination in the natural environment due to anthropogenic activities and excessive industrialization threatens the health of millions of people around the globe. Plant roots exposed to numerous kinds of microbes in the environment play a vital role in the preservation of soil structure, nutrient recycling, detoxification of hazardous chemicals, and management of pathogens and pests. The interaction between plant roots and microbes is either nonspecific or specific. In cases of spe-cific interactions, plant species deliver a source of carbon that boosts the bacterial ability to decrease the toxicity of polluted water, and in cases of nonspecific interac-tions, various kinds of metabolic reactions in plants facilitate the growth of microbes, which causes a decrease in the HMs in contaminants present in the water. To over-come the impact of phytotoxicity, the microorganisms found in the rhizosphere of plant species (bacteria and mycorrhizae) can actively contribute to the change in metal speciation and the effective regeneration of used plant species. Various meth-ods, including proper application of fertilizers, pH adjustment, and chelators, can be used to further enhance the capability of rhizofiltration. The conventional method of remediation of HMs from water is a nonspecific and expensive process, whereas rhi-zofiltration is a successful method for decontaminating different kinds of HMs from water and wastewater. The enzymes, or microorganisms, play an integral role in the growth of plant species through interaction with various environmental factors. The use of genes that encode enzymes can change the oxidation state of bacteria. A gene can encode HgO reductase. It can also change the toxic metals into species of comparatively less toxic forms. For example, enzymes or microorganisms can methylate the metal Se into the less toxic form dimethylselenate. Multiple biotechnological approaches, such as genetic alteration of plant species, may be adopted to strengthen the capability of phytoremediation or rhizofiltration. Plant genetic engineering can be applied to rhizofiltration, which can be treated as a potential way to exploit the appropriate active genes that are responsible for the uptake of HMs, translocation, vacuolar sequestration, reduction, volatilization, and complexation. Genetic modification may be applied to express the enzymes or microbes that participated in the metabolic pathway of existing plant species or suggest new pathways for plant species. The various kinds of plant species and microorganisms possess a variety of mechanisms to protect against HM poisoning in water bodies. The genes introduced in rhizofiltration promote the detoxification of HM contaminants and subsequently improve the rate of degradation of the contaminants, including HMs and organic contaminants. Transgenic plants specifically containing bacterial genes can convert Se and Hg into comparatively fewer toxic species. Some organic acids and glutathione (GHS) metabolism play a significant role in the metal tolerance level of plant species. Glutathione appears in plant species, basically as reduced GSH. Glutathione can be synthesized by glutathione synthetase and glutamylcysteine synthetase enzymes. The metabolism of glutathione is linked with sulfur and cysteine metabolism in plants.

The concentration of cysteine in plants restricts glutathione biosynthesis. Thiol peptide phytochelatins (PCs) of low molecular weight are always termed class III metallothionein and can be synthetized in plant species from glutathione induced by HM ions. Glutathione is the cause of the synthesis of peptides via the α-glutamylcysteine transferase enzyme, also termed phytochelatin synthase (PCS), which catalyzes the transfer of the α-Glu-Cys group from the donor molecules of glutathione to glutathione, which is an acceptor molecule. PCS is an important cytosolic enzyme that is activated through metal ions such as Cd^{2+}, Au^{2+}, Zn^{2+}, Pb^{2+}, Hg^{2+}, $Ag1^+$, Bi^{3+}, Cu^{2+}, and PCs, synthesizing chelate HMs and ultimately producing complexes. The complexes formed are then moved via the cytosol in an ATP-dependent pathway via the tonoplast into the vacuole. The impact of genetic engineering on phytoremediation is illustrated in Figure 10.1. The present remediation technique for decontaminating HMs in water bodies is environmentally hostile, highly expensive, and requires more labor. The use of genetic engineering in rhizofiltration for the removal of HMs from water and wastewater effluents makes the remediation method more environmentally friendly and cost-effective. This approach has attracted attention in the past few years. The use of plant species in genetic engineering provides a novel opportunity for rhizofiltration and phytoremediation of water contaminated by HMs. But this technique or method can be completely exploited only if the mechanisms of translocation, metal tolerance, and accumulation of HMs are completely understood. In some cases, the decrease in pH value of the soil noticeably improves the phytoavailability of HMs by mixing bacteria like *Alyssum murale*, *Sphingomonas macroscopica*, and *Microbacterium arabinogalactanolyticum*, resulting in an improvement in phytostabilization. *Brassica napus* plants were injected to improve their tolerance level to different toxic HMs against varying kinds of bacterial species like *Microbacterium*

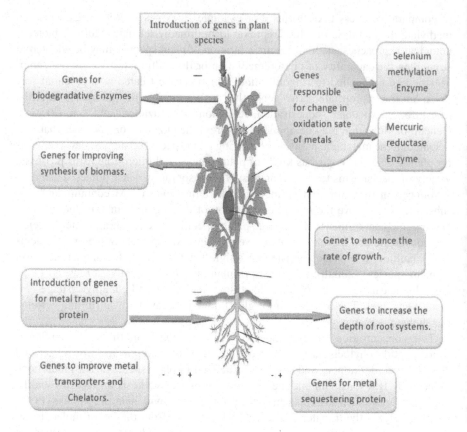

FIGURE 10.1 Use of genetic engineering in phytoremediation and rhizofiltration.

lacticum, Pseudomonas chlororaphis, Azotobacter vinelandii, and *Bacillus muci-laginosus* [34–36].

Significance of Rhizofiltration

- Rhizofiltration is used for remediation or decontamination of groundwater, surface water, storm waters, industrial effluents, household effluents, down-washes released from power lines, effluents from acid mine drainage, water containing radionuclides, runoff from agricultural fields, and sludges in dilute forms. Ultimately, we can say that most of the categories of polluted water can be remediated through rhizofiltration.
- The root system of the plants used appropriately for application in rhizofiltration grows at a rapid rate, resulting in the effective remediation of toxic HMs.
- Although many aquatic plants were found to be suitable for rhizofiltration, different terrestrial plants have been recognized to be efficient for the remediation of toxic heavy metals like Cu^{2+}, Zn^{2+}, Cr^{6+}, Cd^{2+}, Ni^{2+}, and Pb^{2+} from water bodies or in aqueous solutions. It was observed that the presence of

radioactive pollutants at low levels in water resources can be successfully remediated through rhizofiltration. The system to attain rhizofiltration contains a "feeder layer" of soil, which is suspended over the polluted stream in which plant species grow and their roots enter the water bodies.

- The feeder layer permits the plant species to absorb fertilizers through their root systems with the simultaneous removal of HMs without contaminating the water bodies. The tree or plant species of various kinds utilized for the remediation of water bodies must have low costs and long lifespans and must grow effectively in soil of marginal quality.
- The maintenance costs for rhizofiltration are very low, and the most popular plant species include poplars and willows that exhibit high flood tolerance capabilities. In cases of deep toxic pollutants, hybrid poplars with root systems extending up to 30 feet deep have been applied. Their root systems can enter microscopic pores in the matrix system of soil and can recycle 100 L of water per day per tree.
- These plant species behave like a pump for HMs from water bodies and cause the efficient remediation of contaminants in a green pathway from water bodies. Terrestrial plant species such as *Brassica* species and sunflower species can be effective rhizofiltrators.
- It is a low-cost green technology with low labor costs, minimum maintenance costs, and low operational costs. Hence, it is an appropriate alternative to the conventional approach for the remediation of wastewater and HM-contaminated water [37, 38].

FACTORS AFFECTING THE TECHNIQUE OF RHIZOFILTRATION

Numerous factors are responsible for effective rhizofiltration for the remediation of organic contaminants, HMs, and nutrients present in water (Figure 10.2). Various kinds of plant species in different ecosystems possess variable capabilities for remediating HMs and organic contaminants from water systems; therefore, not all categories of plant species are appropriate for remediating pollutants from water systems. For example, the use of maize and wheat can accumulate comparatively more phthalic acid esters (PAEs) from water and soil than only a single maize or wheat species. The maize and wheat plants are not suitable for remediating other pollutants like organochlorine pesticides (OCPs) such as HCHs, DDT, and their degradation products. Since rhizofiltration is specially needed for the remediation of groundwater, aquatic plants, particularly macrophytes, have been suitably used for remediating organic contaminants from groundwater. Plant species with fibrous roots or root systems have higher effectiveness in rhizofiltration because fibrous roots possess comparatively more surface area and are exposed to more contaminants in water systems, which enhances the accumulation of contaminants over the root systems of plants. Besides biological factors, the concentration of microorganisms present in plants or trees also significantly affects rhizofiltration. Furthermore, the multispecies plant system is less sensitive to variations in meteorological factors, which boosts the efficiency of the remediation of contaminants. It was observed that in the process of rhizofiltration, the mixed species of plants facilitated the treatment of

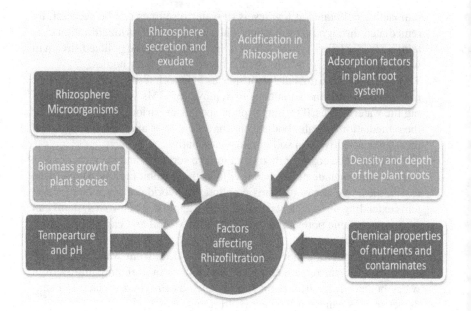

FIGURE 10.2　Various factors affecting rhizofiltration.

wastewater or heavily contaminated water. The interaction between plant species and microbes influences the quality of water in the rhizosphere. In the rhizosphere, the underground zone of the plant species is continuously exposed to water. It is an extremely active zone associated with diversified microorganisms. The microbes need energy, carbon, and oxygen to boost the immunity of the plants and to stimulate the exchange of minerals in plants. Therefore, nutrient consumption and protection from phytotoxicity are enhanced by the interaction between plant species and microorganisms [38, 40]. Various important factors that affect the efficiency of the rhizosphere are as follows (Figure 10.2).

TIME OF RHIZOFILTRATION

The method of rhizofiltration is dependent on time because of the mechanisms of bioaccumulation and biosorption. A greater number of active sites on plant roots (the surface of the biosorbent) in the first hour during biosorption resulted in a higher rate of adsorption of HMs from the water sources. The process of biosorption will be comparatively higher between one and three hours into process, and after that, the number of active sites decreases and is exhausted because of the transfer of adsorbate (HMs) from the water bodies to the active sites.

Example: The plant species *Hydrilla vercillatta* eliminates Cu in 10 days (efficiency of 89.2%), whereas Fe elimination occurs with an efficiency of 91.2% in five days. Hence, the rate of elimination or remediation of HMs from a water sample is measured by knowing the variety of plant species used, the optimum values of contact times, and the kinds of metals that must be remediated [41].

TEMPERATURE AND pH DURING RHIZOFILTRATION

Temperature is one of the key factors that results in an increase in the efficiency of HM removal at the time of rhizofiltration. The rate of biosorption of HMs is highly dependent on the temperature maintained during that process. The rise in temperature results in an increase in contact time between the absorber and the HMs present in the water sample, which therefore increases the overall performance of absorption by the plant roots. But with a decrease in temperature, the number of active sites at the surface of the plant roots deactivates, resulting in a decrease in the rate of absorption of HMs. Furthermore, the rise in temperature reduces the time required to attain the state of equilibrium in the biosorption of HMs. The variation of temperature also causes a variation in the pH of groundwater, which influences the efficiency of rhizofiltration. At some specific pH, the growth of hyperaccumulators is not appropriate. If the growth of hyperaccumulators occurs in an unsuitable pH condition, then the uptake capability of organic pollutants by the hyperaccumulator will decrease or even die, triggering the failure of rhizofiltration. The pollutant concentration in the groundwater also significantly influences the efficiency of the removal of HMs through rhizofiltration. If the organic pollutant concentration within groundwater is low, then in that situation, the hyperaccumulators may not be capable of executing rhizofiltration effectively because the organic contaminants cannot approach the roots of the hyperaccumulator. Therefore, some kinds of organic toxins may travel across the biofilter, which is made of hyperaccumulators. In the process of rhizofiltration, pH influences biosorption by modifying the bioavailability of the metals present in the water medium and the specifications of the plants used. With an increase in pH from a highly acidic medium to a slightly acidic medium or even a neutral medium, different hydrolyzed species are identified by the variation of active sites. Hence, there must be a good correlation between the number of active sites and the pH of the medium. The cations of the metal will attach to the active sites in the cell walls of the plant roots if the pH value of the medium is increased. Hence, we cannot ignore the effect of pH in the process of rhizofiltration for the removal of HMs [42, 43].

CHEMICAL PROPERTIES OF THE POLLUTANTS AND NUTRIENT

The nutrient concentration in the water bodies shows both a negative and a positive effect on the biosorption of metals by the plant roots. If the availability of nutrients for plants is much higher, then the adsorbed metals can bond to the anions that are present in nutrients. Hence, the opposition between the cations of the nutrient and HMs results in the uptake of HMs. The growth of plants is stimulated by the presence of nutrients in the water bodies. Subsequently, more active sites for the absorption of metals are produced that assist the process of rhizofiltration, and its efficiency in the removal of HMs and other toxic contaminants increases. However, the acceptance of HMs and other organic contaminants depends on the response of the plant species to metals and the growth rate of the plant species. Some organic pollutants are hydrophobic in nature and can easily dissolve in groundwater or surface water. But still, the problem of absorbing organic pollutants from the root systems of plant species is challenging. In association with surfactants, the process of rhizofiltration efficiency for the removal of organic substances will be noticeably improved [44].

Density and Depth of the Plant Roots

The density of the plant root depends on the depth of the water bodies, and the absorption of metal by the root systems of the plant is also dependent on the depth. A major part of the plant roots is found near the surface of the soil, and the root density declines with an increase in depth. Therefore, only plants whose roots reach adequate depths are suitable for rhizofiltration. The method of acceptance of contamination from water by the root region of the plant species and subsequent translocation to the leaves and stems is termed transpiration. Many studies have reported that the translocation of metal from the roots through transpiration has been well recognized. The higher concentrations of metals absorbed by the roots cause water consumption by facilitating the migration of metals through mass flow to the surface of the root system, where precipitation occurs. The rate of transpiration also depends on the topographic or climatic conditions of the environment. In mid-latitude climates, plant roots are basically dormant during the winter season and consume very little water. The reverse is true in the summer, when more water is available to transpire for photosynthesis [45].

Absorption Factor of Plant root System

The total quantity of metals accumulated by the root system of the plant species normally equals the concentration of the metal in mg/L in its soluble state. The metal must be absorbed by the roots before translocation, through either the apoplastic or symplastic pathway, where a passive or active filter system may be created. The factors related to root absorption are influenced by rhizobiological activity, moisture, temperature, pH, concentration of competing ions, root exudate, and the concentration of metal in the soil solution [46].

Acidification of Rhizosphere

The rhizosphere is considered the interface between the root system and the soil. In this region, numerous interactions between the solution of the soil and the root system of the plant species take place, which is the cause of the significant modification of the physicochemical characteristics of this region. Plants need huge quantities of nutrients for their optimum development and growth, and the absorption of such a huge amount of nutrients through the root system of plants may cause modification of the rhizosphere, specifically acidification, which is a continuous phenomenon in many soils throughout the world. The mechanism behind it is the proton extrusion at the plasma membrane that improves cation influx, which is a key route utilized by the roots of plants to remove excess nutrients from the water or soil solution. The acidification of the rhizosphere plays a crucial role in the acquisition of nutrients by plant species and stimulates the activity of H^+-ATPase present at the plasmalemma. The lower availability of nutrients proves the release of protons due to nutrient acquisition by the plant roots. Hence, a shortage of nutrients in the water bodies or fewer nutrients in the water or wastewater is an indication of high activity of H^+-ATPase. Acidification in the rhizosphere is a significant natural process in which plant species consume many nutrients. Proton liberation usually promotes the consumption of one cation. Hence, the release of protons (H^+) is essential to facilitate nutrient transport into the plant roots. The secretion of H^+ ions by the root systems of plants is the cause of acidification in the rhizosphere and therefore improves the dissolution of metals. The secretion of H^+ ions from the roots of plants is regulated by triphosphatase (H^+-ATPase) along with H^+ pumps [47, 48].

RHIZOSPHERE SECRETIONS AND EXUDATES

The root systems of several plant species exude organic acids, including malic, oxalic, and malonic acids, which is the cause of the decrease in the pH value of the rhizosphere that converts the cations of to make the metal bioavailable. This secretion of acids significantly impacts the rhizosphere and the biotic community living within it. The nature of the exudates and quantity of the material released usually depend on several factors, such as the architecture of the root system, the level of toxicity of metals, and the existence of harmful microbes. Histidine is an amino acid that is accountable for the hyperaccumulation and transport of Ni in *Alyssum*. There are numerous kinds of organic acids of low molecular weight, like succinic acid and acetic acid, which are found in the rhizosphere of a Cd-accumulating genotype such as Kyle but never in a nonaccumulating genotype such as Arcola. In dicotyledons, the root reductases decrease the concentration of Cu^{2+} and $Fe3^+$ under less Cu and Fe to enhance the uptake of plant species of Fe, Mn, Cu, and Mg. The bioavailability of many metals may be increased through the secretion of phytosiderophores by plants into the rhizosphere. Siderophores are formed by a multitude of classes of microbes, from plant pathogens to animal pathogens and free-living organisms, which are symbiotic N-fixing microorganisms. Siderophores are low molecular weight Fe-chelators produced by microorganisms to obtain Fe from the soil, which is usually not readily available to plants and microorganisms, such as lack of nutrients and the presence of high concentrations of HMs. The siderophore bacteria can bind a variety of metals, hence improving phytoavailability in the zone of contamination. Phytosiderophores, including avenic and mugenic acids, are released because of their action on the deficiency of nutrients and metals. This causes an improvement in the bioavailability of metals in the case of Fe, Cu, Pb, Cd, Zn, and Pt, which facilitates carrying these metals into the plant tissues. The complex based on metal chelates can be transported through the plasma membrane for iron–phytosiderophore in cereals. Soil immunization with the help of *Pseudomonas aeruginosa* basically improves the bioavailable concentrations of Pb and Cr compared to uninoculated controls [49, 50].

RHIZOSPHERE MICROORGANISMS

The rhizosphere of plant species is highly populated by a huge number of microorganisms, mostly mycorrhizal fungi and bacteria. The presence of these microorganisms is responsible for increasing the production of biomass and the tolerance level of plants to HMs and is considered a significant component in the pathway of phytoextraction. These root-colonizing mycorrhizae and bacteria remarkably enhance the bioavailability of different HM ions. The microbes within the rhizosphere function as catalysts for exudate organic material and redox transformations, which promote the bioavailability and root absorption of Cd^{2+}, Mn^{2+}, and some other metal ions. It was observed that microorganisms of the rhizosphere, such as *Bacillus* and *Pseudomonas*, enhance the quantity of Cd accumulation in *Brassica juncea*. The biosurfactant developed by *Pseudomonas aeruginosa* can be used for the remediation of HMs like Pb and Cd [51–53].

BIOMASS GROWTH OF PLANT SPECIES

The biomass of the plant species above the ground is one of the key factors, in addition to the bioavailability of the metal ions, which impacts phytoextraction efficiency. The efficiency of phytoextraction must be improved if the biomass of the plant above the land is increased, which leads to an increase in the concentration of total metal in shoots. HM contamination is the cause of growth retardation in plant species because of the deficiency of organic substances and essential nutrients, including N, K, and P, in association with them. Irrigation and the use of fertilizers are two vital factors that can stimulate the growth of plant species by modifying the presence of essential nutrients and organic substances. Fertilizers not only promote the growth of plant species but also decrease the pH of the soil, therefore improving the bioavailability of metals present in the water or soil [54]. Chou et al. (2005) studied the impact of K-fertilizer on plant species and found that rape (*Brassica campestris*) shows the maximum production of biomass in aboveground parts and the maximum accumulation of contaminants in shoots. The impact of nitrate fertilizers on the development and growth of *Brassica juncea* (Indian mustard) on the accumulation of Zn was studied, and it was found that by applying the required quantity of nitrate fertilizers, the formation of shoot biomass remarkably increases, which is the cause of more accumulation of Zn [55]. If plant species take 90% of the NO_3^- and 10% of the total N as NH^+, then the capability of phytoextraction by Indian mustard reaches its maximum. The increase in metal bioavailability and biomass of plant species is the cause of stimulating microbial-induced phytoextraction. We observed four kinds of bacterial strains from the rhizosphere contaminated by HMs that exhibit the activity of ACC (1-aminocyclopropane-1-carboxylate) deaminase. Two strains, *Pseudomonas fluorescens* ACC and *Pseudomonas tolaasii* ACC, form acetic acid, and in that case, the activity of siderophores becomes greater even under Cd stress [56]. These strains can not only influence the Cd accumulation in the shoot and root systems of *Brassica napus* but can also cause an increase in the biomass formation of the plant species, leading to a higher accumulation of total Cd present in the water bodies and soil. It was reported that more *Salix caprea* biomass and bioavailability were achieved in the accumulation of Zn and Cd on the shoot when inoculated with *Streptomyces* sp. AR17 and *Agromyces* sp. AR33 [57].

CONCLUSION

The rapid development of industrial and mining activity has introduced excessive amounts of toxic contaminants into freshwater bodies and groundwater, which is the cause of significant health hazards for humans and degradation of the environment.

Among the various toxic substances present in water resources, HMs are more dangerous and pose a severe threat to animals and human beings. Hence, researchers throughout the globe have focused on developing an ecologically safe, low-cost, green technology that is viable for remediating toxic contaminants, especially HMs, from water bodies. Rhizofiltration and phytostabilization are the most commonly adopted plant-based methods for remediating toxins and HMs from the water bodies of industrial and mining zones. Rhizofiltration offers an excellent ecological approach for the removal of HMs from water systems in a cost-effective and eco-friendly way, reducing metal toxicity. Plant species selection, the action plan of rhizofiltration, and innovative strategies for effective removal of toxic heavy metallic contaminants are the challenges for researchers. The rhizofiltration method can use plants whose root systems are capable of accumulating HMs, and it is a good, cost-effective, eco-friendly, sustainable, and environmentally safe approach for remediation of heavy metals from water systems. Furthermore, strategies must be developed to use microorganisms for developing HM bioavailability to facilitate more HM deposition in the root systems of plants and to make the method more efficient. Extensive research work must be undertaken to develop plant species that can transfer rhizofiltration to commercial settings.

REFERENCES

1. Dushenkov, V., Kumar, P. N., Motto, H., & Raskin, I. (1995). Rhizofiltration: The use of plants to remove heavy metals from aqueous streams. *Environmental Science & Technology, 29*(5), 1239–1245.
2. Rawat, K., Fulekar, M. H., & Pathak, B. (2012). Rhizofiltration: A green technology for remediation of heavy metals. *International Journal of Innovations in Bioscience, 2*(4), 193–199.
3. Verma, P., George, K. V., Singh, H. V., Singh, S. K., Juwarkar, A., & Singh, R. N. (2006). Modeling rhizofiltration: Heavy-metal uptake by plant roots. *Environmental Modeling & Assessment, 11*, 387–394.
4. Yadav, B. K., Siebel, M. A., & van Bruggen, J. J. (2011). Rhizofiltration of a heavy metal (lead) containing wastewater using the wetland plant *Carex pendula*. *CLEAN–Soil, Air, Water, 39*(5), 467–474.
5. Abubakar, M. M., Ahmad, M. M., & Getso, B. U. (2014). Rhizofiltration of heavy metals from eutrophic water using *Pistia stratiotes* in a controlled environment. *IOSR Journal of Environmental Science, Toxicology and Food Technology, 8*(6), 27–34.
6. Yadav, A. K., Pathak, B., & Fulekar, M. H. (2015). Rhizofiltration of heavy metals (cadmium, lead and zinc) from fly ash leachates using water hyacinth (*Eichhornia crassipes*). *International Journal of Environment, 4*(1), 179–196.
7. Awa, S. H., & Hadibarata, T. (2020). Removal of heavy metals in contaminated soil by phytoremediation mechanism: A review. *Water, Air, & Soil Pollution, 231*(2), 47–61.
8. Woraharn, S., Meeinkuirt, W., Phusantisampan, T., & Chayapan, P. (2021). Rhizofiltration of cadmium and zinc in hydroponic systems. *Water, Air, & Soil Pollution, 232*(5), 204–220.
9. Mithembu, M. S. (2012). Nitrogen and phosphorus removal from agricultural wastewater using constructed rhizofiltration in Durban, South Africa. *Journal of Agricultural Science and Technology*, (2), 1142–1148.
10. Mthembu, M. S. (2016). *Removal of organic and inorganic nutrients in a constructed rhizofiltration system using macrophytes and microbial biofilms* (Doctoral dissertation).

11. Kristanti, R. A., Ngu, W. J., Yuniarto, A., & Hadibarata, T. (2021). Rhizofiltration for removal of inorganic and organic pollutants in groundwater: A review. *Biointerafce Research in Applied Chemistry, 4*, 12326–12347.

12. Akhtar, M. S., Chali, B., & Azam, T. (2013). Bioremediation of arsenic and lead by plants and microbes from contaminated soil. *Research in Plant Sciences, 1*(3), 68–73.

13. Thayaparan, M., Iqbal, S. S., Chathuranga, P. K. D., & Iqbal, M. C. M. (2013). Rhizofiltration of Pb by *Azolla pinnata*. *International Journal of Environmental Sciences, 3*(6), 1811–1823.

14. Dürešová, Z., Šuňovská, A., Horník, M., Pipíška, M., Gubišová, M., Gubiš, J., & Hostin, S. (2014). Rhizofiltration potential of *Arundo donax* for cadmium and zinc removal from contaminated wastewater. *Chemical Papers, 68*, 1452–1462.

15. Kamusoko, R., & Jingura, R. M. (2017). Utility of Jatropha for phytoremediation of heavy metals and emerging contaminants of water resources: A review. *CLEAN–Soil, Air, Water, 45*(11), 1–17.

16. Leblebici, Z., Dalmiş, E., & Andeden, E. E. (2019). Determination of the potential of *Pistia stratiotes* L. in removing nickel from the environment by utilizing its rhizofiltration capacity. *Brazilian Archives of Biology and Technology, 62*, e19180487.

17. Mohanty, M., Dhal, N. K., Patra, P., Das, B., & Reddy, P. S. R. (2010). Phytoremediation: A novel approach for utilization of iron-ore wastes. *Reviews of Environmental Contamination and Toxicology, 206*, 29–47.

18. Pérez-Palacios, P., Agostini, E., Ibáñez, S. G., Talano, M. A., Rodríguez-Llorente, I. D., Caviedes, M. A., & Pajuelo, E. (2017). Removal of copper from aqueous solutions by rhizofiltration using genetically modified hairy roots expressing a bacterial Cu-binding protein. *Environmental Technology, 38*(22), 2877–2888.

19. Galal, T. M., Eid, E. M., Dakhil, M. A., & Hassan, L. M. (2018). Bioaccumulation and rhizofiltration potential of *Pistia stratiotes* L. for mitigating water pollution in the Egyptian wetlands. *International Journal of Phytoremediation, 20*(5), 440–447.

20. Yazdanpanah, S., & Rajaei, P. (2014). A mini review: Metal remediation by microbes and plants. *Journal of Biodiversity and Environmental Sciences (JBES), 5*(2), 222–226.

21. Raj, D., & Maiti, S. K. (2019). Sources, toxicity, and remediation of mercury: An essence review. *Environmental Monitoring and Assessment, 191*, 1–22.

22. Malaviya, P., Singh, A., & Anderson, T. A. (2020). Aquatic phytoremediation strategies for chromium removal. *Reviews in Environmental Science and Bio/Technology, 19*, 897–944.

23. El-Aassar, M. R., Fakhry, H., Elzain, A. A., Farouk, H., & Hafez, E. E. (2018). The Rhizofiltration system consists of chitosan and natural *Arundo donax* L. for removal of basic red dye. *International Journal of Biological Macromolecules, 120*, 1508–1514.

24. Tomé, F. V., Rodríguez, P. B., & Lozano, J. C. (2008). Elimination of natural uranium and 226Ra from contaminated waters by rhizofiltration using *Helianthus annuus* L *Science of the Total Environment, 393*(2–3), 351–357.

25. Bakshe, P., & Jugade, R. (2023). Phytostabilization and rhizofiltration of toxic heavy metals by heavy metal accumulator plants for sustainable management of contaminated industrial sites: A comprehensive review. *Journal of Hazardous Materials Advances, 10*, 1–16.

26. Belle, G. N., Oberholster, P. J., Fossey, A., Esterhuizen, L., & Moodley, R. (2022). Using pollution indices to develop a risk classification tool for gold mining contaminated soils. *Journal of Environmental Science and Health, Part A, 57*(12), 1047–1057.

27. Laghlimi, M., Baghdad, B., El Hadi, H., & Bouabdli, A. (2015). Phytoremediation mechanisms of heavy metal contaminated soils: A review. *Open Journal of Ecology, 5*(08), 375–388.

28. Castro-Castellon, A. T., Hughes, J. M. R., Read, D. S., Azimi, Y., Chipps, M. J., & Hankins, N. P. (2021). The role of rhizofiltration and allelopathy on the removal of

cyanobacteria in a continuous flow system. *Environmental Science and Pollution Research, 28,* 27731–27741.

29. Lee, M., & Yang, M. (2010). Rhizofiltration using sunflower (*Helianthus annuus* L.) and bean (*Phaseolus vulgaris* L. var. vulgaris) to remediate uranium contaminated groundwater. *Journal of Hazardous Materials, 173*(1–3), 589–596.

30. Ghosh, M., & Singh, S. P. (2005). A review on phytoremediation of heavy metals and utilization of it's by products. *Asian J Energy Environ, 6*(4), 214–23.

31. Gaur, N., Flora, G., Yadav, M., & Tiwari, A. (2014). A review with recent advancements on bioremediation-based abolition of heavy metals. *Environmental Science: Processes & Impacts, 16*(2), 180–193.

32. Ignatius, A., Arunbabu, V., Neethu, J., & Ramasamy, E. V. (2014). Rhizofiltration of lead using an aromatic medicinal plant *Plectranthus amboinicus* cultured in a hydroponic nutrient film technique (NFT) system. *Environmental Science and Pollution Research, 21,* 13007–13016.

33. Koźmińska, A., Hanus-Fajerska, E., & Muszyńska, E. (2014). Possibilities of water purification using the rhizofiltration method. *Woda Środowisko-Obszary-Wiejskie, 14*(47), 89–98.

34. Eapen, S., & D'souza, S. F. (2005). Prospects of genetic engineering of plants for phytoremediation of toxic metals. *Biotechnology Advances, 23*(2), 97–114.

35. Kensa, V. M. (2011). Bioremediation-an overview. *Journal of Industrial Pollution Control, 27*(2), 161–168.

36. Raskin, I. (1996). Plant genetic engineering may help with environmental cleanup. *Proceedings of the National Academy of Sciences, 93*(8), 3164–3166.

37. Veselý, T., Tlustoš, P., & Száková, J. (2011). The use of water lettuce (*Pistia stratiotes* L.) for rhizofiltration of a highly polluted solution by cadmium and lead. *International Journal of Phytoremediation, 13*(9), 859–872.

38. Banerjee, A., & Roychoudhury, A. (2022). Rhizofiltration of combined arsenic-fluoride or lead-fluoride polluted water using common aquatic plants and use of the 'clean' water for alleviating combined xenobiotic toxicity in a sensitive rice variety. *Environmental Pollution, 304,* 1–18.

39. Sheoran, V., Sheoran, A. S., & Poonia, P. (2016). Factors affecting phytoextraction: A review. *Pedosphere, 26*(2), 148–166.

40. Malaviya, P., & Singh, A. (2012). Phytoremediation strategies for remediation of uranium-contaminated environments: A review. *Critical Reviews in Environmental Science and Technology, 42*(24), 2575–2647.

41. Odinga, C. A., Kumar, A., Mthembu, M. S., Bux, F., & Swalaha, F. M. (2019). Rhizofiltration system consisting of *Phragmites australis* and *Kyllinga nemoralis*: Evaluation of efficient removal of metals and pathogenic microorganisms. *Desalination and Water Treatment, 169,* 120–132.

42. Sikhosana, M. L. M., Botha, A., Mpenyane-Monyatsi, L., & Coetzee, M. A. (2020). Evaluating the effect of seasonal temperature changes on the efficiency of a rhizofiltration system in nitrogen removal from urban runoff. *Journal of Environmental Management, 274,* 1–16.

43. Singh, R., Ahirwar, N. K., Tiwari, J., & Pathak, J. (2018). Review of sources and effect of heavy metal in soil: Its bioremediation. *International Journal of Research in Applied, Natural and Social Sciences, 2018,* 1–22.

44. Das, P., & Paul, K. K. (2023). Hydroponic rhizofiltration of dairy wastewater by *Coleus scutellarioides & Portulaca oleracea. Journal of Water Process Engineering, 52,* 103589.

45. Douglas, J. A., Douglas, M. H., Lauren, D. R., Martin, R. J., Deo, B., Follett, J. M., & Jensen, D. J. (2004). Effect of plant density and depth of harvest on the production and quality of licorice (*Glycyrrhiza glabra*) root harvested over 3 years. *New Zealand Journal of Crop and Horticultural Science, 32*(4), 363–373.

46. Bloom, A. J., Meyerhoff, P. A., Taylor, A. R., & Rost, T. L. (2002). Root development and absorption of ammonium and nitrate from the rhizosphere. *Journal of Plant Growth Regulation, 21*, 416–431.

47. Carrillo, A. E., Li, C. Y., & Bashan, Y. (2002). Increased acidification in the rhizosphere of cactus seedlings induced by *Azospirillum brasilense*. *Naturwissenschaften, 89*, 428–432.

48. Petersen, W., & Böttger, M. (1991). Contribution of organic acids to the acidification of the rhizosphere of maize seedlings. *Plant and Soil, 132*, 159–163.

49. Walker, T. S., Bais, H. P., Grotewold, E., & Vivanco, J. M. (2003). Root exudation and rhizosphere biology. *Plant Physiology, 132*(1), 44–51.

50. Vives-Peris, V., De Ollas, C., Gómez-Cadenas, A., & Pérez-Clemente, R. M. (2020). Root exudates: From plant to rhizosphere and beyond. *Plant Cell Reports, 39*, 3–17.

51. Salt, D. E., Blaylock, M., Kumar, N. P., Dushenkov, V., Ensley, B. D., Chet, I., & Raskin, I. (1995). Phytoremediation: A novel strategy for the removal of toxic metals from the environment using plants. *Biotechnology, 13*(5), 468–474.

52. Dimitroula, H., Syranidou, E., Manousaki, E., Nikolaidis, N. P., Karatzas, G. P., & Kalogerakis, N. (2015). Mitigation measures for chromium-VI contaminated groundwater-the role of endophytic bacteria in rhizofiltration. *Journal of Hazardous Materials, 281*, 114–120.

53. Khan, A. U., Khan, A. N., Waris, A., Ilyas, M., & Zamel, D. (2022). Phytoremediation of pollutants from wastewater: A concise review. *Open Life Sciences, 17*(1), 488–496.

54. Nedelkoska, T. V., & Doran, P. M. (2000). Characteristics of heavy metal uptake by plant species with potential for phytoremediation and phytomining. *Minerals Engineering, 13*(5), 549–561.

55. Chou, F. I., Chung, H. P., Teng, S. P., & Sheu, S. T. (2005). Screening plant species native to Taiwan for remediation of 137Cs-contaminated soil and the effects of K addition and soil amendment on the transfer of 137Cs from soil to plants. *Journal of Environmental Radioactivity, 80*(2), 175–181.

56. Arshad, M., Saleem, M., & Hussain, S. (2007). Perspectives of bacterial ACC deaminase in phytoremediation. *Trends in Biotechnology, 25*(8), 356–362.

57. De Maria, S., Rivelli, A. R., Kuffner, M., Sessitsch, A., Wenzel, W. W., Gorfer, M., & Puschenreiter, M. (2011). Interactions between accumulation of trace elements and macronutrients in *Salix caprea* after inoculation with rhizosphere microorganisms. *Chemosphere, 84*(9), 1256–1261.

11 Plant–Microbe Nexus in Sustainable Environmental Management and Waste Mitigation

Poulomi Sarkar and Sanjay Swarup

INTRODUCTION

Environmental pollution has become a major issue in recent times, primarily stemming from the growing discharge of xenobiotic and recalcitrant compounds into terrestrial ecosystems. High amounts of both organic and inorganic wastes are released into the environment annually (Thakare et al., 2021). Pollutants enter the ecosystem through various anthropogenic activities. These pollutants can be classified into more than 20 different groups, mostly coming from sources such as agricultural runoff, household activities, combustion, nuclear plants, and biomedical, industrial, municipal, petrochemical, and agrochemical waste (Nasr, 2019; Kumar et al., 2022). Prolonged exposure to and accumulation of these hazardous wastes can prove detrimental to living organisms and natural ecosystems (Singh and Pant, 2023). Due to their distinctive chemical and physical properties, they remain persistent in nature. Additionally, they invade the food chain in the biosphere. These contaminants can cause congenital disabilities, cancer, diabetes, and allergies. They can harm the nervous system and disrupt the immune, reproductive, and endocrine systems in living organisms (Xiang et al., 2022). Decontamination of these wastes is a major challenge. Various physical as well as chemical methods are being used to degrade these wastes. However, application of these methods often proves to be ineffective because they are associated with the generation of toxic by-products (Sarkar et al., 2017; Pant et al., 2021; Singh and Pant, 2023). On the other hand, biological waste remediation methods are environmentally benign and low-cost alternatives for the efficient removal of hazardous substances (Sarkar et al., 2017).

The biological methods make use of microorganisms (bacteria, fungi, and microalgae), microbial enzymes, or plants and their associated phytomicrobiomes for bioremediation of wastes (Kumar et al., 2022). Although it currently has limitations for full-scale applications, phytoremediation has emerged as an innovative technology for biodegradation of recalcitrant environmental pollutants. This technology relies upon the potential of plants and their microbiomes to biotransform and

DOI: 10.1201/9781003442295-11

degrade xenobiotic substances (Augusta et al., 2022). The challenges associated with plant–microbe-assisted remediation can be minimized by promoting rhizosphere microbiomes which provide growth-promoting benefits to the host plants and facilitate bioremediation (Augusta et al., 2022).

In the process of phytoremediation, the plant and microbe coexist or compete for nutrient resources, and their synergistic associations play a crucial role in biomineralization of the toxic substances (Kumar et al., 2022). The plant–microbe nexus within plant rhizospheres functions in a synergistic way in environmental remediation. The plant exudates and metabolites act as sources of carbon for the microbes, and they in return increase the bioavailability of different metals, ions, or nutrients for the plant. In this way the overall metabolic processes decrease the pollutant load from the contaminated matrix. The rhizospheric microbial communities also degrade many organic pollutants by making use of the contaminants as sources of carbon or energy, consequently providing nutrients to the host plants and helping them grow (Kumar et al., 2022). Additionally, plant roots secrete exudates and enzymes which aid in the degradation of various contaminants in soils. Moreover, plants can scavenge obnoxious gases, such as carbon dioxide or carbon monoxide, from the atmosphere. Therefore, due to their ability to revive the ecosphere and degrade pollutants, green plants are equally used for sustainable environmental management (Singh and Pant, 2023).

Besides being vital drivers of phytoremediation, plants and microbes are also valuable players in terrestrial soil carbon sequestration. Recently, carbon farming and carbon sequestration using plants and their rhizospheric microbiomes have attracted intense scientific attention. It has also been recognized as an important strategy for combating climate change (Mattila et al., 2022). Carbon farming is the adoption of methods to capture large pools of atmospheric carbons and to store them in plant or soil organic matter. Moreover, when carbon storage exceeds the rate of carbon loss, carbon farming becomes successful (Debnath et al., 2022). Microbes and plants participate in cycling of many elements and greenhouses gases in the environment. Hence, they are crucial controllers of global climate change (Ooi et al., 2022). Considering all the parameters of environmental sustainability and waste mitigation, this chapter delves into the plant–microbe nexus in combating environmental pollution. We describe the different sources as well as the hazards of conventional and classic pollutants in nature. We further describe the aspects of sustainable environmental management via plant–microbe interactions and carbon farming.

SOURCES AND CATEGORIES OF ENVIRONMENTAL POLLUTANTS

CONVENTIONAL CONTAMINANTS

Industrial effluents, agricultural runoff, and municipal run-off are prominent sources of waste in the environment. The term "conventional wastes" is used because there are numerous methods and techniques applied worldwide to effectively manage these types of pollutants (Babar et al., 2022). Petroleum hydrocarbons, heavy metals, pesticides, textile dyes, and industrial chemicals are some of the conventional contaminants in nature.

Petroleum Hydrocarbons

Petroleum hydrocarbons and their various derivatives serve as primary energy sources in both everyday activities and industrial settings. Therefore, petroleum exploration, processing, and transport become the major sources of hydrocarbon contamination in nature (Das and Chandran, 2011). It is estimated that more than 60 million tons of petroleum wastes are generated per annum, a challenge to the existing remediation strategies (Sarkar et al., 2017). Petroleum oil spills in aquatic environments can have especially devastating consequences. These spills disrupt the aquatic ecosystem, with lighter fractions of petroleum floating on the water's surface and heavier particles settling at the bottom, adversely impacting fish and other aquatic organisms. Additionally, some petroleum fractions evaporate into the atmosphere, while the majority seeps into groundwater, eventually being transported to distant locations (ASTDR, 1999).

Heavy Metals

Anthropogenic activities related to urbanization and industrialization have contributed largely to the release of heavy metals into the ecosystem. These metals mostly get distributed within the soil matrix and aquatic environments. Small fractions of these metals even get into the atmosphere in the form of aerosols and particulate matter (Sharma and Agrawal, 2005). Recent experiments reveal the importance of certain metals like nickel, copper, and zinc for human well-being, as they are naturally abundant. Nevertheless, the environmental impact of heavy metals is substantial. For instance, the transformation of mercury into methylmercury in water results in highly toxic sediments. Chromium, widely employed in industry, poses carcinogenic risks. Despite these drawbacks, specific heavy metals play a crucial role in regulating physiological functions. Naturally occurring essential heavy metals enter the body through food, air, and water, influencing various biological processes (Mitra et al., 2022).

Pesticides

In the past century, pesticides have significantly contributed to sustaining the growing global population by enhancing agricultural yields through effective pest control. They have gained widespread adoption as a straightforward, rapid, and efficient approach to minimize crop losses, improve cosmetic appeal, and occasionally enhance nutritional yield (Parra-Arroyo et al., 2022). But these chemicals have high adsorption rates and can travel within air, soil, or water matrices (Lofrano et al., 2020). The World Health Organization (WHO) indicated that pesticide poisoning resulted in more than 193,460 unintentional deaths in 2012 (Parra-Arroyo et al., 2022). Certain pesticides, such as atrazine (approximate half-life of 72 days), Diuron (302 days), and chlorsulfuron (66.9 days) exhibit prolonged persistence. Besides the detrimental impact on the environment, pesticides also raise concerns for human health. Residues of toxic compounds contact the human body through domestic animals, livestock, marine food, fisheries, and aquaculture. Furthermore, pesticides accumulate over time, and their effects are amplified up the trophic level. They can induce neurotoxic, mutagenic, teratogenic, and carcinogenic effects on human health (Parra-Arroyo et al., 2022).

EMERGING CONTAMINANTS

Emerging pollutants encompass both synthetic and naturally occurring chemicals and microbial particles that persist in the environment yet have been studied relatively little by the research community. These contaminants can stay in nature for longer times because of their ability to transform/biotransform from one state to another (Abdulrazaq et al., 2020). Due to their altered characteristics, these contaminants require different treatment methods from the conventional approaches. The emerging pollutants are increasingly associated with effluents from agricultural, industrial, and municipal resources (Gavrilescu et al., 2015). The emerging contaminants can be classified as follows.

Agricultural Waste

The world population is projected to reach to 9 billion by 2050. Thus, for the sake of food security it is certain there will be intense demand for agricultural and livestock production. But advances in agricultural and livestock production will be accompanied by the generation of enormous amounts of wastes (Koul et al., 2022).

Crop Residues, Stubbles, and Biomass Burning

Crop residues comprising leaf litter, stubble, seed pods, stems, husks, weeds, and straw are one of the major categories of agricultural wastes (Koul et al., 2022). India ranks as the second largest generator of agricultural wastes and accounts for about 130 million metric tons per annum. It is estimated half this waste is utilized as fodder, whereas the other half is discarded, aggravating the problem of solid waste generation and management. The burning of stubble or parali and plant biomass is a prime cause of environmental pollution (Koul et al., 2022). Stubbles are leftover cut stalks of crops that remain after harvest. Stubble burning is the incineration of these leftover residues by farming. These incineration processes generate huge amounts of obnoxious gases (Abdurrahman et al., 2020). The uncontrolled burning processes cause emission of greenhouse gases (GHG). The GHG primarily comprises CO_2 and other gaseous contaminants such as CO, CH_4, NO_x, and lesser quantities of SO_x (Siddiqi et al., 2021; Koul et al., 2022). Apart from GHG emission, these fires also emit carcinogenic polyaromatic hydrocarbons and dioxins (Siddiqi et al., 2021). The open-air burning of agricultural waste has a wide range of harmful and multifaceted consequences. It has been highlighted by researchers and environmental experts that the formation of intense haze over South Asia in winter is linked to biomass and stubble burning since it coincides with the crop harvest seasons (Abdurrahman et al., 2020). The process of rice stubble burning impacts the environment more severely—the rice stubble burning season coincides with winter, and lower ambient temperatures help the pollution to stay longer in the atmosphere, thereby increasing the severity of the contaminants (Abdurrahman et al., 2020). Stubble and biomass burning is an emergent source of environmental pollution in agriculturally based countries such as India. These open-air burning processes not only degrade the air quality, they are also potential causes of reduction in soil fertility and agricultural productivity (Abdurrahman et al., 2020). The burning process depletes the natural soil nutrients (phosphates, nitrates, potassium, etc.) and microbial communities which are essential for soil productivity (Jain et al., 2014; Singh et al., 2018). All these processes in turn cause damage to overall agricultural productivity since air and soil quality are prime factors for thriving agriculture. The burning processes produce huge amounts of volatile organic compounds (VOCs) and NO_x gases that harm the atmospheric ozone layer, further contributing to climate change (Sharma et al., 2019) The increased air pollution caused by burning of stubbles also contributes to high mortality rates due to atmospheric contamination. India stands as the second most polluted nation in the world with its diminished air quality standards and particulate acquisition in air. Particulate pollution in India has increased by 61.4% since 1998 and has further reduced the average life expectancy by 2.1 years. It is reported that if the current pollution persists in India, over 40% of the population will lose 7.6 years of their life expectancy in the coming years (Energy Policy Institute at the University of Chicago [EPIC], 2023). Air quality and environmental pollution also affect a country's economy. High environmental pollution in India led to a decline in tourism in many places such as Delhi. Air pollution also affected the productivity of workers, who suffer from poor vision and various other heath disorders. The Indian economy suffered around 4.5% to 7.7% loss in gross domestic product (GDP) in 2018, which is projected to increase by 15% in 2060 if the same pollution rate prevails (Ghosh et al., 2019; Abdurrahman et al., 2020).

Agricultural Wastewaters

Agriculture uses over 70% of the fresh water which is present in the world; hence environmental pollution due to agricultural wastewater is a critical problem. Agricultural wastewater origins have been classified as point sources and nonpoint sources. The point sources for pollution include agricultural field runoff, animal feeding operations including and animal feedlots, whereas nonpoint sources are hard to precisely identify and can simply be categorized as pesticides and fertilizers originating from agricultural lands (Zahoor and Mushtaq, 2023). The estimation by the United Nations Food and Agriculture Organization reveals that there is a massive demand for nitrogen fertilizers in agriculture (FAO 2019). About 30% to 50% of the applied nitrogenous fertilizer is used by the plant, and the rest is dissipated into water and air, causing damaging impacts on the environment as well as on human health (Usman et al., 2022). Literature suggests that crop farming and livestock rearing are prime factors that introduce nitrates, phosphates, antibiotics, heavy metals, and *ex situ* microbes into waterways and other ecosystems, thereby causing an imbalance in their functioning (Beattie et al., 2020). Microbes such as *Acinetobacter*, *Arcobacter*, and *Staphylococcus* that are prevalent in livestock manures are major causes of human health complications. Moreover, the massive discharge of nitrate or phosphates into water systems cause freshwater eutrophication (Beattie et al., 2020). Overuse of chemical fertilizers and pesticides contribute to an abundance of different harmful heavy metals and metalloids. Prolonged exposure to these elements is dangerous to the biosphere as well as the environment. A lack of regulation and management practices in developing countries also aggravates the problem of environmental pollution. Reports have noted that indiscriminate use of neonicotinoids as insecticides has affected the biodiversity of terrestrial insects in many countries (Zahoor and Mushtaq, 2023). Improper management of enormous amounts of livestock wastewater in Brazil, Indonesia, and Nigeria have contributed to increased water pollution (Zahoor and Mushtaq, 2023). Stringent regulations and legislation as well as proper farmer awareness can help in combating environmental pollution caused by agricultural or livestock farming wastewater effluents.

Nanomaterials (NMs)

The agricultural sector, one of humankind's oldest industries, confronts numerous challenges in today's world, owing to the burgeoning global population creating an increased demand for both the quality and quantity of food. To cope with this demand, the use of chemical fertilizers and pesticides has increased. However, these chemical compounds easily leach into the environment and bring about serious contamination. Thus, to minimize their hazardous effects, the use of nanobiofertilizers and nanopesticides has increased in recent years (Jha et al., 2023). Additionally, nanotechnology has helped in the manufacturing of nanobiosensors for precision agriculture (Jha et al., 2023; Wang et al., 2023).

Nanomaterials (NMs) have been considered a boon for agriculture because they aid in reducing nonpoint source nitrogen or sulfur contamination. However, at the same time there have also been concerns about NM pollution due to increased dissemination of these materials in the environment in the form of nanopesticides or fertilizers (Wang et al., 2023). Nanomaterials, when released into the environment, undergo aging via chemical transformations. Due to their colloidal stability and interaction ability with most particulate or dissolved organic and inorganic matter, they remain highly persistent in the ecosystem. The ionic composition of the media is another factor which governs the stability of NMs in soil or aqueous media (Bundschuh et al., 2018). Researchers have found that prolonged application of NMs has adverse effects on many biological processes within soil (Jha et al., 2023).

Microorganisms which reside within the phyllosphere and rhizosphere of a plant are beneficial for their growth and survival. They play crucial roles in nutrient uptake and cycling of nitrogen, phosphorus, and other nutrients in terrestrial niches (Wang et al., 2023). Depending on the size, dose, duration, exposure, and soil type, nanomaterials can pose a serious threat to these microbial communities. Several studies have depicted the toxic effects of metal NMs (such as ZnO, TiO, Fe_3O_4, and AgO) on microorganisms such as *Azotobacter, Bacillus, Bradyrhizobiaceae, Clostridium, Folsomia*, and *Rhodospirillaceae*, predominantly affecting nitrogen fixation activity and exogenous nitrogen release into the atmosphere (Ge et al., 2012; Chai et al., 2015; Hsueh et al., 2015; Bundschuh et al., 2018; Wang et al., 2023). Studies have shown that overuse of TiO_2- and ZnO-NPs affected the microbial assemblages which are required for nitrogen fixation, methane production, and organic matter degradation in legume plants (Ge et al., 2012). Further analysis of the ecotoxicology of nanomaterials has shown that they greatly hamper the symbiotic association of nitrogen-fixing bacteria with the root nodules. The overdose of NMs can lead to root damage, which can affect nodulation and the nitrogen fixation potential of the legumes. Nanoparticles such as Ag, TiO_2, and ZnO attach to the root cell walls, thereby damaging polysaccharide synthesis and causing premature root senescence. They can also alter the genetic regulation of the root nodulation process (Wang et al., 2023). For example, the arbuscular mycorrhiza population, enzyme activity, and biofilm formation decreased in the soybean rhizosphere community due to high doses of ZnO-NPs (Burke et al., 2015; Jha et al., 2023). Another persistent form of environmental pollution is micro- or nanoplastics. They are formed as products of the partial degradation of plastics. They mostly occur in aquatic environments and various treated waste effluents. Due to their small particle size, they easily pass through filtration units and contaminate the surrounding environment. Extended exposure to these nanoplastics is a major health risk because nanoplastics tend to easily accumulate in biological molecules (Kik et al., 2020). Hence, the use of various nanomaterials for environmental sustainability should be governed by proper risk assessment strategies (Wang et al., 2023).

FLAME RETARDANTS

Fire can cause substantial loss of property and life. Thus, to avoid such accidents, flame retardants are widely used in items such as electronics, furniture, and textiles to reduce their flammability and eventual fire spread. The most-used flame retardants are hexabromocyclododecanes (HBCDs), polybrominated diphenyl ethers (PBDEs), tetrabromobisphenol A (TBBPA), and polybrominated biphenyls (PBBs) (Okeke et al., 2022). In 2013, HBCDs were listed in Annex A of the Stockholm Convention as one of the persistent organic pollutants in the environment (Okeke et al., 2022). Brominated flame retardants (BFRs) are a major concern for environmental toxicity and bioaccumulation. Brominated flame retardants such as HBCDs and PBDEs are found in higher concentrations in the environment because they are produced in massive amounts. Due to their chemical and physical characteristics, BFRs tend to bioaccumulate easily into ecosystems. Given their widespread applications, they are mostly categorized as point-source contaminants. They are usually found to pollute places such as construction sites and waste landfills (Rani et al., 2022). Over long periods of time, these compounds travel through different matrices such as air, sediments, soil, and sewage sludge. Following this they enter the aquatic system and accumulate in aquatic organisms. Due to their high prevalence, BFRs have been detected in fishes, marine organisms, human samples, birds, and plants. The adverse effects of these compounds on biological systems are caused by damage to cellular metabolism due to the generation of reactive oxygen species within living cells. Studies have summarized that these compounds can induce carcinogenicity and cause alteration of immune systems, allergic reactions, chronic kidney disorders, and neurological problems. They can disrupt the endocrine system and can cause premature cell apoptosis in living organisms (Feiteiro et al., 2021, Tian et al., 2023).

ELECTRONIC WASTE

The Digital 2023 report states that 5.44 billion people in the world use mobile phones and 5.16 billion use the internet (Kemp, 2023). The rise in e-commerce and the worldwide availability of different e-products have contributed to the accumulation of electronic wastes. E-waste can range from simple home appliances to laptops, computers, batteries, mobile phones, and electronic gadgets. These products make modern-day living more comfortable but unfortunately add to environmental pollution problems. Reports suggest that the total amount of e-waste produced in 2019 was more than 53.6 million metric tons. About 18% of these wastes are properly collected and recycled, whereas rest 82% could not be accounted for (Rajesh et al., 2022). These electronic wastes include various deadly chemicals such as heavy metals (arsenic, barium, cadmium, chromium, etc.), flame retardants, chlorofluorocarbons, and polychlorinated biphenyls. The recycling and remediation of these wastes is a critical problem due to their complex nature. Many of the e-wastes end up being incinerated and dumped in landfills, which can cause leaching of toxic compounds into the environment. If not handled properly, they can cause serious health and environmental hazards (Hsu et al., 2021; Ilankoon et al., 2018).

PHARMACEUTICAL AND BIOMEDICAL WASTE

Pharmaceutical and biomedical devices, drugs, and equipment are used for lifesaving purposes. However, improper usage and disposal of these materials create environmental contaminants. The major pharmaceutical products that act as environmental hazards are antibiotics, antiseptics, analgesics, contraceptives, syringes, needles, hormones, antivirals, and psychotherapeutics (Patneedi and Prasadu, 2015). Additionally, the devastating COVID-19 pandemic also led to the use of masks and sanitizers that aggravated the pollution problem in nature (Sarkar et al., 2023). Biomedical devices and equipment also are major concerns as they substantially contribute to e-wastes (Ilankoon et al., 2018). Literature suggests that more than 95% of antibiotics enter the sewage system unaltered, and the persistence of these chemicals poses the threat of developing potent antibiotic-resistant microbes. Nonsteroidal drugs such as ibuprofen, diclofenac, naproxen, and aspirin are also discharged into sewage systems, and local water bodies eventually contaminate the groundwater systems (Patneedi and Prasadu, 2015). The pharmaceutical compounds can escape wastewater treatment processes and enter the food chain via drinking water. They also increase the biological and chemical oxygen demand within water bodies. Additionally, the microbial diversity of aquatic systems, soil, and the human gut are changed due to the impact of these pharmaceutical chemicals (Patneedi and Prasadu, 2015). The long-term effects of pharmaceutical and biomedical waste accumulation in nature adversely affects human health because they can cause reproductive damages, behavioral defects, accumulation in tissue, chronic diseases, and disruption in cell proliferation (Patneedi and Prasadu, 2015).

SYNTHETIC DYES

Textile industries play a crucial role in the economy worldwide; however, they are also substantial contributors to environmental pollution. Textile manufacturing processes generate large volumes of highly colored wastewater that contains a diverse array of persistent pollutants. Globally more than 7×10^7 tons of synthetic dyes are produced, and textile industries use over 10,000 tons of these synthetic dyes (Al-Tohamy et al., 2022). Textile industries discharge wastewater that contains large amounts of heavy metals (mercury, chromium, lead, cadmium, and arsenic) required for producing color pigments. Another major source of synthetic dyes is household substances such as textiles, soaps, shampoos, hair colorants, and cosmetic products. Pharmaceutical industries and biological research also release considerable amounts of dyes into the environment in the form of medicine colors, capsule cover dyes, and biological regents (Sudan black, bromophenol blue, etc.) (Hashemi and Kaykhaii, 2022). These dyes remain in the water and soil systems for a long time, where they pose serious threats to the environment. They reduce soil fertility and block the photosynthetic activity of aquatic flora, creating anoxic conditions. Moreover, these dyes also diminish the aesthetic quality of water as they increase the biological and chemical oxygen demands (Al-Tohamy et al., 2022). Synthetic dyes can be classified into two broad categories, nitro and azo groups (Hashemi and Kaykhaii, 2022). Azo dyes contain one or more azo functional groups (-N=N-) as the chromophore in the chemical structure of these groups imparts stability to these dyes. Hence, they are resilient to light or microbial degradation (Sun et al., 2022). Some of the by-products of azo dyes form harmful aromatic substances because of oxidative and reductive reactions in the environment (Ngo and Tischler, 2022). Some of the azo dyes are also used in food industries as coloring agents to enhance the visual appeal of the foods. Microbes residing within the gastrointestinal tract, liver, and stomach metabolize some of these dyes. However, the aromatic amines which are formed as by-products of azo dye degradation remain in a sulfonated state, and their toxic effects are not clearly known (Ramos-Souza et al., 2023). Dyes are hazardous to human health since they have carcinogenic, neurotoxic, and physiological effects in humans (Garg and Chopra, 2022).

PLANT–MICROBE INTERACTIONS FOR WASTE MANAGEMENT

Plants live in close association with different biotic and abiotic components of the environment. The biotic factors which play a crucial role in plant survival are the microorganisms that reside within the rhizosphere and phyllosphere. In positive plant–microbe synergy, the microbes impart various growth-promoting factors and environmental stress resilience to the plant, and in exchange they extract nutrients from it (Enerijiofi and Ikhajiagbe, 2021). The assembly of the rhizospheric microbiomes are shaped by the plant root exudates and secondary metabolites. The rhizospheric communities and their interactions with the host plants are major drivers of phytoremediation (Giri et al., 2023). The use of phytoremediation for the mitigation of emergent contaminants is an attractive waste cleanup strategy. However, the extent of bioattenuation depends on several factors such as age, concentration, complexity of the microbial community, and plant root exudates (Giri et al., 2023). Although plants and microbes are independently capable of degrading many pollutants, phytoremediation without the microbes will not be sufficient for the attenuation of a wide variety of recalcitrant organic wastes (Ubogu and Akponah, 2021). Plant–microbe interactions for waste removal can be stimulated by the application of nutrient sources or exogenous microbial formulations (Giri et al., 2023). The process of phytoremediation can be categorized as phytodegradation, phytoextraction, phytostabilization, phytovolatilization, phytofiltration, and rhizodegradation (Veerapagu et al., 2023) (Figure 11.1). A variety of the microorganisms, viz. *Actinomycetes*, *Azospirillum*, *Azotobacter*, *Bacillus*, *Clostridium*, *Cellulomonas*, *Flavobacterium*, *Klebsiella*, *Micrococcus*, and *Mycobacterium*, as well as plant

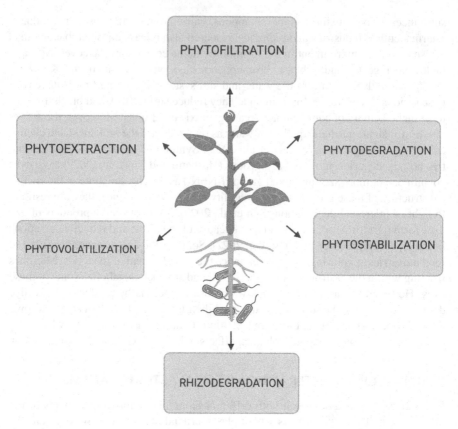

FIGURE 11.1   The different methods of phytoremediation.

species such as *Arabidopsis, Brassica, Nicotiana*, and *Triticum* have been identified as important drivers of phytoremediation (Giri et al., 2023, Veerapagu et al., 2023). Phyla such as *Actinobacteria, Bacteroidetes, Firmicutes, Nocardioides* and *Proteobacteria* are well-known degraders of aliphatic as well as aromatic petroleum hydrocarbons (Singha and Pandey, 2021). Combined use of *Medicago sativa* and an *Alphaproteobacteria*-predominated bacterial consortium showed enhanced biodegradation of petroleum hydrocarbons (Eze et al., 2022). Similarly, endophytic bacteria *Pseudomonas* when used in combination with *Lolium perenne* and *Arabidopsis thaliana* for phytoremediation of total petroleum hydrocarbons showed promising results (Iqbal et al., 2019). Plant growth–promoting microbes help the plants to grow under stressed conditions. These microbes help the plant to sustain growth in environments polluted by such substances as heavy metals, pesticides, polychlorinated biphenyls, and chemical dyes. Strains such as *Alcaligenes, Azotobacter, Bacillus, Paenibacillus, Pseudomonas, Pseudoalcaligenes*, and *Staphylococcus* showed promising results in the remediation of heavy metals (As, Co, Mn, Ni, and Zn), pesticides, and petroleum hydrocarbons (Veerapagu et al., 2023). *Enterobacter* in association with hibiscus helped in the remediation of Cd in multimetal polluted soil. The study further showed that production of phytohormones, siderophore, and enterobactin increased during the remediation process (Chen et al., 2017). Mutualistic relationships of *Bacillus* with *Vallisneria*, and PGPR (*Rhizobacterium, Pseudomonas, Luteibacter, Variovorax*) with *Lathyrus* helped in significant absorption of As, Pb, and Cd respectively (Abdelkrim et al., 2020; Irshad et al., 2020). Plant–microbe association is an efficient remediation process in the mitigation of persistent organic pollutants such as polychlorinated biphenyls, triphenyls, trichloroethylene, and pesticides (Veerapagu et al., 2023). During the process of bioremediation, different compounds are produced by the plants and the microbes. Such compounds range from phytohormones to microbial metabolites. Indole acetic acid, ACC deaminase, siderophores, organic acids, biosurfactants, and exopolysaccharides are some of the compounds which facilitate plant growth promotion as well as phytoremediation of hazardous pollutants. Phytoremediation of pharmaceutical wastes, synthetic dyes, and endocrine-disrupting chemicals (flame retardants) have also been studied extensively. Remediation of sulfamethoxazole, carbamazepine, ibuprofen, and their derivatives were successfully achieved by different plant–microbe associations (Giri et al., 2023). The removal of different pharmaceutical and personal care products (PPCPs) have gained popularity in the form of constructed wetlands and hydroponic systems that are used for treating PPCP-contaminated wastewaters (Nguyen et al., 2019). It is interesting to note that the mean degradation rates of PPCPs such as sulfadimethoxine, caffeine, and salicylic acids were more than 75%, however, very low efficiency (<25%) was observed for triclocarban, sulfapyridine, sulfamethoxazole, sotalol, gemfibrozil, sulfapyrine, and triclosan in constructed wetlands (Nguyen et al., 2019). Authors have demonstrated the applicability of *Phragmites australis, Typha angustifolia, Scirpus validus, Iris pseudacorus, Berula erecta, Juncus effusus*, and *Heliconia* sp. in efficient phytoremediation of diclonac, ibuprofen, naproxen, bisphenol, galaxolide, carbamazepine, and caffeine (Matamoros et al., 2007; Ávila et al., 2010; Hijosa-Valsero et al., 2010; Zhang et al., 2013; Toro-Vélez et al., 2016; Zhang et al., 2017). Uptake and degradation of paracetamol by *Spinacia* and its rhizospheric communities in a hydroponic setup was also reported (Badar et al., 2022). Additionally, another study demonstrated the use of a hydroponic plant–microbe system for removal of ibuprofen and iohexol. The plants which were used for the study included *Typha, Phragmites, Iris,* and *Juncus* along with the microbiome for degradation of the pharmaceutical wastes (Zhang et al., 2016).

Plant–microbe interactions have also gained immense popularity in the remediation of synthetic chemicals which are designated as endocrine disrupting compounds, mostly found in flame retardants, furniture, and electronic equipment. Many aquatic plants (*Spirodela, Lemna,* or *Phragmites*) and their rhizosphere microbes (*Acinetobacter, Stenotrophomonas,* and *Sphingobium*) are capable of degrading endocrine-disrupting chemicals (bisphenol A, B, C, 4-nonylphenol). These aquatic plants have a strong ability to release phenolic root exudates, which helps in the biotransformation of these toxic chemicals (Giri et al., 2023). Synthetic dyes are another hazardous environmental waste which can be biodegraded using plants and microbes. A beneficial synergistic interaction between *Petunia* and its rhizospheric *Bacillus* could degrade Navy Blue RX dyes up to over 90% within 36 hours (Watharkar et al., 2013). Another study reported removal of 53% and 92% of methyl red, a class of azo dye, using a synergistic plant-microbe-mediated aerobic-anaerobic biodegradation regime using *Vetiveria* and a bacterial consortium (Jayapal et al., 2018). For sustainable environmental management, waste to wealth recovery has evolved as a promising field. Hence the concept of waste "a material which has no use" has changed over years, and now it is seen as a resource for the recovery of valuable products (Ezejiofor et al., 2014). The use of agrowaste as a valuable resource is an interesting avenue which has long been explored. A case study shows the development of Nutrigel from soya waste, which is also known as okara. Yet another study also demonstrated the recovery of usable phosphate from wastewater using the synergy between struvite and microbial consortia (Zhu et al., 2022). Published literature have indicated the recovery of valuable metals from fly ash from municipal waste incineration processes (Bakalár et al., 2021). All these processes of waste-to-resource recovery support a circular bioeconomy for a sustainable future.

CARBON FARMING AS A STRATEGY FOR SUSTAINABLE ENVIRONMENTAL MANAGEMENT

Industrialization and several anthropogenic activities have increased atmospheric carbon dioxide emission levels as well as overall carbon accumulation in the environment. At present, the global carbon emission has increased about 400 Gt, of which CO_2 emissions in the atmosphere account for 400 ppm (Giri et al., 2023). The carbon pools in nature can be classified as: (a) atmosphere (875 Gt), (b) ocean (38000 Gt), (c) soil (2500 Gt), and (d) vegetation (450 Gt) (Giri et al., 2023). Soil organic matter is an important reserve for terrestrial carbon (Bhattacharyya et al., 2022). The soil organic matter is primarily composed of plant and microbial organic inputs. Hence, the plant–microbe interactions within soil systems are important parameters for defining terrestrial carbon neutrality and sustainable environmental management. A rise in CO_2 levels in the atmosphere interferes with global climate change patterns; therefore, managing the carbon losses becomes imperative (Tang et al., 2016). Agricultural activities, croplands, and peatlands are the hot spots of carbon emission as well as greenhouse gas emissions (GHG) into the ecosystem (Paul et al., 2023). To mitigate carbon emission and global climate change, carbon farming has evolved as an important alternative. The strategies that deal with carbon capture and minimization of GHG emissions are collectively termed *carbon farming* (Tang et al., 2016). It aims at reducing carbon emissions and enhancing carbon storage in the soil. Many countries have passed legislation for minimizing carbon losses. For example, the European Union adopted a new EU Soil Strategy 2030, which aims at designing strategies for increasing carbon storage in soil organic matter in order to achieve climate neutrality in land surfaces (Paul et al., 2023).

PLANT–MICROBE INTERACTIONS GOVERNING CARBON TRANSACTIONS IN NATURE

The soil carbon balances are primarily governed by metabolic processes of the plant and microbial communities. The overall carbon balance in soil can be represented by the following equation (Jansson et al., 2021).

$$C_i - C_o = C_s$$

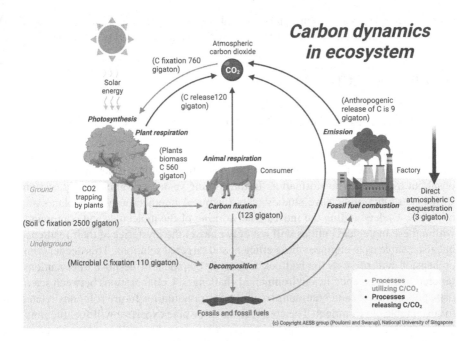

FIGURE 11.2 Terrestrial ecosystem carbon dynamics.

where C_i is the carbon input, C_o is the carbon output, and C_s is the net storage. The carbon inputs come from photosynthesis, microbial metabolisms, root exudates, different soil amendments, plant litter, and microbial masses, whereas the carbon escape occurs through plant–microbe respirations and the emission of volatile organic compounds (Jansson et al., 2021). On the other hand, cellular biomass, metabolites, and exudates act as reserves for carbon (Babu et al., 2023). An overall carbon cycle and sequestration process in nature is described in Figure 11.2. Increasing atmospheric CO_2 levels to speed up plant photosynthesis thereby fixes more carbon in plant systems. The allocation of carbon within the plant body is an important physiological process. According to functional equilibrium principles, plants store optimum levels of carbon in forms of photosynthate among its different organs (leaf, shoot, and root). Interestingly, the allocation process within a plant body also varies with the environmental conditions (Jansson et al., 2021). For example, in light-deficient conditions, shoots accumulate more carbon substrates than roots, whereas in water or nutrient-limited condition roots will store more photosynthate. Reports suggest that the plant accumulates about 30% of newly synthesized photosynthate in belowground biomass. New root growth utilizes almost 50% of the allocated carbon, and nearly 30% is released into the atmosphere as plant volatile organic carbons (Jansson et al., 2021). Plants along with their phytobiome (both phyllosphere microbes and rhizosphere microbes) play a remarkable role in carbon cycling. An estimated 20% to 30% of the plant-allocated carbons are released in the environment in the forms of root exudates and rhizodeposits, which act as carbon sources for the microbes. The arbuscular fungi, endo- and epiphytic bacteria feed on these released carbon substrates and sequester them in forms of cell biomass. Hence, the soil organic matter acts as the largest pool of environmentally fixed carbons (Giri et al. 2023).

FUTURE PERSPECTIVES

One promising avenue for a greener tomorrow lies in waste mitigation strategies that capitalize on the remarkable capabilities of plants and microbes. By tapping into the intricate biochemical interactions within these organisms, we can unlock innovative approaches to reduce, reuse, and recycle waste materials. The integration of plant-based systems and microbial interventions holds the promise of not just waste disposal but true waste transformation. In the coming years, we anticipate a paradigm shift toward comprehensive studies of biochemical interactions within plants and microbes. Understanding the intricate networking of the molecules and compounds within these biological entities will not only unravel the mysteries of their resilience but also enable us to engineer more efficient and targeted solutions. This deeper comprehension will serve as the bedrock for the development of bespoke interventions, tailored to address specific environmental challenges. Collaborations between scientists, policymakers, and communities will be paramount for future solutions related to sustainable environment. The convergence of diverse expertise will fuel the innovation needed to turn these futuristic visions into tangible realities. Governments, industries, and individuals alike must join hands to implement and support initiatives that promote the sustainable cohabitation of humanity and the environment.

REFERENCES

Abdelkrim S, Jebara SH, Saadani O, Abid G, Taamalli W, Zemni H, Mannai K, Louati F, Jebara M. In situ effects of *Lathyrus sativus*-PGPR to remediate and restore quality and fertility of Pb and Cd polluted soils. Ecotoxicology and Environmental Safety. 2020 Apr 1;192:110260.

ASTDR. An Overview of Total Petroleum Hydrocarbons. In: Toxicological Profile for Total Petroleum Hydrocarbons (pp. 9–16). 1999. Atlanta, Georgia: U.S. Department of Health and Human Services, Public Health Service.

Abdulrazaq Y, Abdulsalam A, Rotimi AL, Abdulbasit AA, Clifford O, Abdulsalam OA, Racheal ON, Joy AA, Victor FO, Johannes ZM, Bilal M. Classification, Potential Routes and Risk of Emerging Pollutants/Contaminant. In: Emerging Contaminants. 2020 Dec 11. London, UK: IntechOpen.

Abdurrahman MI, Chaki S, Saini G. Stubble burning: Effects on health & environment, regulations and management practices. Environmental Advances. 2020 Dec 1;2:100011.

Al-Tohamy R, Ali SS, Li F, Okasha KM, Mahmoud YA, Elsamahy T, Jiao H, Fu Y, Sun J. A critical review on the treatment of dye-containing wastewater: Ecotoxicological and health concerns of textile dyes and possible remediation approaches for environmental safety. Ecotoxicology and Environmental Safety. 2022 Feb 1;231:113160.

Augusta AC, Bertha EE, Eromosele AS. Plant-microbe Interaction: Prospects and Applications in Sustainable Environmental Management. In: Plant Hormones: Recent Advances. New Perspectives and Applications. 2022 May 25:43. London, UK: IntechOpen.

Ávila C, Pedescoll A, Matamoros V, Bayona JM, García J. Capacity of a horizontal subsurface flow constructed wetland system for the removal of emerging pollutants: An injection experiment. Chemosphere. 2010 Nov 1;81(9):1137–1142.

Babar ZB, Haider R, Sattar H. Conventional and Emerging Practices in Hazardous Waste Management. In: Hazardous Waste Management: Advances in Chemical and Industrial Waste Treatment and Technologies (pp. 57–93). 2022. Cham: Springer International Publishing.

Babu S, Singh R, Avasthe R, Kumar S, Rathore SS, Singh VK, Ansari MA, Valente D, Petrosillo I. Soil carbon dynamics under organic farming: Impact of tillage and cropping diversity. Ecological Indicators. 2023 Mar 1;147:109940.

Badar Z, Shanableh A, El-Keblawy A, Mosa KA, Semerjian L, Mutery AA, Hussain MI, Bhattacharjee S, Tsombou FM, Ayyaril SS, Ahmady IM. Assessment of uptake, accumulation and degradation of paracetamol in spinach (*Spinacia oleracea* L.) under controlled laboratory conditions. Plants. 2022 Jun 21;11(13):1626.

Bakalár T, Pavolová H, Hajduová Z, Lacko R, Kyšeľa K. Metal recovery from municipal solid waste incineration fly ash as a tool of circular economy. Journal of Cleaner Production. 2021 Jun 15;302:126977.

Beattie RE, Bandla A, Swarup S, Hristova KR. Freshwater sediment microbial communities are not resilient to disturbance from agricultural land runoff. Frontiers in Microbiology. 2020 Oct 15;11:539921.

Bhattacharyya SS, Ros GH, Furtak K, Iqbal HM, Parra-Saldívar R. Soil carbon sequestration–An interplay between soil microbial community and soil organic matter dynamics. Science of The Total Environment. 2022 Apr 1;815:152928.

Bundschuh M, Filser J, Lüderwald S, McKee MS, Metreveli G, Schaumann GE, Schulz R, Wagner S. Nanoparticles in the environment: Where do we come from, where do we go to?. Environmental Sciences Europe. 2018 Dec;30(1):1–7.

Burke DJ, Pietrasiak N, Situ SF, Abenojar EC, Porche M, Kraj P, Lakliang Y, Samia AC. Iron oxide and titanium dioxide nanoparticle effects on plant performance and root associated microbes. International Journal of Molecular Sciences. 2015 Oct 5;16(10): 23630–23650.

Chai H, Yao J, Sun J, Zhang C, Liu W, Zhu M, Ceccanti B. The effect of metal oxide nanoparticles on functional bacteria and metabolic profiles in agricultural soil. Bulletin of Environmental Contamination and Toxicology. 2015 Apr;94:490–495.

Chen Y, Yang W, Chao Y, Wang S, Tang YT, Qiu RL. Metal-tolerant *Enterobacter* sp. strain EG16 enhanced phytoremediation using *Hibiscus cannabinus* via siderophore-mediated plant growth promotion under metal contamination. Plant and Soil. 2017 Apr;413:203–216.

Das N, Chandran P. Microbial degradation of petroleum hydrocarbon contaminants: An overview. Biotechnology Research International. 2011;2011:941810.

Debnath N, Nath AJ, Majumdar K, Das AK. Carbon Farming with Bamboos in India: Opportunities and Challenges. International Journal of Ecology and Environmental Sciences. 2022 Jul 31;48(5):521–531.

Energy Policy Institute at the University of Chicago (EPIC). 2023. https://epic.uchicago.in/aqli-tool-shows-pollutions-impact-in-india/ (Accessed on 3rd September 2023).

Enerijiofi KE, Ikhajiagbe B. Plant–Microbe Interaction in Attenuation of Toxic Wastes in Ecosystem. In: Rhizobiont in Bioremediation of Hazardous Waste (pp. 291–315). 2021. Singapore: Springer.

Eze MO, Thiel V, Hose GC, George SC, Daniel R. Enhancing rhizoremediation of petroleum hydrocarbons through bioaugmentation with a plant growth-promoting bacterial consortium. Chemosphere. 2022 Feb 1;289:133143.

Ezejiofor TI, Enebaku UE, Ogueke C. Waste to wealth-value recovery from agro-food processing wastes using biotechnology: A review. British Biotechnology Journal. 2014 Apr 1;4(4):418.

FAO. World fertilizer trends and outlook to 2022, 2019. http://www.fao.org/3/ca6746en/ca6746en.pdf.

Feiteiro J, Mariana M, Cairrão E. Health toxicity effects of brominated flame retardants: From environmental to human exposure. Environmental Pollution. 2021 Sep 15;285:117475.

Garg A, Chopra L. Dye waste: A significant environmental hazard. Materials Today: Proceedings. 2022 Jan 1;48:1310–1315.

Gavrilescu M, Demnerová K, Aamand J, Agathos S, Fava F. Emerging pollutants in the environment: Present and future challenges in biomonitoring, ecological risks and bioremediation. New Biotechnology. 2015 Jan 25;32(1):147–156.

Ge Y, Schimel JP, Holden PA. Identification of soil bacteria susceptible to TiO_2 and ZnO nanoparticles. Applied and Environmental Microbiology. 2012 Sep 15;78(18):6749–6758.

Ghosh, P., Sharma, S., Khanna, I., Datta, A., Suresh, R., Kundu, S., Goel, A., Datt, D., 2019. Scoping study for South Asia air pollution. New Delhi: The Energy and Resources Institute (TERI). https://www.gov.uk/dfid-research-outputs/scoping-study-for-south-asia-air-pollution.

Giri A, Pant D, Srivastava VC, Kumar M, Kumar A, Goswami M. Plant-microbe assisted emerging contaminants (ECs) removal and carbon cycling. Bioresource Technology. 2023 Jun 26:129395.

Hashemi SH, Kaykhaii M. Azo Dyes: Sources, Occurrence, Toxicity, Sampling, Analysis, and their Removal Methods. In: Emerging Freshwater Pollutants (pp. 267–287). 2022. Elsevier.

Hijosa-Valsero M, Matamoros V, Sidrach-Cardona R, Martín-Villacorta J, Bécares E, Bayona JM. Comprehensive assessment of the design configuration of constructed wetlands for the removal of pharmaceuticals and personal care products from urban wastewaters. Water Research. 2010 Jun 1;44(12):3669–3678.

Hsu E, Durning CJ, West AC, Park AH. Enhanced extraction of copper from electronic waste via induced morphological changes using supercritical CO_2. Resources, Conservation and Recycling. 2021 May 1;168:105296.

Hsueh YH, Ke WJ, Hsieh CT, Lin KS, Tzou DY, Chiang CL. ZnO nanoparticles affect Bacillus subtilis cell growth and biofilm formation. PloS One. 2015 Jun 3;10(6):e0128457.

Ilankoon IM, Ghorbani Y, Chong MN, Herath G, Moyo T, Petersen J. E-waste in the international context–A review of trade flows, regulations, hazards, waste management strategies and technologies for value recovery. Waste Management. 2018 Dec 1;82:258–275.

Iqbal A, Mukherjee M, Rashid J, Khan SA, Ali MA, Arshad M. Development of plant-microbe phytoremediation system for petroleum hydrocarbon degradation: An insight from alkb gene expression and phytotoxicity analysis. Science of the Total Environment. 2019 Jun 25;671:696–704.

Irshad S, Xie Z, Wang J, Nawaz A, Luo Y, Wang Y, Mehmood S. Indigenous strain Bacillus XZM assisted phytoremediation and detoxification of arsenic in *Vallisneria denseserrulata*. Journal of Hazardous Materials. 2020 Jan 5;381:120903.

Jain N, Bhatia A, Pathak H. Emission of air pollutants from crop residue burning in India. Aerosol and Air Quality Research. 2014 Jan;14(1):422–430.

Jansson C, Faiola C, Wingler A, Zhu XG, Kravchenko A, De Graaff MA, Ogden AJ, Handakumbura PP, Werner C, Beckles DM. Crops for carbon farming. Frontiers in Plant Science. 2021 Jun 4;12:636709.

Jayapal M, Jagadeesan H, Shanmugam M, Murugesan S. Sequential anaerobic-aerobic treatment using plant microbe integrated system for degradation of azo dyes and their aromatic amines by-products. Journal of Hazardous Materials. 2018 Jul 15;354:231–243.

Jha A, Pathania D, Damathia B, Raizada P, Rustagi S, Singh P, Rani GM, Chaudhary V. Panorama of biogenic nano-fertilizers: A road to sustainable agriculture. Environmental Research. 2023 Jun 19:116456.

Kemp, S., 2023. Digital 2023: Global Overview Report. https://datareportal.com/reports/digital-2023-global-overview-report.

Kik K, Bukowska B, Sicińska P. Polystyrene nanoparticles: Sources, occurrence in the environment, distribution in tissues, accumulation and toxicity to various organisms. Environmental Pollution. 2020 Jul 1;262:114297.

Koul B, Yakoob M, Shah MP. Agricultural waste management strategies for environmental sustainability. Environmental Research. 2022 Apr 15;206:112285.

Kumar V, Agrawal S, Bhat SA, Américo-Pinheiro JH, Shahi SK, Kumar S. Environmental impact, health hazards, and plant-microbes synergism in remediation of emerging contaminants. Cleaner Chemical Engineering. 2022 Jun 1;2:100030.

Lofrano G, Libralato G, Meric S, Vaiano V, Sacco O, Venditto V, Guida M, Carotenuto M. Occurrence and Potential Risks of Emerging Contaminants in Water. In: Visible Light Active Structured Photocatalysts for the Removal of Emerging Contaminants (pp. 1–25). 2020. Elsevier.

Matamoros V, Arias C, Brix H, Bayona JM. Removal of pharmaceuticals and personal care products (PPCPs) from urban wastewater in a pilot vertical flow constructed wetland and a sand filter. Environmental Science & Technology. 2007 Dec 1;41(23):8171–8177.

Mattila TJ, Hagelberg E, Söderlund S, Joona J. How farmers approach soil carbon sequestration? Lessons learned from 105 carbon-farming plans. Soil and Tillage Research. 2022 Jan 1;215:105204.

Mitra S, Chakraborty AJ, Tareq AM, Emran TB, Nainu F, Khusro A, Idris AM, Khandaker MU, Osman H, Alhumaydhi FA, Simal-Gandara J. Impact of heavy metals on the environment and human health: Novel therapeutic insights to counter the toxicity. Journal of King Saud University-Science. 2022 Apr 1;34(3):101865.

Nasr M. Environmental Perspectives of Plant-Microbe Nexus for Soil and Water Remediation. In: Microbiome in Plant Health and Disease: Challenges and Opportunities (pp. 403–419). 2019. Singapore: Springer.

Ngo AC, Tischler D. Microbial degradation of azo dyes: Approaches and prospects for a hazard-free conversion by microorganisms. International Journal of Environmental Research and Public Health. 2022 Apr 14;19(8):4740.

Nguyen PM, Afzal M, Ullah I, Shahid N, Baqar M, Arslan M. Removal of pharmaceuticals and personal care products using constructed wetlands: Effective plant-bacteria synergism may enhance degradation efficiency. Environmental Science and Pollution Research. 2019 Jul 1;26:21109–21126.

Okeke ES, Huang B, Mao G, Chen Y, Zhengjia Z, Qian X, Wu X, Feng W. Review of the environmental occurrence, analytical techniques, degradation and toxicity of TBBPA and its derivatives. Environmental Research. 2022 Apr 15;206:112594.

Ooi QE, Nguyen CT, Laloo A, Bandla A, Swarup S. Urban Soil Microbiome Functions and their Linkages with Ecosystem Services. In: Soils in Urban Ecosystem (pp. 47–63). 2022. Singapore: Springer.

Pant G, Garlapati D, Agrawal U, Prasuna RG, Mathimani T, Pugazhendhi A. Biological approaches practised using genetically engineered microbes for a sustainable environment: A review. Journal of Hazardous Materials. 2021 Mar 5;405:124631.

Parra-Arroyo L, González-González RB, Castillo-Zacarías C, Martínez EM, Sosa-Hernández JE, Bilal M, Iqbal HM, Barceló D, Parra-Saldívar R. Highly hazardous pesticides and related pollutants: Toxicological, regulatory, and analytical aspects. Science of The Total Environment. 2022 Feb 10;807:151879.

Patneedi CB, Prasadu KD. Impact of pharmaceutical wastes on human life and environment. Rasayan Journal of Chemistry. 2015;8(1):67–70.

Paul C, Bartkowski B, Dönmez C, Don A, Mayer S, Steffens M, Weigl S, Wiesmeier M, Wolf A, Helming K. Carbon farming: Are soil carbon certificates a suitable tool for climate change mitigation?. Journal of Environmental Management. 2023 Mar 15;330:117142.

Rajesh R, Kanakadhurga D, Prabaharan N. Electronic waste: A critical assessment on the unimaginable growing pollutant, legislations and environmental impacts. Environmental Challenges. 2022 Apr 1;7:100507.

Ramos-Souza C, Bandoni DH, Bragotto AP, De Rosso VV. Risk assessment of azo dyes as food additives: Revision and discussion of data gaps toward their improvement. Comprehensive Reviews in Food Science and Food Safety. 2023 Jan;22(1):380–407.

Rani M, Sillanpää M, Shanker U. An updated review on environmental occurrence, scientific assessment and removal of brominated flame retardants by engineered nanomaterials. Journal of Environmental Management. 2022 Nov 1;321:115998.

Sarkar P, Banerjee S, Saha SA, Mitra P, Sarkar S. Genome surveillance of SARS-CoV-2 variants and their role in pathogenesis focusing on second wave of COVID-19 in India. Scientific Reports. 2023 Mar 22;13(1):4692.

Sarkar P, Roy A, Pal S, Mohapatra B, Kazy SK, Maiti MK, Sar P. Enrichment and characterization of hydrocarbon-degrading bacteria from petroleum refinery waste as potent bioaugmentation agent for in situ bioremediation. Bioresource Technology. 2017 Oct 1;242:15–27.

Sharma R, Kumar R, Sharma DK, Son LH, Priyadarshini I, Pham BT, Tien Bui D, Rai S. Inferring air pollution from air quality index by different geographical areas: Case study in India. Air Quality, Atmosphere & Health. 2019 Nov;12:1347–1357.

Sharma RK, Agrawal M. Biological effects of heavy metals: An overview. Journal of Environmental Biology. 2005 Jun 1;26(2):301–313.

Siddiqi H, Mishra A, Kumari U, Maiti P, Meikap BC. Utilizing agricultural residue for the cleaner biofuel production and simultaneous air pollution mitigation due to stubble burning: A net energy balance and total emission assessment. ACS Sustainable Chemistry & Engineering. 2021 Nov 12;9(47):15963–15972.

Singh H, Pant G. Phytoremediation: Low input-based ecological approach for sustainable environment. Applied Water Science. 2023 Mar;13(3):85.

Singh J, Singhal N, Singhal S, Sharma M, Agarwal S, Arora S. Environmental implications of rice and wheat stubble burning in north-western states of India. In: Advances in Health and Environment Safety: Select Proceedings of HSFEA 2016 (pp. 47–55). 2018. Singapore: Springer.

Singha LP, Pandey P. Rhizosphere assisted bioengineering approaches for the mitigation of petroleum hydrocarbons contamination in soil. Critical Reviews in Biotechnology. 2021 Jul 4;41(5):749–66.

Sun L, Mo Y, Zhang L. A mini review on bio-electrochemical systems for the treatment of azo dye wastewater: State-of-the-art and future prospects. Chemosphere. 2022 May 1; 294:133801.

Tang K, Kragt ME, Hailu A, Ma C. Carbon farming economics: What have we learned?. Journal of Environmental Management. 2016 May 1;172:49–57.

Thakare M, Sarma H, Datar S, Roy A, Pawar P, Gupta K, Pandit S, Prasad R. Understanding the holistic approach to plant-microbe remediation technologies for removing heavy metals and radionuclides from soil. Current Research in Biotechnology. 2021 Jan 1;3:84–98.

Tian YX, Chen HY, Ma J, Liu QY, Qu YJ, Zhao WH. A critical review on sources and environmental behavior of organophosphorus flame retardants in the soil: Current knowledge and future perspectives. Journal of Hazardous Materials. 2023 Jun 15;452:131161.

Toro-Vélez AF, Madera-Parra CA, Peña-Varón MR, Lee WY, Bezares-Cruz JC, Walker WS, Cárdenas-Henao H, Quesada-Calderón S, García-Hernández H, Lens PN. BPA and NP removal from municipal wastewater by tropical horizontal subsurface constructed wetlands. Science of the Total Environment. 2016 Jan 15;542:93–101.

Ubogu M, Akponah E. Plant–Microbe Interactions in Attenuation of Toxic Waste in Ecosystem. In: Rhizobiont in Bioremediation of Hazardous Waste (pp. 131–150). 2021. Singapore: Springer.

Usman M, Sanaullah M, Ullah A, Li S, Farooq M. Nitrogen pollution originating from wastewater and agriculture: Advances in treatment and management. Reviews of Environmental Contamination and Toxicology. 2022 Dec;260(1):9.

Veerapagu M, Jeya KR, Sankaranarayanan A. Role of Plant Growth-Promoting Microorganisms in Phytoremediation Efficiency. In: Plant-Microbe Interaction-Recent Advances in Molecular and Biochemical Approaches (pp. 45–61). 2023. Elsevier.

Wang Q, Gao L, Li Y, Shakoor N, Sun Y, Jiang Y, Zhu G, Wang F, Shen Y, Rui Y, Zhang P. Nano-agriculture and nitrogen cycling: Opportunities and challenges for sustainable farming. Journal of Cleaner Production. 2023 Aug 17:138489.

Watharkar AD, Rane NR, Patil SM, Khandare RV, Jadhav JP. Enhanced phytotransformation of Navy Blue RX dye by *Petunia grandiflora* Juss. with augmentation of rhizospheric *Bacillus pumilus* strain PgJ and subsequent toxicity analysis. Bioresource Technology. 2013 Aug 1;142:246–54.

Xiang L, Harindintwali JD, Wang F, Redmile-Gordon M, Chang SX, Fu Y, He C, Muhoza B, Brahushi F, Bolan N, Jiang X. Integrating biochar, bacteria, and plants for sustainable remediation of soils contaminated with organic pollutants. Environmental Science & Technology. 2022 Oct 27;56(23):16546–66.

Zahoor I, Mushtaq A. Water Pollution from Agricultural Activities: A Critical Global Review. International Journal of Chemical and Biochemical Science. 2023;23:164–176.

Zhang DQ, Hua T, Gersberg RM, Zhu J, Ng WJ, Tan SK. Carbamazepine and naproxen: Fate in wetland mesocosms planted with *Scirpus validus*. Chemosphere. 2013 Mar 1; 91(1):14–21.

Zhang Y, Lv T, Carvalho PN, Arias CA, Chen Z, Brix H. Removal of the pharmaceuticals ibuprofen and iohexol by four wetland plant species in hydroponic culture: Plant uptake and microbial degradation. Environmental Science and Pollution Research. 2016 Feb;23:2890–2898.

Zhang Y, Lv T, Carvalho PN, Zhang L, Arias CA, Chen Z, Brix H. Ibuprofen and iohexol removal in saturated constructed wetland mesocosms. Ecological Engineering. 2017 Jan 1;98:394–402.

Zhu J, Tan WK, Song X, Zhang Z, Gao Z, Wen Y, Ong CN, Swarup S, Li J. Synthesis and characterization of okara-poly (acrylic acid) superabsorbent hydrogels for enhancing vegetable growth through improving water holding and retention properties of soils. ACS Food Science & Technology. 2022 Jul 25;3(4):553–61.

12 Phytoremediation of Environmental Pollutants Using Ornamental Plants in Terrestrial and Aquatic Environments

Abanti Pradhan, Swayam Prakash Nanda,
Bibhu Prasad Panda, and Aditya Kishore Dash

INTRODUCTION

The growing world population needs an adequate amount of produce and facilities to maintain their livelihood in the biosphere (Shang et al., 2019). Due to this, the pressures on natural resources and unhealthy industrial practices lead to catastrophic environmental events for living organisms (Cazalis et al., 2018). Aggressive economic growth, unsustainable human development, and compromises on ecological sustainability result in environmental pollution and biosphere contamination (Vita et al., 2019). Environmental contaminations have been increasing significantly in the past decade. Due to unhealthy industrial practices, huge amounts of heavy metals (HMs) and other pollutants illegally enter the environment. These heavy metals contaminate the natural environment due to both human-made and natural activities. The continuous inflow of pollutants poses a serious threat to the sustainability of the earth's three spheres—hydrosphere, lithosphere, and atmosphere.

Pollutants enter the food chain through the lower trophic levels with biomagnification concentrating them at higher levels (Arshad et al., 2017; Majed et al., 2016). Some heavy metals are very important for plant growth as micronutrients and are required in minute quantities, but some have harmful and toxic effects on plant, animal, and human health even at lower concentrations (Alirzayeva et al., 2017). The xenobiotic substances create further problems in the environment due to their toxic, persistent, and nonbiodegradable nature (El-Naggar et al., 2018). Heavy metals like chromium (Cr), zinc (Zn), nickel (Ni), and iron (Fe) are essential nutrients for plant metabolic activities leading to growth and development, but their presence above threshold limits can have many adverse effects on the plant development process (Lajayer et al., 2017). The higher concentrations can affect enzymatic activity, photosynthesis, membrane transport of water and nutrients, and genetic structure and function (Drazkiewicz and Baszyński, 2010; Goswami and Das, 2016). Apart from these essential heavy metal nutrients, there are

 DOI: 10.1201/9781003442295-12

some nonessential heavy metals like lead (Pb), arsenic (As), cadmium (Cd), and mercury (Hg) in the environment that have no biological functions (at least, none that are currently known). They show high toxicity at low concentrations, and they alter the biochemical and physiological processes of plants (Alloway, 2012; Lajayer et al., 2017).

Various conventional methods have been employed to remove heavy metals from contaminated sites, such as adsorption, oxidation, precipitation and reduction (Venkatachalam et al., 2017), electrochemical methods, ion exchange, coagulation and flocculation (Fu and Wang, 2011), leaching/acid extraction, and soil washing. As an alternative to these methods, choosing plants that can absorb the HMs with a low risk of biomagnification in the food chain is highly preferable to scientists. Further, all the chemical processes are more expensive and consume more energy, whereas plant-based remediation using ornamental plants is environmentally friendly and cost effective (Feng et al., 2017; Mahar et al., 2016). The ornamental plants have high importance in Indian culture and traditions compared to other countries in the world. Every family uses flowers for offerings to the gods on many occasions. Ornamental plants are also planted by the government to enhance the aesthetic beauty of cities, to beautify landscapes, and for the development of mining area wasteland (Zeng et al., 2018). The use of ornamental plants for bioremediation is highly economical and beneficial, since these species produce large amounts of aboveground biomass, have great efficacy in removing contaminants, and have rapid life cycles. The ornamental plants can also be used for phytoextraction and phytostabilization of toxic metals in mining areas to check the outflow of heavy metals during the rainy season. Plants accumulate the metals and hold them in their structures in less toxic form (Zeng et al., 2018). The plants can be used as natural filters in contaminated areas with landscape values. They can remove contaminants from soil, water, and air (Chen et al., 2009). In highly populated and poor countries where the government cannot afford to apply heavy metal cleanup processes, the local people may adopt this technique for reducing heavy metal contamination from the environment using ornamental plants, and the generated by-product can enhance their economic status (Lajayer et al., 2019).

FEATURES FOR SPECIES SELECTION

Ornamental plants of various shapes, sizes, and colors are being used for phytoremediation by different workers. The rate of growth, biomass production, growth rate, canopy coverage, and reproductive growth are important parameters for selection of ornamental plant species. Similarly, market analysis, feasibility, and acclimatization to the environment are equally important for choosing species for bioremediation using ornamental plants (Lajayer et al., 2019). Physiological and ecological characteristics like pollutant degradation and growth and reproduction of microorganisms in the rhizosphere are also very important selection criteria. The leaf area index, plant height, and stem diameter are equally responsible for accumulation of contaminants and tolerance as these parameters influence the photosynthesis rate and pollutant detoxification (Liu et al., 2018).

Species are also selected for their biotic tolerance factors like disease resistance and tolerance for abiotic factors like temperature, drought, salinity, and other contaminants, as these are the indicators of cellular defense mechanisms. Plants generally secrete reactive oxygen species (ROS) under stressful conditions, which leads to oxidative damage to cells (Liu et al., 2018). During these situations, plants activate their antioxidant defense mechanisms and produce nonenzymatic antioxidants like glutathiones, phytoquelatins, ascorbate, tocopherols, and others phenolic compounds (Gomes et al., 2016). The peroxidases are the scavenging enzymes to protect plants from internal cellular damage. These properties of plants can be used as biomarkers for contamination tolerance and for evaluating their phytoremediation efficiency (Gomes et al., 2017).

CLASSIFICATION OF ORNAMENTAL PLANTS

Various classes of ornamental plants are generally used for phytoremediation. Some plants are cultivated for their flowers and some are cultivated for their leaves, fruits, stem, and fragrance. They can be broadly classified as terrestrial, aquatic, herbaceous, woody, and higher and lower plants. The classification of ornamental plants can be made based on their suitability to different types of contaminants.

a. **Heavy metal accumulators**
   i. *Hyperaccumulators:* These plants have the inherent ability to bioaccumulate high concentrations of heavy metals in their body tissues. Examples include *Thlaspi caerulescens*, which is known for accumulating zinc, nickel, and cadmium. *Arabidopsis halleri* can tolerate high cadmium and zinc (Liang et al., 2009; Liu et al., 2018b).
   ii. *Moderate accumulators:* These plants accumulate moderate levels of heavy metals and are used for remediating areas with lower levels of contamination. Examples include sunflowers (*Helianthus annuus*), which are used for removing arsenic, lead, and other heavy metals. Indian mustard (*Brassica juncea*) is used for removing selenium and sulfur.
b. **Organic pollutant degraders**
   Poplar (*Populus* spp.) trees are known for their ability to break down organic pollutants like trichloroethylene and benzene in contaminated groundwater.
c. **Nutrient uptake plants**
   The water hyacinth (*Eichhornia crassipes*) is popularly used for absorbing excesses of inorganic nutrients like phosphorus and nitrogen from water bodies, preventing eutrophication.
d. **Salt-tolerant plants**
   Saltbush (*Atriplex* spp.) can tolerate high salt concentrations and remediate saline soil.
e. **Air pollutant absorbers**
   The spider plant (*Cholorophytum comosum*) can help remove indoor air pollutants like formaldehyde and xylene.
f. **Aesthetic ornamental plants with phytoremediation potential**
   Lavender (*Lavandula* spp.) can grow for its fragrance and beauty and also help in phytoremediation by absorbing heavy metals.

Rose (*Rosa rubiginosa*) can be grown in contaminated soil and is effective at removing air pollutants.

g. **Wetland plants**

Wetland ornamental plants are also very important for removing contaminants from water. Examples *Typha, Nymphaea, Iris,* etc

SELECTION CRITERIA OF ORNAMENTAL PLANTS

While selecting ornamental plants for phytoremediation, it is essential to consider various criteria to ensure effectiveness in cleaning up the environmental contaminants while maintaining the aesthetic beauty of the site. Some of the selection criteria are discussed below.

i. **Contaminant-specific traits**

Identification of the specific contaminants is essential in the target site. Plants should be chosen considering the natural affinity of the plants to the target contaminant.

ii. **Tolerance to site conditions**

The soil physicochemical characteristics, such as pH, alkalinity, soil type, and moisture conditions of the site are very important for phytoremediation, as the plants have to thrive in these conditions.

iii. **Phytoremediation potential**

The phytoremediation potential of the candidate plant is very important. The heavy metal accumulation efficiency, pollutant degradation, and transformation efficiency are essential to study before species selection.

iv. **Adaptability and growth rate**

The growth rate of the plant is important as fast-growing plants can compete with other undesirable species. The pollutant-resistant behavior is also important for species selection.

v. **Root depth and structure**

Root depth and structure of the candidate plant are important considerations for extraction of pollutants from subsurface soil. The deep-rooted species are more effective at extracting contaminants from soil than the surface-growing plants.

vi. **Biomass production**

High biomass–producing plants with more canopy can absorb and store more pollutants from air and soil.

vii. **Aesthetic value**

Ornamental plants with attractive foliage, flowers, and other aesthetic qualities suitable for landscaping purposes are also given importance, so that they will enhance the visual appeal of the bioremediation site.

viii. **Maintenance of the plant**

The maintenance needs of the chosen plants, including pruning, watering, and pest control are important to maintain the plants for long periods of time. Hence, low-maintenance plants are preferable for this purpose.

ix. **Local availability**

Native species are well suited to local conditions and need less maintenance, and they are locally available in plentiful numbers.

x. **Cost effectiveness**

Cost of the plants for procurement and maintenance should be considered for phytoremediation. High-cost plant are not preferred for this purpose.

xi. **Plan design and regular monitoring**

Comprehensive planning and design along with space requirements for plant growth are very important. Regular monitoring and assessment of the site is essential to know whether or not the pollution load is being reduced over time.

xii. **Safety and environmental impact**

Plants which are environmentally safe and also show no toxic properties for humans and other animals should be preferred for phytoremediation.

xiii. **Long-term viability and regulatory compliance**

The selected plant should meet the local regulation permit for phytoremediation project. It should be viable for long term and should be thrived for years for remediation.

Considering the above factors, many workers have chosen plants with specific criteria. Pradhan et al. (2015) have used *Tagetes erecta* for remediation of fly ash and observed that 5% to 10% of fly ash amendment is enhancing plant growth. Phytoremediation of fly ash was also studied by many workers using different species of plants, who recorded encouraging results (Dash et al., 2015).

MECHANISM OF PHYTOREMEDIATION BY ORNAMENTAL PLANTS

Basically, ornamental plants are classified into aquatic and terrestrial plants. The ornamental plants are generally used in gardens for decorative purposes for landscape beautification. They are commonly harvested for leaves, flowers, scent, and foliage. The useful by-products like flowers and leaves are used for decoration and production of fragrances. Many of the flowering plants are not edible, hence there is low risk of heavy metal contamination through the food chain (Nakbanpote et al., 2016). Ornamental plants, often used for aesthetic purposes, can play a vital role in cleaning the environment by absorbing and sequestering heavy metals from soil and water. This process involves various mechanisms, including intracellular and extracellular uptake, chelation, diffusion, and ion exchange (Soudzilovskaia et al., 2010). Ornamental plants possess specialized transport proteins, such as metal transporters, that facilitate the uptake of heavy metals from the surrounding environment. This intracellular uptake can occur through root cell membranes. The process typically involves the metal transporters. These proteins are embedded in the root cell membrane and have a high affinity for specific heavy metal ions. They help transport heavy metals across the root cell membrane and into the plant's vascular system. In some cases, the plant uses active transport to take up heavy metals with ATP (adenosine triphosphate) energy against the concentration gradient (active uptake). This ensures that the plant can accumulate even trace amounts of heavy metals (Mahapatra et al., 2019). Some heavy metals can passively diffuse through the plant cell following the concentration gradient without using any form of energy.

Extracellular uptake mechanisms involve the adsorption of heavy metals onto the root cell surface and subsequent transport into the plant. Heavy metals can adhere to the cell walls of plant roots. This extracellular binding prevents them from entering the plant's interior. Some heavy metals form insoluble precipitates with substances in the root exudates or in the soil solution, reducing their bioavailability for uptake. Chelation is a common mechanism employed by ornamental plants to reduce the toxicity of heavy metals (Lajayer et al., 2019). It involves the formation of metal-organic complexes within plant tissues. The ornamental plants produce some chelating compounds, such as phytochelatins and organic acids, which have a high affinity for binding with heavy metal ions. When heavy metals are taken up by plant roots or absorbed from the environment, these chelating compounds bind to the metal ions, forming stable complexes. This prevents the toxic metals from interfering with essential cellular processes. The beneficial soil microbes can also release some chelating agents that help the plants in heavy metal uptake and detoxification (Franzetti et al., 2020). Diffusion is a passive transport mechanism that allows heavy metals to move through plant tissues based on concentration gradients. It can occur between cells, within a plant's organs, or along the plant's vascular system. The plant's roots possess cation exchange sites that can exchange essential nutrient ions (e.g., calcium, magnesium) with toxic heavy metal ions present in the rhizosphere. The diffusion process facilitates the distribution of heavy metals within the plant. The ion exchange method involves the replacement of one type of ion with another ion at the exchange sites within plant tissues. The heavy metal ions compete with essential nutrient ions for binding sites on these exchange sites. In some cases, heavy metal anions can also compete with essential nutrient anions on exchange sites, although this is a less common mechanism. This competition can reduce the uptake of toxic metals and promote the acquisition of essential nutrients. The detailed mechanism of heavy metal uptake and remediation is shown in Figure 12.1.

DIFFERENT TECHNIQUES OF PHYTOREMEDIATION

Phytoremediation can be achieved by different techniques, such as phytoextraction, phytostabilization, phytovolatilization, and rhizofiltration. The details of the techniques are described below.

PHYTOEXTRACTION

This method implies the uptake of contaminants by roots of the plant from terrestrial and aquatic media. The contaminants accumulate in the aboveground biomass of the plant. Soil contaminated with different types of heavy metals is generally treated with this technique (Pandya et al., 2022). The cost of this technique is very low compared to other traditional methods. Nowadays more people are attracted to phytoextraction, as this is a long-lasting and sustainable technique (Ali et al., 2013; Sarwar et al., 2017). The procedural steps for phytoextraction are:

  i. *Site assessment:* The extent and type of contamination are determined at the site, including the specific contaminants and their concentration.

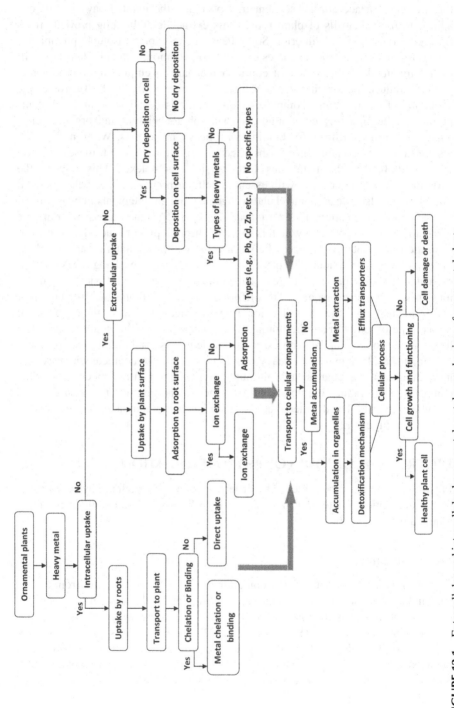

FIGURE 12.1 Extracellular and intracellular heavy metals uptake mechanism of ornamental plants.

ii. *Plant selection:* Appropriate plant species should be chosen that can grow in that contaminated area with high accumulating ability (hyperaccumulators) in their biomass.

iii. *Soil preparation:* Soil quality may be improved by adjusting pH, adding organic matter, and providing essential nutrients for plant growth.

iv. *Planting:* Hyperaccumulating plant species are planted in the contaminated area. Proper spacing and planting should be ensured with depth according to the plant species requirements. Plants can help stabilize the contaminants through their root systems.

v. *Irrigation and maintenance:* Adequate irrigation should be provided to support plant growth. Regular monitoring and maintenance are required to ensure that the plants are healthy and thriving.

vi. *Monitoring:* The concentration of contaminants should be regularly monitored in the soil or water to assess the progress of phytoextraction. This may involve periodic soil or water sampling and analysis.

vii. *Harvesting:* After maturity, plants may be harvested as they accumulate sufficient amounts of contaminant. This can be done by cutting the aboveground biomass.

viii. *Contaminant removal:* The harvested plant material should be removed from the site. The plant biomass can be disposed of as hazardous waste or processed to recover the accumulated contaminants, depending on the nature of the contamination and regulations.

ix. *Replanting (optional):* Depending on the level of contamination and desired remediation goals, one option may be to replant the site with hyperaccumulating plants for additional cycles of phytoextraction.

x. *Final soil/water testing:* After one or more cycles of phytoextraction, final assessment should be conducted to examine the soil or water quality to determine if the remediation goals have been met.

xi. *Postremediation monitoring:* After phytoextraction, the site should be monitored to ensure that contaminants do not return and that the site remains in a healthy condition.

xii. *Reporting and documentation:* Maintain detailed records of the phytoremediation process, including plant growth, contamination levels, and any actions taken. This documentation is essential for regulatory compliance and reporting.

It's important to note that the success of phytoextraction as a phytoremediation technique depends on various factors, including the choice of appropriate plant species, soil conditions, and the specific contaminants involved. Additionally, it may take several years to achieve significant reductions in contaminant levels through phytoextraction, so patience and long-term commitment are often required.

PHYTOSTABILIZATION

Phytostabilization is a phytoremediation technique generally used to immobilize environmental contaminants in soil systems in order to prevent their migration and to reduce their bioavailability to enter the food chain (Marques et al., 2009; Wong, 2003). It is particularly effective for heavy metal and metalloid-contaminated sites. The steps involved in phytostabilization are:

i. *Site assessment:* Evaluation of types of contaminants with their concentration levels. This assessment will help to reveal whether phytostabilization is a suitable option for this site or not.

ii. *Contaminant characterization:* Identification of contaminants with their properties; different plants have efficient stabilizing properties for different contaminants.

iii. *Plant selection:* Same as phytoextraction.

iv. *Soil preparation:* Same as phytoextraction.

v. *Planting:* Same as phytoextraction.

vi. *Irrigation and maintenance:* Same as phytoextraction.

vii. *Root zone management:* The focus should be on the root zone of the plants, as this is where most of the contaminant immobilization occurs. Hence, root growth and colonization should be encouraged.

viii. *Monitoring:* Continuous monitoring of the concentration of contaminants should be done to assess the effectiveness of phytostabilization.

ix. *Maintenance and plant monitoring:* Management and trimming of the plant are needed to prevent overgrowth and to optimize the contaminant-stabilizing capacity. Dead and diseased plants should be removed.

x. *Long-term management:* Phytostabilization is often a long-term remediation strategy; hence periodic assessment is required to monitor the contaminant levels in soil.

xi. *Reporting and documentation:* Records should be kept about the growth, soil conditions, and contaminant levels for regulatory compliance.

Phytostabilization is a sustainable and environmentally friendly approach to managing contaminated sites, as it reduces the risk of contaminant migration and exposure to the surrounding environment. The mycorrhiza and rhizobacteria present in the rhizosphere help to increase the surface area of the root to ease the phytostabilization process (Gohre and Paszkowski, 2006).

PHYTOVOLATILIZATION

Phytovolatilization is a phytoremediation technique used to remediate soil water contaminated with volatile organic compounds (VOCs) or certain contaminants that can be converted into volatile forms by plants (Pandya et al., 2022; Rocha et al., 2022). The primary goal of phytovolatilization is to capture and release these contaminants into the atmosphere, where they can be dispersed harmlessly (Dela et al., 2014). Here are the steps involved in phytovolatilization.

i. *Site assessment:* The extent and type of contamination should be assessed at the site, and the specific volatile contaminants and their concentrations should be identified.

ii. *Contaminant characterization:* Determination of the properties of the contaminants to understand their volatility and the potential for phytovolatilization should be done as a remediation option

iii. *Plant selection:* Plant species should be chosen for their ability to take up volatile and target contaminants. Plants are known for their capacity to absorb and release VOCs or inorganic volatiles.

iv. *Soil preparation:* Same as phytoextraction.

v. *Planting:* Same as phytoextraction.

vi. *Irrigation and maintenance:* Provision of adequate irrigation to support plant growth. Regular maintenance of plants is needed to ensure healthy growth to facilitate phytovolatilization.

vii. *Contaminant uptake:* The target contaminants will be absorbed by the plants from soil and water through the root system. The process occurs through active and passive transport mechanisms.

viii. *Contaminant volatilization:* The contaminants taken up by the plants will move through the plant tissues and eventually be released into the atmosphere through transpiration or metabolic processes. This conversion into a volatile form is crucial for phytovolatilization

ix. *Monitoring:* Continuous monitoring is required to evaluate the pollutant concentration in the contaminated sites.

x. *Harvesting:* Harvesting of the plant biomass is necessary to remove accumulated contaminants from the site and to check the contaminants from reentering the soil and water.

xi. *Postremediation monitoring:* Continuous monitoring is required to ensure that contaminants do not return to the site.

xii. *Reporting and documentation:* Same as phytoextraction.

Phytovolatilization can be an effective method for addressing volatile contaminant issues in soil and water, but its success depends on different factors, such as the choice of suitable plant species, site-specific conditions, and the nature of the contaminants involved (Rock et al., 2000). Not all contaminants can be effectively remediated through phytovolatilization, so a thorough site assessment and feasibility study are critical before implementing this technique.

RHIZOFILTRATION

Rhizofiltration is a phytoremediation technique in which plant roots are used to remove environmental contaminants, primarily heavy metals and metalloids, from water. It can be an effective method for treating industrial wastewater, agricultural runoff, and other water sources contaminated with toxic elements (Ishak and Hum, 2022; Khan et al., 2021). Here are the steps involved in rhizofiltration:

i. *Site assessment:* Same as phytoextraction.

ii. *Contaminant characterization:* Same as phytoextraction.

iii. *Plant selection:* Same as phytoextraction.

iv. *System design:* The rhizofiltration system should be designed with a container or beds filled with an appropriate growth medium (soil, sand, or hydroponic substrates), into which the selected plants will be planted.

v. *Planting:* Same as phytoextraction.

vi. *Water circulation:* A water circulation system should be set up that allows contaminated water to flow through the rhizofiltration system. This can be achieved with a pump or with gravity flow.

vii. *Contaminated water flow:* Contaminated water will be directed into the rhizofiltration system, allowing it to percolate through the growth medium to interact with the plant roots.

viii. *Contaminant uptake:* The contaminants will be absorbed by the plant root system and accumulate in the tissue.

ix. *Monitoring:* Same as phytoextraction.

x. *Maintenance:* Same as phytoextraction.

xi. *Harvesting:* Depending on the extent of contamination and accumulation of contaminants in the plant roots, periodic harvesting of the plants may be done to remove hazardous waste.

xii. *Postremediation monitoring:* After rhizofiltration treatment, continuous monitoring of water quality is required to measure the pollutant level in the contaminated water or soil.

xiii. Reporting and documentation: Same as phytoextraction.

Rhizofiltration can be an efficient and sustainable method for removing heavy metals and metalloids from contaminated water sources, but its success depends on factors such as the choice of suitable plant species, system design, and the nature of the contaminants (Van der Ent et al., 2013). It is essential to perform regular maintenance and monitoring to achieve optimal results in rhizofiltration projects.

PHYTOREMEDIATION OF HEAVY METALS BY TERRESTRIAL PLANTS

The accumulation, extraction, transformation, and stabilization of pollutants depends upon various environmental factors like their bioavailability, nutrients, plant species, growth stage of plant, optimum temperature, pH, and many more factors (Prasad and Freitas, 2003). Both phytostabilization of heavy metals and beautification of contaminated environments can be done using some plants like *Tagetes patula* and *Tagetes erecta* (Chaturvedi et al., 2014; Chintakovid et al., 2008; Pradhan et al., 2015), *Erica australis* and *Erica andevalensis* (Pérez-López et al., 2014), *Nerium oleander* (Trigueros et al., 2012), and *Calendula officinalis* (Liu et al., 2010). In Thailand, people are using the *Tagetes patula* (French marigold) ornamental plant for remediation of heavy metals as well as for income generation (Nakbanpote et al., 2016). Likewise, many workers have used different terrestrial ornamental plants for phytoremediation of different heavy metals (Liu et al., 2017; Zhi et al., 2011). Table 12.1 lists some ornamental plants along with remediation strategies of various heavy metals like Cd, Mn, Al, As, Pb, Cr, and Zn. This study describes the plant species in different system with providing heavy metal stress to the plant. The selected terrestrial ornamental plants have a higher potential to remove heavy metals from contaminated soil sites.

TABLE 12.1
Different Ornamental Plants with Their Heavy Metal Phytoremediation Potential

Sl No	Plant Species Botanical Name	Heavy Metals	Results	References
1	*Mirabilis jalapa*	Cd	Cd extraction in shoot.	Wang and Liu, 2014
2	*Erica Australis and E. andevalensis*	Al, As, Fe, Mn	Hyperaccumulators of Mn and tolerant to Al⁻, Fe⁻, and As⁻.	Pérez et al., 2014
3	*Antirrhinum majus, Quamoclit pennata, Celosia cristata pyramidalis*	Pb	*Celosia cristata pyramidalis* could be considered an accumulator of Pb.	Cui et al., 2013
4	*Lonicera japonica*	Cd	Accumulation of Fe and Cd in plant tissue and the concentration of Zn, Cd, and Cu showed a significant negative correlation.	Liu et al., 2009
5	*Chrysanthemum maximum*	Cd, Cu, Ni, Pb	Roots accumulated Cd, Cu, and Ni, excluding Pb. Cu and mycorrhizal plants accumulate lower concentration of Pb than non-mycorrhizal plants in both shoot and root.	González-Chávez and Carrillo-González, 2013
6	*Nerium oleander*	Pb	Phytostabilization of Pb. Pb concentration increased in malondialdehyde content in leaves and inhibited plant growth.	Trigueros et al., 2012
7	*Amaranthus hypochondriacus*	Cd	Increased in accumulation of Cd in plant.	Li et al., 2013
8	*Calendula officinalis*	Cd	Cd accumulation increases due to application of EGTA and SDS.	Liu et al., 2010
9	*Althaea rosea, Calendula officinalis*	Cd	Cd uptake was enhanced. These plants can considered hyperaccumulators of Cd.	Liu et al., 2008a
10	*Calendula officinalis*	Cd	EGTA alone enhances Cd uptake.	Liu et al., 2010
11	*Helianthus annuus*	Cd	Cd uptake increased due to application of EDTA and electrical field (10 and 30 V).	Tahmasbian and Safari Sinegani, 2014
12	*Calendula alata*	Cs, Pb	Plants remediate Pb and Cs. Cs uptake is possible by the presence of Pb. This plant has a very high growth rate.	Borghei et al., 2011
13	*H. annuus*	Cr	Decreased stomata activities due to Cr(III) and Cr(VI) accumulation and also affected plant growth. Arbuscular micorrhiza increase chromium toxicity.	Davies et al., 2002
14	*H. annuus, Celosia cristata, Tagetes patula*	Ca, Cr, Mn, Fe, Cu, Zn, Pb	The heavy metals accumulated in different parts of the plants in the order of root > leaves > stem > flower.	Chatterjee et al., 2012

(Continued)

TABLE 12.1 *(Continued)*

Different Ornamental Plants with Their Heavy Metal Phytoremediation Potential

Sl No	Plant Species Botanical Name	Heavy Metals	Results	References
15	*H. annuus*	Zn, Cd, Cu	These metals are responsible for a decrease in chlorophyll concentration and enhanced shoot biomass.	Nehnevajova et al., 2012
16	*H. annuus*	Cd, Zn	Cd and Zn uptake increased by application of KCl.	Hao et al., 2012
17	*H. annuus*	Zn, Cd	*Chrysiobacterium humi* and *Ralstonia eutropha* control the weight loss in metal-contaminated plants. *C. humi* prevented the loss of biomass and increased short-term stabilization.	Marques et al., 2013
18	*H. annuus*	Hg	Hg uptake and translocation enhanced due to the treatment with ammonium thiosulfate and cytokinine.	Cassina et al., 2012
19	*Impatiens balsamin, Tagetes erecta, Mirabilis jalapa*	Cr	*M. jalapa* had the greatest potential to tolerate and extract Cr.	Miao and Yan, 2013
20	*Cosmos sulphureus, Panicum maximum, Tagetes erecta, H. annuus*	Cd	Cd accumulation was increased due to numerous roots and high biomass.	Rungruang et al., 2011
21	*Ricinus communis, Tagetes erecta*	Ni, Pb	*R. communis* is more efficient than *T. erecta* for Ni accumulation in roots; further Ni accumulation was increased by application of farmyard manure.	Malarkodi et al., 2008
22	*Althaea rosea, Impatiens balsamina, Calendula officinalis*	Cd, Pb	Roots and shoots of *C. officinalis* had higher potential to tolerate, accumulate, and stabilize Cd contamination.	Liu et al., 2008b
23	*H. annus, Salvia splendens, Tagetes erecta*	Cd	The uptake capacity of these plants is increased with increased dose of Cd with a lower yield value.	Bosiacki, 2008
24	*Gladiolus grandiflorus, T. erecta, Chrysanthemum indicum*	Cd	The Cd removal efficiency was *C. indicum* > *G. grandiflorus* > *T. erecta*	Lal et al., 2008
25	*Tagetes patula*	Cd, Cu, Pb, and benzo[a] pyrene (B[a]P)	Marigold plant considered a good accumulator of benzo[a]pyrene.	Sun et al., 2011

(Continued)

TABLE 12.1 *(Continued)*
Different Ornamental Plants with Their Heavy Metal Phytoremediation Potential

Sl No	Plant Species Botanical Name	Heavy Metals	Results	References
26	*Tagetes erecta*	Cd	*A. mycorrhiza* improved the efficiency of reactive oxygen species scavenging the Cd metal.	Liu et al., 2011
27	*Tagetes patula*	Fe	Marigold considered the best plant for phytostabilization and had high tolerance in excess heavy metal stress from Fe ore tailings.	Nakbanpote et al., 2016
28	*Nugget marigold* (triploid hybrid between *T. erecta* L. and *T. patula*)	As	Trivalent As and pentavalent As found in roots, shoots, and leaves of plants.	Chintakovid et al., 2008
29	*Tagetes erecta*	Cd, Pb	Accumulation of Cd in marigold plant as leaves > stems > inflorescence and Pb accumulated in stems > leaves > inflorescence.	Bosiacki, 2008
30	*Tagetes erecta*	Cu	Efficient Cu accumulation was done by the colonization of marigold and *A. mycorrhiza* and they phytostabilized Cu contamination.	Castillo et al., 2011

PHYTOREMEDIATION OF HEAVY METALS BY AQUATIC PLANTS

Freshwater bodies are generally contaminated with a variety of environmental pollutants, thereby increasing the immediate need for effective eco-friendly remediation. Hence, in order to take up, degrade, and sequester environmental pollutants in water bodies, aquatic phytoremediation is adopted. It is a phytotechnology which is applied for pollutant removal from surface water bodies. In this technique, freshwater angiosperms, ferns, and pteridophyte groups of aquatic plants are used to degrade and remove aquatic environmental pollutants. This type of phytoremediation is popularly termed green remediation, green technology, agro remediation, or vegetative remediation technology (Kushwaha et al., 2018; Rahman and Hasegawa, 2011; Sarwar et al., 2017). This is one of the noninvasive, versatile, and cost-effective methods for the removal of trace contaminants and nutrients from different water bodies (Sinha et al., 2018). In phytoremediation, both *in situ* and *ex situ* remediation methods are employed. However, *in situ* phytoremediation technique is more commonly applied as it reduces contaminant proliferation in aquatic systems in a way that minimizes risks to the surrounding environment (Ensley, 2000). Bioremediation techniques for the treatment of heavy metals in wastewater has attracted the attention of researchers all over the globe. Physicochemical methods for treating water and wastewater are expensive and also generate secondary pollutants (Sinha et al., 2017; Wang et al.,

2018). Hence, biological treatment methods, more particularly aquatic microphytes, are the best option for treatment of wastewater from a techno-economic viewpoint (Dash, 2012; Dash and Mishra, 1996a, 1996b, 1997, 1998, 1999; Dash and Pradhan, 2013). Wetland species like *Z. aethiopica* are being employed in wastewater treatment and offer the dual benefits of water reclamation and flower production with financial returns (Zurita et al., 2008).

The use of aquatic macrophytes in the field of phytoremediation demonstrates high pollution removal due to their capacity to sequester pollutants in large amounts and in less time (Bartucca et al., 2016; Pradhan et al., 2013). Aquatic macrophytes can easily take up various substances from the water, thereby reducing the pollution load (Dhote and Dixit, 2009). Treatment of wastewater containing heavy metals using certain plant-based systems has revealed their metal hyperaccumulation properties (Ensley, 2000). The ability of aquatic macrophytes like *Phragmites* sp., *Leersia* sp., and *Spartina* sp. for bioremediation/phytoremediation of metals in contaminated water has been well demonstrated (Deng et al., 2004; Neralla et al., 1999; Shankers et al., 2005). N, P, K, micropollutants, heavy metals, trace metals, both organic and inorganic forms of nitrogen, and fluoride can be removed from water by employing different types of aquatic macrophytes (Liu et al., 2017; Sinha et al., 2018). Further, the use of aquatic macrophytes like *Lemms* sp. and *Eichhornia* sp. in treating wastewater has also been well studied (Perry and Robinson, 1997; Reed et al., 1988). However, higher concentration of metals in water has deleterious effects on the chlorophyll content of aquatic plants (Broadley et al., 2007) by causing problems with pigment synthesis (Kupper et al., 1996; Prasad and Prasad, 1987) and also oxidative damage to pigments (Oláh et al., 2010). One of the important tasks in the area of phytoremediation is to discover hyperacuumulator which can adapt to various environmental conditions. Ornamental plants are ideally selected and screened as they tolerate different abiotic stresses like drought, heat, and salt concentrations and biotic stresses like pests and diseases (Cay, 2016). Plants perform better in areas with less contamination because high levels of contaminants potentially restrict the growth and development of the plants and require more time to decontaminate the environment. Metal phytoremediation is restricted to metal uptake followed by efficiency of translocation in various plant parts. It is best to cultivate plants that are nonedible in order to prevent the entry of contaminants into the human body through the food chain (Ramana et al., 2009).

PHYTOREMEDIATION MECHANISM USING AQUATIC PLANTS

Plant species used for phytoremediation should have the following characteristics: they should be native with a fast growth rate, biomass yield should be high, they should be able to take up large amounts of metal from the water and transport it into the aboveground plant parts, and they should be able to tolerate high levels of metal toxicity, among other features (Ali et al., 2013; Arslan et al., 2017; Burges et al., 2018). Like terrestrial plants, the aquatic plants also perform different types of phytoremediation, including phytostabilization, rhizofiltration, phytoextraction, and phytovolatilization (Rahman and Hasegawa, 2011; Sarwar et al., 2017). Various mechanisms of phytoremediation used by plants are shown in Table 12.2.

TABLE 12.2

Phytoremediation Mechanism under Different Growth Media with Their Process of Extraction. Species Wise Accumulation of Different Pollutants in Various Plant Parts Is Also Shown.

Mechanism of Phytoremediation	Growth Medium	Category of Pollutant	Process of Extraction	Accumulation Part	Species
Phytofiltration	Water	Pollutants of inorganic, organic, and types of heavy metals	Absorption and/or adsorption from polluted water	Root and/or shoot	*Eichhornia, Lemna, Hydrocharis*
Phytoextraction	Water/soil	Inorganic pollutants and types of heavy metals	Absorption through roots followed by translocation to different parts	Shoots	*Juncus, Schoenoplectus*
Phytostabilization	Sediment/soil	Inorganic pollutants and types of heavy metals	Immobilization of pollutants in soil matrix because of action of plant root	Rhizosphere reduction	*Chenopodium*
Phytovolatilization	Water/soil/sediment	Organic pollutants	Pollutant conversion into volatile form	Release from atmosphere	*Phragmites*
Phytodegradation	Water/soil/sediment	Inorganic, organic, and microbial pollutants	Microbial degradation of rhizosphere or by plant metabolism	Rhizosphere degradation/degradation of pollutants in plant	*Myriophyllum, Typha, Phragmites*

Source: Dhir, 2013; Rezania et al., 2016.

AQUATIC PHYTOREMEDIATION AND MICROBIAL ACTIVITY

Microbial activities in water play a significant role in the removal of contaminants from water and wastewater. However, the independent role of different microbial communities in the process is now being given more attention (Houda et al., 2014). Many microorganisms are found to be associated with the roots of aquatic macrophytes, influencing the degradation and removal of pollutants from the aquatic environment (Faulwetter et al., 2009). The nitrifying and denitrifying bacteria, which are responsible for mineralization of nitrogen and phosphorous products need to be characterized to find out their efficiency. With adequate characterization of different microbial communities along with their association with both plant and abiotic environments, new opportunities may arise to increase the use of microbes to promote

TABLE 12.3

Aquatic Plant Species and Their Effects on Heavy Metals for Phytoremediation. Different Heavy Metals Like Lead, Cadmium, Chromium, Zinc and Iron Were Studied.

Sl No	Botanical Name	Heavy Metal		References
1	*Iris tectorum and Iris lactea*	Pb	Shoot accumulated more Pb than root in both *Iris lactea* and *Iris tectorum*.	Han et al., 2008
2	*Nymphaea*	Cd	In the mature leaf lamina during daytime, maximum Cd accumulation occurred.	Lavid et al., 2001
3	*Iris pseudacorus*	Cr and Zn	Cr and Zn were removed by *I. pseudacorus* plant through rhizofiltration and phytoextraction.	Caldelas et al., 2012
4	*Iris tectorum and Iris lactea*	Cd	Cd metal was spotted in cytoplasm and cell wall of both plants.	Han et al., 2007
5	*Nymphaea odorata*	Pb	Pb accumulated in shoots.	Outridge, 2000
6	*Iris pseudacorus*	Pb	Pb accumulation was enhanced with provided dose of Fe.	Zhong et al., 2010
7	*Zantedeschia aethiopica*	Fe	*Z. aethiopica* accumulated high Fe and has less tolerance to high Fe toxicity, but this plant can help in rehabilitation and phytoremediation of Fe-contaminated water bodies.	Casierra-Posada et al., 2014
8	*Nuphar variegata*	Cd	*N. variegata* leaves accumulated more Cd. There is no effect of organic content in the sediment.	Thompson et al., 1997
9	*Nymphaea spontanea*	Cr(VI)	Hexavalent Cr was accumulated in *N. variegata* in the order of roots > leaves > petioles	Choo et al., 2006

the removal of different environmental pollutants. Toxic substances like pesticides, herbicides, and chemical fertilizers may enter the food chain and have adverse effects on animal and plant health. Like terrestrial ornamental plants, aquatic ornamental plants are also being used for wastewater remediation and financial returns (Nakbanpote et al., 2016). To improve water quality through environmentally friendly processes, some special types of ornamental plants are being used to effectively clean the heavy metal contamination. Table 12.3 depicts the potential of these ornamental plants to remove heavy metal contamination from aquatic environments.

CONCLUSION

Heavy metals which are a persistent pollutant in the environment require complete removal. Phytoremediation using ornamental plants is a promising biotechnology for removing pollutants/toxicants from both the terrestrial and aquatic environments. It also offers other benefits, like beautification of contaminated sites, providing a source of income, and negligible threat to the food chain since the plants are not edible and thus do not have direct links to the food chain. Since physicochemical methods of pollutant removal are costlier, phytoremediation can be easily employed as a low-cost technique for pollutant removal. The metal uptake efficiency of plants can be enhanced by using genetic engineering and by modifying other environmental factors. It can fit into the sustainable development model for its eco-friendly approach. There is a need for future study of the remediation of pharmaceutical waste, radioactive waste, and hazardous contaminants using ornamental plants. Aquatic ornamental plants can also be used for the successful removal of contaminants from industrial, agricultural, and domestic wastewater.

ACKNOWLEDGEMENT

The authors would like to thank the authority of Siksha O Anusandhan (Deemed to be University) for providing financial and moral support for the study.

REFERENCES

Ali H, Khan E, Sajad MA (2013) Phytoremediation of heavy metals—Concepts and applications. Chemosphere 91(7):869–881.

Alirzayeva E, Neumann G, Horst W, Allahverdiyeva Y, Specht A, Alizade V (2017) Multiple mechanisms of heavy metal tolerance are differentially expressed in ecotypes of *Artemisia fragrans*. Envrion Pollut 220:1024–1035.

Alloway BJ (2012) Sources of Heavy Metals and Metalloids in Soils. In: Alloway. B.J., Ed., Heavy Metals in Soils: Trace Metals and Metalloids in Soils and Their Bioavailability, Environmental Pollution, vol. 22:11–50, Springer, Dordrecht.

Arshad M, Khan AHA, Hussain I, Anees M, Iqbal M, Soja G, Linde C, Yousaf S (2017) The reduction of chromium (VI) phytotoxicity and phytoavailability to wheat (*Triticum aestivum* L.) using biochar and bacteria. Appl Soil Ecol 114:90–98.

Arslan M, Imran A, Khan QM, Afzal M (2017) Plant–bacteria partnerships for the remediation of persistent organic pollutants. Environ Sci Pollut Res 24:4322–4336.

Bartucca ML, Mimmo T, Cesco S, Del Buono D (2016) Nitrate removal from polluted water by using a vegetated floating system. Sci Total Environ 542:803–808.

Borghei M, Arjmandi R, Moogouei R (2011) Potential of *Calendula alata* for phytoremediation of stable cesium and lead from solutions. Environ Monitor Assess 18:63–68. https://doi.org/10.1007/s10661-010-1813-9

Bosiacki M (2008) Accumulation of cadmium in selected species of ornamental plants. Acta Sci Pol Hortorum Cultus 7:21–31.

Broadley MR, White PJ, Hammond JP, Zelko I, Lux A (2007) Zinc in plants. New Phytol 173:677–702.

Burges A, Alkorta I, Epelde L, Garbisu C (2018) From phytoremediation of soil contaminants to phytomanagement of ecosystem services in metal contaminated sites. Int J Phytoremediat 20:384–397.

Caldelas C, Araus J, Febrero A, Bort J (2012) Accumulation and toxic effects of chromium and zinc in *Iris pseudacorus* L. Acta Physiol Plant 34:1217–1228. https://doi.org/10.1007/s11738-012-0956-4

Casierra-Posada F, Blanke M, Guerrero-Guío JC (2014) Iron tolerance in Calla lillies (*Zantedeschia aethiopica*). Gesunde Pflanzen 66:63–68. https://doi.org/10.1007/s10343-014-0316-y

Cassina L, Tassi E, Pedron F, Petruzzelli G, Ambrosini P, Barbafieri M (2012) Using a plant hormone and a thioligand to improve phytoremediation of Hg-contaminated soil from a petrochemical plant. J Hazard Mater 231:36–42. https://doi.org/10.1016/j.jhazmat.2012.06.031

Castillo OS, Dasgupta-Schubert N, Alvarado CJ, Zaragoza EM, Villegas HJ (2011) The effect of the symbiosis between *Tagetes erecta* L. (marigold) and *Glomusintraradices* in the uptake of copper(II) and its implications for phytoremediation. New Biotech 29(1): 156–164. https://doi.org/10.1016/j.nbt.2011.05.009

Cay S (2016) Enhancement of cadmium uptake by *Amaranthus caudatus*, an ornamental plant, using tea saponin. Environ Monit Assess 188(6): 320.

Cazalis V, Loreau M, Henderson K (2018) Do we have to choose between feeding the human population and conserving nature? Modelling the global dependence of people on eco-system services. Sci Total Environ 634:1463–1474.

Chatterjee S, Singh L, Chattopadhyay B, Datta S, Mukhopadhyay S (2012) A study on the waste metal remediation using floriculture at East Calcutta Wetlands, a Ramsar site in India. Environ Monit Assess 184:5139–5150. https://doi.org/10.1007/s10661-011-2328-8

Chaturvedi AK, Patel MK, Mishra A, Tiwari V, Jha B (2014) The *SbMT-2* Gene from a halophyte confers abiotic stress tolerance and modulates ROS scavenging in transgenic tobacco. PLoS ONE 9(10): e111379. https://doi.org/10.1371/journal.pone.0111379

Chen Y, Bracy RP, Owings AD, Merhaut DJ (2009) Nitrogen and phos-phorous removal by ornamental and wetland plants in a green- house recirculation research system. Hort Science 44:1704–1711. https://doi.org/10.21273/hortsci.44.6.1704

Chintakovid W, Visoottiviseth P, Khokiattiwong S, Lauengsuchonkul S (2008) Potential of the hybrid marigolds for arsenic phytoremediation and income generation of remedia-tors in Ron Phibun District, Thailand. Chemosphere 70(8):1532–1537.

Choo T, Lee C, Low K, Hishamuddin O (2006) Accumulation of chromium (VI) from aque-ous solutions using water lilies (*Nymphaea spontanea*). Chemosphere 62:961–967. https://doi.org/10.1016/j.chemosphere.2005.05.052

Cui S, Zhang T, Zhao S, Li P, Zhou Q, Zhang Q, Han Q (2013) Evaluation of three ornamen-tal plants for phytoremediation of Pb-contamined soil. Int J Phytoremed 15:299–306. https://doi.org/10.1080/15226514.2012.694502

Dash AK, Pradhan A, Das S, Mohanty SS (2015) Fly ash as a potential sources of soil amend-ment in agriculture and a component of integrated plant nutrient supply system. J Ind Poll Cont 31(2):249–257.

Dash, AK (2012) Impact of domestic wastewater on seed germination and physiological parameters of wheat and rice. Int J Res Rev Appl Sci 12(2):280–286.

Dash, AK and Mishra, PC (1996a) Changes in pigment and protein content of *Westiellopsis prolifica*, a blue-green alga, grown in paper mill waste water. Microbio 85:257–266.

Dash AK and Mishra PC (1996b) Changes in biomass, pigment and protein content of *Westiellopsis prolifica*, a blue-green alga, grown in nutrient manipulated paper mill wastewater. Cytobio 88:11–16.

Dash, AK and Mishra PC (1997) Blue –green alga in sewage – Amended paper mills waste water. Int J Environ Stud 53:9–10.

Dash AK and Mishra PC (1998) Role of Cyanobacteria in water pollution abatement. Indian J Environ Ecoplan 1(1 & 2):1–11.

Dash AK and Mishra PC (1999) Role of the blue – Green alga, *Westiellopsis prolifica* in reducing pollution load from paper mill wastewater. Ind J Environ Proct 19(1):1–15.

Dash AK and Pradhan A (2013) Growth and biochemical changes of the blue-green alga, *Anabaena doliolum* in domestic wastewater. Int J Sci Eng Res 4(6):2753–2758.

Davies Jr FT, Puryear JD, Newton RJ, Egilla JN, Saraiva Grossi JA (2002) Mycorrhizal fungi increase chromium uptake by sunflower plants: Influence on tissue mineral concentra-tion, growth, and gas exchange. J Plant Nutri 25(11):2389–2407. https://doi.org/10.1081/PLN-120014702

Dela CM, Christensen JH, Thomsen JD, Müller R (2014) Canorna- mental potted plants remove volatile organic compounds from indoor air? — A review. Environ Sci Pollut Res 21:13909–13928. https://doi.org/10.1007/s11356-014-3240-x

Deng H, Ye ZH, Wong MH (2004) Accumulation of lead, zinc, copper and cadmium by 12 wetland plants species thriving in metal contaminated sites in China. Environ Pollut 132(1):29–40.

Dhir B (2013) Phytoremediation: Role of Aquatic Plants in Environmental Clean-Up. Springer, India. https://doi.org/10.1007/978-81-322-1307-9

Dhote S, Dixit S (2009) Water quality improvement through macrophytes—a review. Environ Monit Assess 152:149–153.

Drazkiewicz M and T Baszyński (2010) Interference of nickel with the photosynthetic apparatus of Zea mays. Ecotoxicol Environ Saf 73(5):982–986. https://doi.org/10.1016/j.ecoenv.2010.02.001

El-Naggar NEA, Hamouda RA, Mousa I, Abdel-Hamid MS, Rabei NH, (2018) Statistical optimization for cadmium removal using Ulva fasciata biomass: Characterization, immobilization and application for almost-complete cadmium removal from aqueous solutions. Sci Rep 8:12456.

Ensley BD (2000) Rationale for Use of Phytoremediation. In: Raskin, I., Ensley, B.D. (Eds.), Phytoremediation of Toxic Metals Using Plants to Clean Up the Environment. John Wiley and Sons, New York, pp. 3–12.

Faulwetter J, Gagnon V, Sundberg C, et al. (2009) Microbial processes influencing performance of treatment wetlands: A review. Ecol Eng 35:987–1004.

Feng NX, Yu J, Zhao HM, Cheng YT, Mo CH, Cai QY, Li YW, Li H, Wong MH (2017) Efficient phytoremediation of organic contaminants in soils using plant–endophyte partnerships. Sci Total Environ 583:352–368. https://doi.org/10.1016/j.scitotenv.2017.01.075

Franzetti A, Gandol I, Bestetti G, et al. (2020) Plant-microorganisms interaction promotes removal of air pollutants in Milan (Italy) urban area. J Hazard Mater 384:121021. https://doi.org/10.1016/j.jhazmat.2019.121021

Fu F, Wang Q (2011) Removal of heavy metal ions from wastewaters: A review. J Environl Manage 92(3):407–418. https://doi.org/10.1016/j.jenvman.2010.11.011

Gohre V, Paszkowski U (2006) Contribution of the arbuscular mycorrhizal symbiosis to heavy metals phytoremediation. Planta 223(6):1115–1122.

Gomes MP, Gonçalves CA, Brito JCM, et al. (2017) Ciprofloxacin induces oxidative stress in duckweed (Lemna minor L.): implications for energy metabolism and antibiotic-uptake ability. J Hazard Mater 328:140–149. https://doi.org/10.1016/j.jhazmat.2017.01.005

Gomes MP, Le Manac'h SG, Moingt M, et al. (2016) Impact of phosphaet on glyphosate uptake and toxicity in willow. J Hazard Mater 304:269–279. https://doi.org/10.1016/j.jhazmat.2015.10.043

González-Chávez MDCA, Carrillo-González R (2013) Tolerance of Chrysantemum maximum to heavy metals: The potential for its use in the revegetation of tailings heaps. J Environ Sci 25:367–375.

Goswami S and Das S (2016) Copper phytoremediation potential of Calandula officinalis L. and the role of antioxidant enzymes in metal tolerance Ecotoxicol Environ Saf 126:211–218. https://doi.org/10.1016/j.ecoenv.2015.12.030

Han YL, Huang SZ, Gu JG, et al. (2008) Tolerance and accumulation of lead by species Iris L. Ecotoxicology 17:853. https://doi.org/10.1007/s10646-008-0248-3

Han YL, Yuan HY, Huang SZ, et al. (2007) Cadmium tolerance and accumulation by two species of Iris. Ecotoxicology 16:557–563. https://doi.org/10.1007/s10646-007-0162-0

Hao XZ, Zhou DM, Li DD, Jiang P (2012) Growth, cadmium and zinc accumulation of ornamental sunflower (Helianthus annuus L.) in contaminated soil with different amendments. Pedosphere 22(5):631–639. https://doi.org/10.1016/S1002-0160(12)60048-4

Houda N, Hanene C, Ines M, et al. (2014) Isolation and characterization of microbial communities from a constructed wetlands system: A case study in Tunisia. Afr J Microbiol Res 8:529–538.

Ishak IR, Hum NNMF (2022) Phytoremediation using ornamental plants in removing heavy metals from waste water sludge. IOP Conference Series: Earth and Environmental Science 1019:012009. https://doi.org/10.1088/1755-1315/1019/1/012009

Khan AHA, Kiyani A, Mirza CR, Butt TA, Barros R, Ali B, Iqbal M, Yousaf S. (2021) Ornamental plants for the phytoremediation of heavy metals: Present knowledge and future perspectives, Environ Res 195:110780. https://doi.org/10.1016/j.envres.2021.110780.

Kupper H, Kupper F, Spiller M (1996) Environmental relevance of heavy metal-substituted chlorophylls using the example of water plants. J Exp Bot 47:259–266.

Kushwaha A, Hans N, Kumar S, Rani, R (2018) A critical review on speciation, mobilization and toxicity of lead in soil-microbe-plant system and bioremediation strategies. Ecotoxicol Environ Saf 147:1035–1045.

Lajayer AB, Khadem Moghadam N, Maghsoodi MR, et al. (2019) Phytoextraction of heavy metals from contaminated soil, water and atmosphere using ornamental plants: Mechanisms and efficiency improvement strategies. Environ Sci Poll Res 26:8468–8484. https://doi.org/10.1007/s11356-019-04241-y

Lajayer BA, Ghorbanpour M, Nikabadi S (2017) Heavy metals in contaminated environment: Destiny of secondary metabolite biosynthesis, oxidative status and phytoextraction in medicinal plants. Ecotoxicol Environ Saf 145:377–390. https://doi.org/10.1016/j.ecoenv.2017.07.035

Lal K, Minhas P, Chaturvedi R, Yadav R (2008) Extraction of cadmium and tolerance of three annual cut flowers on Cd-contaminated soils. Bioresour Technol 99:1006–1011. https://doi.org/10.1016/j.biortech.2007.03.005

Lavid N, Schwartz A, Yarden O, et al. (2001) The involvement of polyphenols and peroxidase activities in heavy metals accumulation by epidermal glands of the water lily (*Nymphaeaceae*). Planta 212:323–331. https://doi.org/10.1007/s004250000400

Li N, Li Z, Fu Q, et al. (2013) Agricultural technologies for enhancing the phytoremediation of cadmium-contaminated soil by *Amaranthus hypochondriacus L.* Water Air Soil Poll 224:1673. https://doi.org/10.1007/s11270-013-1673-3

Liang HM, Lin TH, Chiou JM, Yeh KC (2009) Model evaluation of the phytoextraction potential of heavy metal hyperaccumulators and non-hyperaccumulators. Environ Poll 157:1945–1952. https://doi.org/10.1016/j.envpol.2008.11.052

Liu J, Zhou Q, Wang S (2010) Evaluation of chemical enhancement on phytoremediation effect of Cd-contaminated soils with *Calendula officinalis* L. Int J Phytoremed 12:503–515. https://doi.org/10.1080/15226510903353112

Liu JN, Zhou QX, Sun T, Ma LQ, Wang S (2008a) Growth responses of three ornamental plants to Cd and Cd–Pb stress and their metal accumulation characteristics. J Hazard Mater 151:261–267. https://doi.org/10.1016/j.jhazmat.2007.08.016

Liu JN, Zhou QX, Sun T, Ma LQ, Wang S (2008b) Identification and chemical enhancement of two ornamental plants for phytoremediation. Bull Environ Contam Toxicol 80:260–265. https://doi.org/10.1007/s00128-008-9357-1

Liu LZ, Gong ZQ, Zhang YL, Li PJ (2011) Growth, cadmium accumulation and physiology of marigold (*Tagetes erecta* L.) as affected by arbuscular mycorrhizal fungi. Pedosphere 21(3):319–327. https://doi.org/10.1016/S1002-0160(11)60132-X

Liu Z, Chen W, He X (2018) Evaluation of hyperaccumulation potentials to cadmium (Cd) in six ornamental species (Compositae). Int J Phytorem 20:1464–1469. https://doi.org/10.1080/15226514.2018.1501343

Liu Z, He X, Chen W, Yuan F, Yan K, Tao D (2009) Accumulation and tolerance characteristics of cadmium in a potential hyperaccumulator—*Lonicera japonica* Thunb. J Hazard Mater 169:170–175. https://doi.org/10.1016/j.jhazmat.2009.03.090

Liu J, Xin X, Zhou Q (2017) Phytoremediation of contaminated soils using ornamental plants. Environ Rev 999.

Mahapatra B, Dhal NK, Dash AK, et al. (2019) Perspective of mitigating atmospheric heavy metal pollution: Using mosses as biomonitoring and indicator organism. Environ Sci Pollut Res 26:29620. https://doi.org/10.1007/s11356-019-06270-z

Mahar A, Wang P, Ali A, Awasthi M, Lahori AH, Wang Q, Li R, Zhang Z (2016) Challenges and opportunities in the phytoremediation of heavy metals contaminated soils: A review. Ecotoxicol Environ Saf 126:111–121. https://doi.org/10.1016/j.ecoenv.2015.12.023

Majed N, Real M, Isreq H, Akter M, Azam HM (2016) Food adulteration and bio-magnification of environmental contaminants: A comprehensive risk framework for Bangladesh. Front Environ Sci 4:34.

Malarkodi M, Krishnasamy R, T. Chitdeshwari T (2008) Phytoextraction of nickel contaminated soil using castor phytoextractor. J Plant Nutri 31(2):219–229. https://doi.org/10.1080/01904160701853654

Marques AP, Rangel AO, Castro PM (2009) Remediation of heavy metals contaminated soils: Phytoremediation as a potentially promising clean-up technology. Critical Rev Environ Sci Technol 39(8):622–654.

Marques APGC, Moreira H, Franco AR, et al. (2013) Inoculating *Helianthus annuus* (sunflower) grown in zinc and cadmium contaminated soils with plant growth promoting bacteria effects on phytoremediation strategies. Chemosphere 92:74–83. https://doi.org/10.1016/j.chemosphere.2013.02.055

Miao Q, Yan J (2013) Comparison of three ornamental plants for phytoextraction potential of chromium removal from tannery sludge. J Mat Cycles Waste Manag 15:98–105. https://doi.org/10.1007/s10163-012-0095-4

Nakbanpote W, Meesungnoen O, Prasad MNV (2016) Potential of ornamental plants for phytoremediation of heavy metals and income generation. Bioremed Bioeconomy 179–217. https://doi.org/10.1016/B978-0-12-802830-8.00009-5

Nehnevajova E, Lyubenova L, Herzig R, Schröder P, Schwitzguébel JP, Schmülling T (2012) Metal accumulation and response of antioxidant enzymes in seedlings and adult sunflower mutants with improved metal removal traits on a metal-contaminated soil. Environ Expe Botany 6:39–48. https://doi.org/10.1016/j.envexpbot.2011.10.005

Neralla S, Weaver RW, Varvel TW, Lesikar BJ (1999) Phytoremediation and on-site treatment of septic effluents in subsurface flow constructed wetlands. Environ Technol 20:1139–1146.

Oláh V, Lakatos G, Bertók C, Kanalas P, Szőllősi E, Kis J, Mészáros I (2010) Short-term chromium(VI) stress induces different photosynthetic responses in two duckweed species, *Lemna gibba* L. and *Lemna minor* L. Photosynthetica 48:513–520. https://doi.org/10.1007/s11099-010-0068-6

Outridge PM (2000) Lead biogeochemistry in the littoral zones of south central Ontario lakes, Canada, after elimination of gasoline lead additives. Water Air Soil Poll 118:179–201. https://doi.org/10.1023/A:1005194309413

Pandya K, Rajhans S, Pandya H, Mankad A (2022) Phytoremediation: An innovative perspective for reclaiming contaminated environmental using ornamental plants. World J Adv Res Rev 13(2):1–14. https://doi.org/10.30574/wjarr.2022.13.2.0088

Pérez R, Márquez-García B, Abreu MM, Nieto JM, Córdoba F (2014) *Erica andevalensis* and *Erica australis* growing in the same extreme environments: Phytostabilization potential of mining areas. Geoderma 230–231:194–203. https://doi.org/10.1016/j.geoderma.2014.04.004

Perry DS, Robinson P (1997) Water Gardening: Water Lilies and Lotuses. Timber Press, Oregon.

Pradhan A, Dash AK, Mohanty SS, Das S (2015) Potential use of fly ash in floriculture: A case study on the photosynthetic pigments content and vegetative growth of *Tagetes erecta* (Marigold). Eco Env Cons 21:AS369–AS376.

Pradhan A, Sahu SK, Dash AK (2013) Changes in pigment content (chlorophyll and carotenoid), enzyme activities (catalase and peroxidase), biomass and yield of rice plant

(*Orizasativa*. L) following irrigation of rice mill wastewater under pot culture conditions. Int J Sci Eng Res 4(6):2706–2717.

Prasad DDK, Prasad ARK (1987) Altered delta-aminolevulinic-acid metabolism by lead and mercury in germinating seedlings of bajra (*Pennisetum typhoideum*). J Plant Phys 127:241–249.

Prasad MNV, de Oliaveira Freitas HM (2003) Metal hyperaccumulation in plants - Biodiversity prospecting for phytoremediation technology. Elect J Biotech 6(3):285–321. http://dx.doi.org/10.2225/vol6-issue3-fulltext-6

Rahman MA, Hasegawa H (2011) Aquatic arsenic: Phytoremediation using floating macrophytes. Chemosphere 83:633–646.

Ramana S, Biswas AK, Subba Rao A (2009) Phytoextraction of cadmium by African marigold and chrysanthemum. Natl Acad Sci Lett 32(11–12):333–336.

Reed SC, Middlebrooks EJ, Crites RW (1988) Natural System for Waste Management and Treatment. McGraw Hill, New York.

Rezania S, Taib S, Fadhil M, Dahalan FA, Kamyab H (2016) Comprehensive review on phytotechnology: Heavy metals removal by diverse aquatic plants species from wastewater. J Hazard Mater 318:587–599. https://doi.org/10.1016/j.jhazmat.2016.07.053

Rocha CS, Rocha DC, Kochi LY, Carneiro DNM, dos Resi MV, Gomes MP (2022) Phytoremediation by ornamental plants: A beautiful and ecological alternative. Env Sci Poll Res 29:3336–3354. https://doi.org/10.1007/s11356-021-17307-7

Rock S, Pivetz B, Mandalinski K, Adams N, Wilson T (2000) Introduction to Phytoremediation: EPA/600/R-99/107 (NTISPB2000-106690). U.S. Enviromental Protection Agency.

Rungruang N, Babel S, Parkpian P (2011) Screening of potential hyperaccumulator for cadmium from contaminated soil. Desalin Water Treat 32:19–26

Sarwar N, Imran M, Shaheen MR, Ishaque W, Kamran MA, Matloob A, Hussain S (2017) Phytoremediation strategies for soils contaminated with heavy metals: Modifications and future perspectives. Chemosphere 171:710–721.

Shang C, Wu T, Huang G, Wu J (2019) Weak sustainability is not sustainable: Socioeconomic and environmental assessment of inner Mongolia for the past three decades. Resour Conserv Recycl 141:243–252.

Shankers AK, Cervantes C, Losa-Tavera H, Avdainayagam S (2005) Chromium toxicity in plants. Environ Int 31(5):739–753.

Sinha V, Pakshirajan K, Chaturvedi R (2018) Chromium tolerance, bioaccumulation and localization in plants: An overview. Enviro Manage 206:715–730.

Sinha V, Pakshirajan K, Manikandan NA, Chaturvedi R (2017) Kinetics, biochemical and factorial analysis of chromium uptake in a multi-ion system by *Tradescantia pallida* (rose) DR Hunt. Int J Phytorem 19(11):1007–1016.

Soudzilovskaia NA, Cornelissen JHC, During HJ, et al. (2010) Similar cation exchange capacities among bryophyte species refute a presumed mechanism of peatland acidification. Ecology 91:2716–2726.

Sun Y, Zhou Q, Xu Y, Wang L, Liangb X (2011) Phytoremediation for co-contaminated soils of benzo[a]pyrene (B[a]P) and heavy metals using ornamental plant *Tagetes patula*. J Hazard Mater 186(2–3):2075–2082. https://doi.org/10.1016/j.jhazmat.2010.12.116

Tahmasbian I, Safari Sinegani AA (2014) Chelate-assisted phytoextraction of cadmium from a mine soil by negatively charged sunflower. Int J Environ Sci Tech 11:695–702. https://doi.org/10.1007/s13762-013-0394-x

Thompson E, Pick F, Bendell-Young BL (1997) The accumulation of cadmium by the yellow pond lily, *Nuphar variegatum* in Ontario Peatlands. Arch Environ Contam Toxicol 32:161–165. https://doi.org/10.1007/s002449900169

Trigueros M, Mingorance D, Rossini Oliva S (2012) Evaluation of the ability of *Nerium oleander* L. to remediate Pb-contaminated soils. J Geochem Explor 114:126–133. https://doi.org/10.1016/j.gexplo.2012.01.005

Van der Ent A, Baker AJ, Reeves RD, Pollard AJ, Schat H (2013) Hyperaccumulators of metal and metalloid trace elements: Facts and fiction. Plant Soil 362:319–334.

Venkatachalam T, Natarajan AV, Parvathi K (2017) Effect of trivalent and hexavalent chromium toxicity on biochemical composition of fresh water teleost *Labeo rohita* (Ham.) Int J Curr Sci Res 3(6):1266–1276.

Vita G, Hertwich EG, Stadler K, Wood R (2019) Connecting global emissions to fundamental human needs and their satisfaction. Environ Res Lett 14:014002.

Wang S, Liu J (2014) The effectiveness and risk effectiveness and risk comparison of EDTA with EGTA in enhancing Cd phytoextraction by *Mirabilis jalapa* L. Environ Monit Assess 186:751–759. https://doi.org/10.1007/s10661-013-3414-x

Wang Z, Jiang Y, Awasthi MK, Wang J, Yang X, Amjad A, Wang Q, Lahori AH, Zhang Z (2018) Nitrate removal by combined heterotrophic and autotrophic denitrification processes: Impact of coexistent ions. Bioresour Technol 250:838–845.

Wong MH (2003) Ecological restoration of mine degraded soils, with emphasis on metal contaminated soils. Chemosphere 50(6):775–780.

Zeng P, Guo Z, Cao X, et al. (2018) Phytostabilization potential of ornamental plants grown in soil contaminated with cadmium. Int J Phytoremediation 20:311–320. https://doi.org/10.1080/15226514.2017.1381939

Zhi LL, Qiang GZ, Long ZY, Jun LP (2011) Growth, cadmium accumulation and physiology of marigold (*Tagetes erecta* L.) as affected by arbuscular mycorrhizal fungi. Pedosphere 21(3):319–327. https://doi.org/10.1016/S1002-0160(11)60132-X

Zhong S, Shi J, Xu J (2010) Influence of iron plaque on accumulation of lead by yellow flag (*Iris pseudacorus*) grown in artificial Pb-contaminated soil. Soils Sediments 10:964–970. https://doi.org/10.1007/s11368-010-0213-7

Zurita F, Belmont MA, De Anda J, Cervantes-Martinez J (2008) Stress detection by laser-induced fluorescence in *Zantedeschia aethiopica* planted in subsurface flow treatment wetlands. Ecol Eng 33:110–118.

13 Treatment Approaches to Wastewater Using Aquatic Weeds
A Comprehensive Review Exploring the Potential of Aquatic Weeds for Wastewater Treatment

Pradip Kumar Prusty, Jatindra Nath Mohanty, and Kunja Bihari Satapathy

INTRODUCTION

Environmental pollution, primarily in the aquatic environment, is considered to be a major issue. Rapid industrialization, urbanization, and population growth have deposited massive amounts of contaminants in water supplies. Because of contaminants released into the environment, the atmosphere has become polluted and equally hazardous for human health and welfare. The discharge of municipal sewage water and industrial effluents containing significant amounts of toxins into receiving water bodies can harm aquatic biota as well as coastal ecosystems (Anjuli et al., 2012). Many toxic wastes acquired from industries are related to the manufacture and processing of iron, steel, coke, petroleum, electroplating, pesticides, textile, paper, pulp, paint, solvent, and pharmaceuticals, releasing massive amounts of organic pollutants and thus elevating the toxicity levels. Water pollution is a major concern especially in developing countries such as India (Lokeshwari and Chandrappa, 2007). However, a lack of technology, poor environmental policy implementation, and limited financial resources have created significant challenges. There are several types of water pollutants, but heavy metals are considered to be more hazardous owing to their bioaccumulative and persistent properties (Chang et al., 2009; Yadav et al., 2009). It would be commendable to develop environmentally friendly and cost-effective technology to eradicate pollutants, particularly heavy metals, and thus improve the quality of water.

Therefore, the search for a new, simple remedy for removing these pollutants from wastewater has directed attention toward phytoremediation. Phytoremediation offers a low-cost technology using aquatic weeds as a way to remove contaminants.

DOI: 10.1201/9781003442295-13

Phytoremediation is the process of cleaning up (remediating) polluted soil or water by using plants (phyto). This technology is considered to be advantageous in contrast to conventional physicochemical approaches. It has the potential to be one of the sustainable solutions for treating huge quantities of waste materials before they are discharged to water sources, thus increasing access to decontaminated water. When compared to conventional treatment systems, water treatment systems based on macrophytes have several potential advantages Aquatic macrophytes can remove a wide range of inorganic pollutants from polluted wastewater, including major agricultural pollutants such as excess nitrogen and phosphorus. An objective of this investigation was also to develop an economically viable technology through phytoremediation using aquatic weeds, including *Azolla pinnata* and *Spirodela polyrhiza*.

PHYTOREMEDIATION METHOD

The conventional term *phytoremediation* comprises the Greek prefix *phyto* (plant) joined to the Latin root *remedium* (to address or eliminate a malevolence). The actual term portrays the utilization of plants to remediate impurities by the uptake of polluted water (Figure 13.1). It is a climatically appropriate, savvy, and sun-driven interaction. Plants can be used to hold, eliminate, or corrupt toxins. It is a new name for the development of an old interaction that happens normally in environments as both inorganic

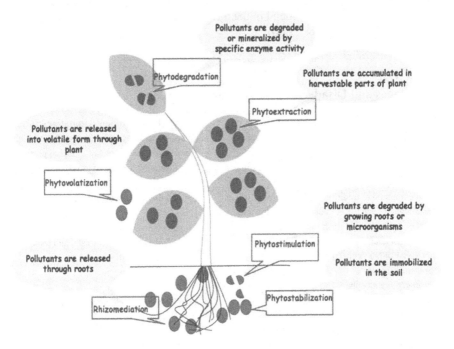

FIGURE 13.1 Various parts of the bioremediation process, which includes rhizomediation, phytostimulation, and phytoextraction.

and natural constituents cycle through plants. Plant roots can eliminate metals from the polluted sources and transport them to leaves and stems for gathering and removal or metal recuperation cycles. In the natural groupings that change pollutants to less toxic mixtures, plants contribute innate enzymatic uptake measures that can reuse or sequester the natural particles they receive. A plant utilized for phytoremediation ought to be poison-tolerant and develop rapidly with an exceptional yield for each hectare. It should be greater than 20 in order to have the option of halving defilement. Plants that aggregate a high concentration of metals are known as hyperaggregators; what's more, they can amass 50 to 100 times more metal than traditional plants (Chaoni et al., 2003).

ADVANTAGES OF PHYTOREMEDIATION

The essential inspiration driving the improvement of phytoremediation innovations is their eco-accommodating and practical nature (Table 13.1). Phytoremediation exploits the exceptional, specific, and normally occurring uptake capacities of plant root frameworks, along with the movement, bioaccumulation, and contamination stockpiling/corruption capacities of the whole plant body.

LIMITATIONS OF PHYTOREMEDIATION

The limitations of this innovation include the long cleaning time, the possibility of introducing foreign substances into the developing ecosystem, and the bioavailability of impurities and toxins found in the formation and maintenance of vegetation in landfills (Salt and Kramer, 2000).

TABLE 13.1
Advantages and Disadvantages of Phytoremediation

Advantages	Disadvantages/Limitations
Adjustable to a variety of organic and inorganic mixtures.	Restricted to areas with minor pollution within the establishment zone of remedial plants.
In situ/ex situ application conceivable with profluent/soil substrate separately.	May take quite a while to remediate a tainted site.
When compared to traditional approaches, *in situ* treatments reduce the amount of soil aggravation.	Limited to locales with low impurity focuses.
Reduces the amount of trash that must be disposed of (up to 95%), and can also be employed as a bio-metal of heavy metals.	Reaped phytoextraction plant biomass may be classified as hazardous waste, and removal must be permitted.
Applications made *in situ* reduce pollution dispersion through the atmosphere and water.	Climate is a limiting factor.
Doesn't need costly gear or exceptionally specific work force.	Introduction of nonlocal species might influence biodiversity.
The potential energy that has been stored can be used to produce nuclear power on a huge scale.	Concerns have been raised about the usage of contaminated plant biomass.

Source: Ghosh and Singh, 2005.

FUTURE OF PHYTOREMEDIATION

Among the vital points to the acceptance of phytoremediation are the selection of items using different samples and getting appropriate results, and its monetary reasonability. While these findings are encouraging, researchers admit that arrangement culture is not the same as soil culture. Many metals are locked in insoluble compounds in natural soil, making them less accessible, which is the most serious issue, according to Kochian (1996). The fate of phytoremediation is still in the early stages, and there are various obstacles that must be addressed. Both agronomic administration practices and plant hereditary capacities should be upgraded to foster monetarily helpful practices. Numerous hyperaccumulator plants still need to be discovered (Raskin and Ensley, 1994).

ROLE OF *AZOLLA* IN PHYTOREMEDIATION

The aquatic weeds implies that the plant dies when it super sider. The sort *weeds* has a place with the sole family Azollaceae (Konar and Kapoor, 1972). The six species that do not exceed our budget are classified into two subgenera. *Azolla filiculoides*, *Azolla caroliniana*, *Azolla mexicana*, and *Azolla microphylla* are the four species that make up the subvariety Euazolla. These four species are thought to have originated in North and South America's warm, subtropical, and tropical climates. Regenerative characters give the most helpful apparatus to ordered, but in a large portion of the examples sporocarps are generally missing, so recognizable proof is lacking. *Azolla* is utilized in transport all around the world, and it meets an assortment of conditions from debilitated water bodies to messy water bodies. It can take in and retain substances straightforwardly from rising waters and has a solid inclination for P, Fe, and K. It gathers these supplements a few times more than its essential requirements, and afterward leisurely conveys these enhancements as it rots. As a result of its ability to focus supplements like heavy metals, it has additionally been utilized in wastewater treatment (Wagner, 1997). *Azolla* has a high potential for use in phytoremediation of waste streams and soils. There are three significant parts to utilizing *Azolla* in phytoremediation. *Azolla* has an amazing capacity to ingest substantial amounts of metals, like different heavy metals. *Azolla's* obstruction and phyto-gathering limits with certain metal particles have been considered in earlier discussions. Scientists from China concentrated on the responses of four *Azolla* species to copper, manganese, iron, zinc, cobalt, cadmium, and other metals in the laboratory and discovered that the fixation limit of *Azolla* for metals influenced its development only minimally, with no impediment or no impact at all. The *Azolla* phytoremediation innovation is better than conventional strategies for metal expulsion from effluents. Metal expulsion happens as *Azolla* plants fill in compartments—for example, tanks or lakes containing wastewater—while in uninvolved (biosorption) conditions the sewage is run through dry weeds stuffed into channels. The two kinds of cycles have been proficient in treating new, harsh, and contaminated water. The chance of utilizing *Azolla* by dynamic cycles for decontaminating wastewaters has been investigated before. It was discovered (Jain et al., 1989) that sewage waste and biomass can remove significant amounts of metals from water and wastewater.

ROLE OF *SALVINIA* IN PHYTOREMEDIATION

These are little, gliding aquatics with crawling stems that spread and bear hairs but have no visible bases. The microgametical follicles, which are small hair-like developments, are not known to have any functional capacity (Schneider and Rubio, 1999; Skinner et al., 2007). Seagoing greeneries, in particular, have a high capacity to remove various foreign chemicals from the environment, such as heavy metals, natural mixes, and radionuclides (Olguin et al., 2002). The presence of specific supplements and chelants, with optional reliance on ecological conditions, is a major component of heavy metal evacuation and compartmentalization in *Salvinia* (Olguin et al., 2003). However, the manner of metal uptake varies depending on the plant species and metal (Maine et al., 2004). The uptake of heavy metals is thought to be driven by auxiliary vehicle proteins, such as channel proteins or H^+-linked transporter proteins, where the negative layer potential inside the plasma film drives the uptake of cations via auxiliary carriers. Metal limiting destinations are provided by free carboxylic groupings present on the surface water (Olguin et al., 2005). The increased convergences of lipids and carbohydrates on the plant surface act as cationic weak exchanger clusters, which contribute to metal sorption via particle responses. The Langmuir isotherm is also used to describe the sorption of heavy metals by dry biomass (Schneider and Rubio, 1999). A few other benefits promote *Salvinia* as a significant species for use in phytoremediation advances.

ROLE OF *SPIRODELA* IN PHYTOREMEDIATION

Spirodela polyrhiza, known as "giant duck grass," is much larger than other algae (*Lemna minor, Lemna gibba*), with leaves up to 1 cm long compared to a few millimeters for most other water lentil species. *Spirodela polyrhiza* has several roots, while *Lemna minor* and *Lemna gibba* have only one. Duckweeds reproduce asexually, but when growing conditions are unfavorable, such as during drought or cold, a dormant stage (known as a "shoot") is produced. Buds are specialized leaves that are smaller and thicker than sinking leaves and remain dormant until favorable growing conditions are resumed. This is about two times of other "fast-flowering" plants and more than three times that of traditional agricultural crops. A significant number of publications discuss and report on the use of duckweed for the biological treatment of wastewater, in particular the treatment of wastewater from agricultural activities with the collection of biomass for feedstock for biofuels.

ROLE OF *PISTIA* IN PHYTOREMEDIATION

Pistia is a seaweed that belongs to the Araceae family. *Pistia stratiotes* L., the single species involved, is also known as water cabbage (Quattrocchi, 2000). It's a hardy monocotyledon with thick, delicate leaves that form a rosette. It floats on the water's surface, its foundations dangling underneath swaying leaves. The leaves can grow to be up to 14 cm long, light green, with equal veins and wavy edges, and are coated in short hairs that create bushel-like structures that trap air bubbles, increasing the plant's lightness. The blossoms are dioecious and are hidden within the leaves in the center of the shrub. After careful preparation, little green berries form. The plant can also undergo agamic generation. A small stolon connects the parent and tiny offspring plants, framing dense mats (Ramey, 2001). *Pistia stratiotes* grows swiftly and has a large root architecture that aids in the removal of heavy metals from its seagoing environment. This plant has demonstrated numerous cases of Pb evacuation and collection; in a constructed wetland it essentially eradicated Pb (99.28%) and Cd (99.28%). *Pistia stratiotes* appears to be a fruitful and low-cost option to be considered in current profluent therapy (Miretzky et al., 2005). The phytoremediation potential of *Pistia stratiotes* L. for addressing lead accumulation from supplement-rich water was investigated. *Pistia* plants grew stronger in a Hoagland supplement arrangement having various Pb fixations. With increasing Pb levels in the development medium, the Pb concentration in plant tissues substantially increased, and the natural fixation factor (BCF) fundamentally decreased. Pb treatment has no negative effects on the total plant biomass. The production of leaves was not substantially different between the control and those generated at greater Pb convergences. The total chlorophyll content fell fundamentally as Pb openness increased. The natural chelates were used to mimic the states of a triggered phytoextraction measure. With increasing Pb fixations in the supplement arrangement, Pb content in both root and leaf tissues grew progressively (Jain et al., 1989). In all drugs, leaves gathered more Pb than establishes. The overall chlorophyll content dropped as Pb fixation and chelate level rose. Total amino corrosive substance grew in leaves, whereas absolute amino corrosive substance decreased in roots as a result of chelate expansion. The use of chelates to expand lead uptake resulted in a reduction in dry biomass weight (Bassi et al., 1990). In sufficiency of major metal uptake from contaminated supplement arrangement by four seagoing macrophytes (*Pistia stratiotes* L., *Salvinia auriculata* L., *Salvinia minima* Baker, and *Azolla filiculoides* Lam.), monitoring chlorophyll content and rate of occurrence for more than 14 days during the trial determined the influence of cadmium (3.5 mg/L and 10.5 mg/L) and lead (25 mg/L and 125 mg/L) on pressure indicators (Miretzky et al., 2005). During the first four days, there were considerable decreases in Cd and Pb fixations in the arrangements.

CONCLUSION

Effluent contamination of the environment, including heavy metals, is of serious concern owing to its impact on human health. In recent years the disposal of wastes from domestic, agricultural, and industrial sources has severely affected our ecosystem. Untreated effluents are released into water bodies, making them unfit for any use. A water crisis seen across the globe has to be dealt with on a priority basis. Reusing water can be a suitable alternative to circumvent this alarming situation. Nevertheless, with the help of several scientific tools wastewater may be converted to usable form, but it is interesting to note that some plants also have the ability to remove pollutants from water bodies. This process of removing pollutants from wastewater with the help of plants is called phytoremediation.

REFERENCES

Anjuli, S., Uniyal, P.L., Prasanna, R. and Ahuluwalia, A.S. (2012). Phytoremediation potential of aquatic macrophyte, *Azolla*. Ambhio, 41: 122–137.

Bassi, M., Grazia-Corradi, M. and Realini, M. (1990). Effect of chromium (VI) in two freshwater plants, Lemna minor and Pistia stratiotes 1 morphological observations. Cytobious, 62: 101–109.

Chang, J.S., Yoon, I.H. and Kim, K.W. (2009). Heavy metal and arsenic accumulating fern species As potential ecological indicators in As-contaminated abandoned mines. Ecological Indicators, 9: 1275–1279.

Chaoni, H.I., Zibilske, L.M. and Ohno, T. (2003). Effects of earthworm casts and compost on soil microbial activity and plant nutrient availability. Soil Biology and Biochemistry, 35(2): 295–302.

Ghosh, M. and Singh, S.P. (2005). A review on phytoremediation of heavy metals and utilization of its byproducts. Applied Ecology and Environmental Research, 3(1): 1–18.

Jain, S., Valsudevan, P. and Jha, N.K. (1989). Removal of some heavy metals from polluted water by aquatic plants: Studies on duckweed and water velvet. Biol. Wastes, 28: 115–126.

Konar, R.N. and Kapoor, R.K. (1972). Anatomical Studies on *Azolla pinnata*. Phytomorphology, 22: 211–223.

Lokeshwari, H. and Chandrappa, G.T. (2007). Effects of heavy metal contamination from anthropogenic sources on Dasarahalli tank, India. Lakes and Reservoirs: Research and Management, 12: 121–128.

Maine, M.A., Suñé, N.L. and Lagger, S.C. (2004). Chromium bioaccumulation: Comparison of the capacity of two floating aquatic macrophytes. Water Research, 38: 1494–1501.

Miretzky, P., Andrea, S. and Fernádez Cirelli, A. (2005). Simultaneous heavy metal removal mechanism by dead macrophytes. Chemosphere, 62: 247–254.

Olguin, E. J., Rodriguez, D., Sanchez, G., Hernandez, E. and Ramirez, M.E. (2003). Productivity, protein content and nutrient removal from anaerobic effluents of coffee wastewater in *Salvinia minima* ponds, under subtropical conditions. Acta Biotechnology, 23: 259–270.

Olguin, J., Hernandez, E. and Ramos, I. (2002). The effect of both different light conditions and the pH value on the capacity of *Salvinia minima* Baker for removing cadmium, lead and chromium. Acta Biotechnology, 22: 121–131.

Olguin, J., Hernandez, E. and Ramos, I. (2005). The effect of both different light conditions and the pH value on the capacity of *Salvinia minima* Baker for removing cadmium, lead and chromium. Acta Biotechnology, 22: 121–131.

Quattrocchi, U. (2000). CRC World Dictionary of Plant Names. Volume III: M-Q. CRC Press. p. 2084. ISBN 978-0-8493-2677-6.

Ramey, V. (2001). Water Lettuce (*Pistia stratiotes*). Center for Aquatic and Invasive Plants, University of Florida. http://plants.ifas.ufl.edu/node/328

Raskin, I. and Ensley. B.D. (1994). Phytoremediation of Toxic Metals: Using Plants to Clean Up the Environment. John Wiley & Sons, Inc., New York.

Salt, D, Kramer, U (2000). Mechanisms of metal hyperaccumulation in plants. In: Raskin I, Ensley B (eds), Phytoremediation of Toxic Metals. John Wiley and Sons Inc., New York, pp. 231–246.

Schneider, I.A.H. and Rubio, J. (1999). Sorption of heavy metal ions by the nonliving biomass of freshwater macrophytes. Environmental Science and Technology, 33: 2213–2217.

Skinner, K., Wright, N. and Porter-Goff, E. (2007). Mercury uptake and accumulation by four aquatic plants. Environmental Pollution, 145: 234–237.

Suñe, N., Sánchez, G., Caffaratti, S. and Maine, M.A. (2007). Cadmium and chromium removal kinetics from solution by two aquatic macrophytes. Environmental Pollution, 145: 467–473.

Wagner, G.M. (1997). *Azolla*: A review of its biology and utilization. The Botanical Review, 63: 1–26.

Yadav, K.K., Trivedi, S.P. (2009). Sublethal exposure of heavy metals induces micronuclei in fish, Channa punctata. Chemosphere, 77(11): 1495–1500.

14 A Combined Use of Plants and Microbes for Management in Wastewater Treatment Plants

Rim Werheni Ammeri, Yassmina Angar, and Najla Sadfi-Zouaoui

INTRODUCTION

Biological remediation, including bioremediation and phytoremediation, employs microorganisms and/or plants to remove, degrade, and/or immobilize contaminants in the environment. With its advantages being low cost, simple operation, and eco-friendliness, biological remediation has wide application in restoring contaminated sites (Cui et al., 2022). Phytoremediation is a sustainable and low-cost remediation approach for treating different kinds of water pollutants, including heavy metals, eutrophication-related components, antibiotics, persistent organic pollutants, and reverse osmosis concentrate (Al-Thani & Yasseen, 2020; Chang et al., 2020; Chen et al., 2021; Hu et al., 2021; Sharma et al., 2021). In recent years, some successful cases have been reported, including sites in cold climates and with co-contamination (Miri et al., 2019; Usmani et al., 2022). However, biological remediation takes time due to issues with its environmental adaptability, nutrient limitation, toxicity resistance, and limited contaminant bioavailability. To overcome these limitations, improvement strategies, such as immobilization of microbial/microalgal cells (Miri et al., 2019; Partovinia & Rasekh, 2018), construction of microbial/microalgal consortia (Li et al., 2022; Usmani et al., 2022), and use of amendments (i.e., chemical agents, nanomaterials, poultry manure, industrial wastes and microbial agents) have been employed (Song et al., 2019; Wang et al., 2020).

Bioaugmentation is a widely known approach to remediate heavy metal from contaminated environments by adding indigenous and exogenous microorganisms that can resist and reduce the toxicity of heavy metals (Hassan et al., 2019; Purwanti et al., 2020). Indigenous microorganisms are isolated from contaminated soil and reinoculated back into the corresponding contaminated soil (Purwanti & Hardiyanti, 2018). By contrast, exogenous microorganisms are isolated outside the contaminated areas and introduced into the desired contaminated area (Huang et al., 2019). Several of

 DOI: 10.1201/9781003442295-14

the processes involved in remediating heavy metals using microbes are biosorption, bioaccumulation, biochelation, biodigestion, biomineralization, and biotransformation (Nwaehiri et al., 2020). Several studies claimed the success of bioaugmentation in treating heavy metal–contaminated soil. Mahbub et al. (2017) reported the successful removal of Hg from artificial contaminated soil using *Sphingobium* SA2 with removal efficiency reaching 50%. Ibarrolaza et al. (2011) also mentioned the removal of hexavalent chromium from artificial contaminated soil by *Sphingomonas paucimobilis* 20006FA with removal efficiency of 90%. Polti et al. (2014) also reported the removal of hexavalent chromium from artificial contaminated soil by using a consortium of *Streptomyces* sp. M7, *Streptomyces* sp. MC1, *Streptomyces* sp. A5, and *Amycolatopsis tucumanensis* with removal efficiency of 86%. Microorganisms are selected based on two main criteria: ability to degrade targeted pollutants and ability to resist and survive in a wide range of environments. Several microorganisms such as bacteria, fungi, yeast, actinomycetes, and algae can resist and survive in a wide range of environments, including the ability to remove heavy metals from contaminated areas. This capability is mostly originated from microbial cell walls made up of polysaccharides, lipids, and protein that play an essential role in attaching metal ions with carboxylate, hydroxyl, and amino and phosphate groups (Girma et al., 2015, Purwanti et al., 2018), resulting in nontoxic complexed compounds (Kurniawan et al., 2022).

This chapter identifies the major role of phytoremediation in restoring wastewater. We discuss the current methods used in aquaculture for the treatment of wastewater, the opportunities to use improved processes, and the knowledge gaps that require further attention.

WASTEWATER TREATMENT

Wastewater is generally defined as "used" water contaminated by human activities (Mateo-Sagasta et al., 2015). Urban wastewater is specifically defined as wastewater mainly from kitchens, bathrooms, and laundry areas. Wastewater from food preparation, dishwashing, trash shredding, toilets, bathrooms, showers, and sinks also contributes to municipal wastewater (Ajibade et al., 2013). Additionally, it includes toilet drains, industrial waste, hospital waste, commercial waste, human waste, rainwater runoff, and all other types of hazardous, toxic, and biological waste (Jayaweera & Kasturiarachchi, 2004; Kumar and Chopra, 2012). Domestic municipal wastewater contains a variety of dissolved and suspended pollutants as well as high levels of biodegradable organic substances (Baldisserotto et al., 2004). Urban wastewater discharges, including domestic, industrial, and agricultural facilities, produce wastewater that can lead to the pollution of many lakes and rivers (Padmapriya & Murugesan, 2012; Salt et al., 1995). Wastewater can be divided into four categories: (1) Domestic: Wastewater discharged from residences, commercial establishments, and similar establishments. (2) Industrial: Wastewater that is mostly industrial waste. (3) Seepage/inflow: Extraneous water that enters the sewer system both indirectly and directly through leaky joints, cracks, porous walls, etc. Inflow water is rainwater that enters the sewer system through gutter connections, roof caps, foundation and underground drains, or manhole covers. (4) Rainwater: Runoff resulting from rainfall flooding (Al-Rekab et al., 2007). Municipal wastewater treatment involves removing various pollutants from municipal wastewater to make it suitable for discharge into the environment or for reuse (Singh et al., 2011). Many treatment methods are used to treat municipal wastewater, including trickling filters, activated sludge processes, oxidation ponds, and upturned sludge blankets (Solt et al., 1998). However, the technology of these processing methods is very expensive and energy-intensive. Phytoremediation of wastewater is being developed as an alternative to these treatment methods (Porra, 2002).

PHYTOREMEDIATION PROCESS

Phytoremediation is a low-cost, environmentally friendly, and long-term technology (Muthusaravanan et al., 2018). Phytoremediation technology is an emerging green approach for the detection, degradation, and removal of various types of pollutants in the environment (Cauwenberghe et al., 2016; Khan et al., 2022). The key concept of phytoremediation is the use of green plants and microorganisms to effectively remove pollutants from the environment. Plants use different methods to purify wastewater, sludge, soil, sediment, etc., which can be broadly called phytoremediation (Figure 14.1) (Jacob et al., 2018). These plants are known for their ability to remove a wide range of inorganic and organic pollutants, heavy metals, pesticides, and nutrients from various sources such as industrial and domestic wastewater, sewage, and agricultural runoff. Furthermore, these plants can grow in contaminated sites with large variations in temperature, pH, and nutrient content (Ali et al., 2020; Javed et al., 2019). Due to its eco-friendly nature, this approach has advantages over conventional techniques that have harmful effects on biological systems and the environment (Jeevanantham et al., 2019; Khan et al., 2022). The use of plant species

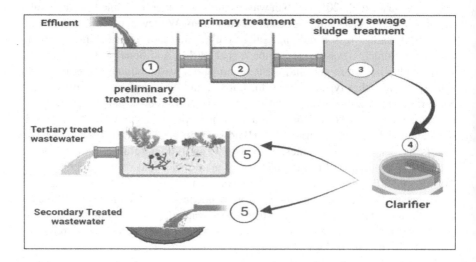

FIGURE 14.1 Graphic description of wastewater treatments.

can remove various types of pollutants that adversely affect human health and other biological systems (Khan et al., 2022). This technology has great potential in tropical regions as it favors plant growth and increases microbial activity due to climatic conditions (Liu et al., 2020).

REMOVAL OF POLLUTANTS FROM WASTEWATER BY PHYTOREMEDIATION PROCESS

The removal of various types of pollution by phytoremediation occurs through a four-step mechanism. Direct absorption of pollutants causes them to accumulate in plant tissues, where they are metabolized. It uses transpiration to remove various volatile organic hydrocarbons from its leaves and releases exudates from other plants to remove various pollutants. These exudates activate plant root–associated microorganisms, such as mycorrhizal fungi and the microorganism *Concortia*, and remove various pollutants (Boroş et al., 2014). Heavy metals pose a serious threat to the environment and ultimately have dangerous effects on human health. Various technologies have been developed to remove heavy metals, but they are expensive compared to plant-based removal of toxic metals in industry (Sumiahadi & Acar, 2018). Much research has been conducted on more cost-effective and environmentally friendly (plant-based) adsorbents for removing heavy metals (Joseph et al., 2019). Biosorption is an effective method to remove heavy metals and other toxic substances in the environment and has been extensively examined in several past studies (Calderón & Nava-Mesa, 2020). Coconut waste and black oak bark can remove metals such as lead, cadmium, and mercury (Alalwan et al., 2020). Due to the large amount of waste generated worldwide, there is a strong need for cost-effective and environmentally friendly adsorbents (Hossain et al., 2020). Plants and microorganisms are deeply involved in nitrogen removal (Tang et al., 2020). Levels of water pollution are increasing due to domestic and industrial wastewater contamination (Khan et al., 2022). The use of pesticide products in agriculture is a major cause of water pollution (Klöppel et al., 1976). Same plants has excellent ability to purify water pollution and is widely used in phytoremediation technology for pesticide treatment (Table 14.1) (Khan et al., 2022). In recent years, the presence of pharmaceutical compounds in the aquatic environment has been recognized as one of the emerging problems in environmental chemistry. Drugs and their metabolites have been detected in wastewater, groundwater, and even drinking water worldwide (Dordio et al., 2010). Petroleum refineries and petrochemical industries emit a variety of hydrocarbons, including polycyclic aromatic hydrocarbons (PAHs), which are major wastewater pollutants. These PAH compounds pose several health problems, mainly mutagenic, carcinogenic, and teratogenic, due to their benzene structure (Dsikowitzky et al., 2011; Khan et al., 2022).

MECHANISTIC APPROACH OF PHYTOREMEDIATION

When removing pollutants from water, several mechanisms are involved to convert the pollutants into nontoxic compounds, resulting in the removal of waste from the water (Cherian & Oliveira, 2005). These mechanisms include phytostabilization,

TABLE 14.1

Example of Plants Used in Phytoremediation Process

Pollutants	Examples of Plants Used	References
Dyes	*Salvinia molesta* plant	Al-Baldawi (2020)
Heavy metals	*Lemna minor*	Ali et al. (2020)
	Nicotiana tabacum	Naseer et al. (2022)
	Arabidopsis thaliana	Shafiq et al. (2018)
	Eichhornia crassipes	Thakur et al. (2016)
		Nazir et al. (2020)
Inorganic elements	*Eichhornia crassipes*	Khan et al. (2022)
	Phragmites australis	
	Pistia stratiotes along with *S. molesta*	
	Lactuca sativa	
	Centaurea cyanus	
Pesticides	*Lemna minor*	Werheni Ammeri et al. (2023)
	Eichhornia crassipes	Khan et al. (2022)
	Elodea canadensis	Olette et al. (2008)
Pharmaceuticals	*Typha* spp. *Phragmites, Iris*, and *Juncus*	Khan et al. 2022
Hydrocarbons	Vetiver and *Phragmites*	Alsghayer et al. 2020
Antibiotics	*Eichhornia crassipes*	Yan et al. 2021

nodule decomposition, phytofiltration (also called root filtration), phytoextraction, phytovolatilization, and phytoaccumulation, as shown in Figure 14.2 (Jeevanantham et al., 2019). Phytostabilization uses an accumulation or adsorption mechanism. In this approach, contaminants in groundwater or soil are adsorbed by roots or accumulate in the rhizosphere, preventing their movement from one location to another in the environment (Raskin et al., 1997). Environmental characteristics

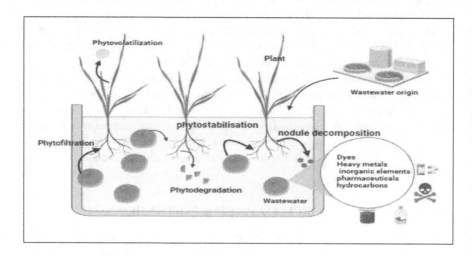

FIGURE 14.2   General diagram of phytoremediation processes in wastewater.

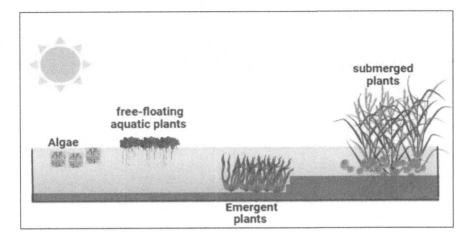

FIGURE 14.3   Different type of plants growth in water.

influence the interaction between water in rhizome sanitation and microorganisms and plants in phytoremediation, but phytoremediation in the case of rhizome decomposition removes pollutants, especially heavy metals and organic wastes, in the rhizosphere. It breaks down and is converted into nontoxic or less toxic substances. This process is also enhanced by using different types of microorganisms (Olaniran et al., 2013). The nodule filtration approach also uses the same mechanism as phytoremediation. However, in this case, pollutants are absorbed through the roots of the plant. Plant decomposition breaks down waste products and pollutants through metabolic pathways. Plants absorb metals and waste products from the environment and wastewater and use a variety of enzymes to break them down into nontoxic compounds.

This process is also known as plant transformation (Dixit et al., 2015). In the plant volatilization approach, plants absorb different types of wastewater and convert them into nontoxic compounds. These nontoxic compounds are then released into the atmosphere through the process of transpiration from the leaves. This process is also called phytoaccumulation when waste products accumulate in different parts of the plant such as roots, shoots, and leaves (Jeevanantham et al., 2018). Different types of plant species have been used to remove different types of heavy metals, organic wastes, and other types of pollutants (Figure 14.3) (Ali et al., 2020; Javed et al., 2019).

TYPES OF AQUATIC PLANTS IN PHYTOREMEDIATION PROCESS

Aquatic plants are plant species visible to the naked eye growing in wetlands, such as lakes, ponds, and streams. They are also found in wastewater streams considered eutrophic media. Various species have been inventoried around the world in various aquatic environments. Their abundance and their distribution depend strongly on the hydraulic regime, nutrients, and water quality (Mishra et al., 2008). Numerous studies have been published on the use of various aquatic plants for phytoremediation of

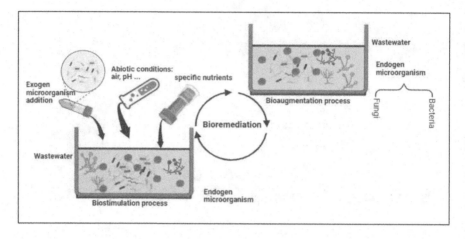

FIGURE 14.4   Different types of plant growth in water.

both organic and inorganic contaminants. Macrophytes, with fixed or free culture (floating, submerged, and emergent) are excellent candidates for cleaning up contaminated sites with toxic or undesirable substances (Figure 14.4). Several species have been tested in the laboratory, and some have been used on a large scale (Khellaf et al., 2022).

TRANSGENIC PLANTS USED IN PHYTOREMEDIATION PROCESS

METAORGANISM AS A STRATEGY TO IMPROVE PHYTOREMEDIATION

Although the competition-based model may appear oversimplified, and incorporating all partners (next to bacteria and fungi, also archaea, arthropoda, protista, and other macro- and mesofauna) will better represent phytoremediation activity, the model shows the inherent interactive character of plant–microbial interactions for contaminant degradation, and it provides a basis for the development of improved and more effective phytoremediation strategies (Wang et al., 2020). In fact, a naturally evolved rhizosphere microbiome may not be optimal for phytoremediation. For degraders to abundantly colonize the niche, there should be a high dominance at the start of the competition (nichecolonization), so that enough degraders are present toward off opportunists that do not contribute to the degradation of bacteria (Table 14.2). In addition to the use of aerobic and anaerobic mixed cultures, microbial isolates have received much attention for decomposing organic pollutants in wastewater (Ahmed et al., 2022). Contaminated water from different locations can be successfully treated using bacterially aided phytoremediation (Bala et al., 2022). Moreover, loss of sensitive taxa can severely alter the way the plant host interacts with the other microbiome members (Bell et al., 2014). Furthermore, the selected host species can in some cases be incompatible for inoculation with certain types of bacteria or fungi, because they are not naturally selected for by the host. With the model in mind, we can see that many new opportunities have arisen to optimize the

TABLE 14.2

Examples of Bacteria Used in Bioaugmentation Process in Wastewater Treatment

Microorganism	Process Target	References
Pseudomonas putida	Transformation of pesticide	Werheni Ammeri et al. (2023)
Bacillus sp.	Metal accumulation	Shin et al. (2012)
Dechloromonas sp. (RCB)	Benzene, toluene, xylene	Chakraborty et al. (2005)
Mycobacterium gilvum	Pyrene, benzo(a)pyrene	Toyama et al. (2018)
Pseudomonas fluorescens	Uptake of Pb and Ni	Wang et al. (2022)
Stenotrophomonas maltophilia	Increase the uptake of heavy metals	
Myceliophthora thermophila	Polycyclic aromatic hydrocarbons	Bulter et al. (2003)

metaorganism in phytoremediation, rather than the plant and microbe part separately (Werheni Ammeri et al., 2023).

Most microorganisms are isolated from contaminated sites (indigenous) because they can easily adapt to the environment compared to exogenous microorganisms (Purwanti & Hardiyanti, 2018). In contrast, exogenous microorganisms face challenges in colonizing contaminated sites and maintaining immobilization effects due to the need to compete with natural microorganisms (Wang et al., 2020).

Four strategies are proposed which are likely to be important in redirecting the microbiome: (1) selecting plants not only for high biomass/rapid growth rate and tolerance, but also for their global interactions with the microbiome, (2) changing the root exudates (microbial diet), (3) taking into account vertical transmission and promoting a higher "immigration" rate (feeding the supply lines, priority effects), and (4) altering the competitive interactions that evolve between host–microbe, microbe–microbe, and microbe–environment (Aliko et al., 2022).

ROLE OF ENDOPHYTE BACTERIA IN PHYTOREMEDIATION PROCESS

Plants are colonized by a variety of microbiota, including bacteria, fungi, yeasts, viruses, and protists, as well as epiphytes such as algae and nematodes (Hegazy et al., 2009). Plant and microbial populations are dynamic, and fluctuations in microbial communities have been observed, influenced by large variations in physical and nutritional conditions and other biotic and abiotic influences (Germaine et al., 2004). Endophytes often benefit host plants through nitrogen fixation, promoting seedling germination, protecting against environmental stressors, improving nutrient availability and vitamin supply, and protecting against and removing pollutants (Mastretta et al., 2006; Ryan et al., 2007). In addition, endophytes can produce bioactive compounds that are associated with increased plant growth and health and provide protection against abiotic and biotic stresses (Krutz et al., 2005; Sun et al., 2014). Moreover, endophytes can be recruited to the host by chemotaxis, electrotaxis, or simply by chance encounter, most commonly from the roots (Lauren & Claudia, 2016). The natural ability of some endophytic fungi to degrade xenobiotics has been studied with the aim of improving the efficiency of phytoremediation (Khan et al., 2014; Liang et al., 2014). The advantage of using endoparasitic degradants for remediation is that toxic xenobiotics can be degraded within the plant, reducing the effects of phytotoxicity and reducing the effects of toxicity on herbivores located at or near the contaminated site (Cauwenberghe & Roote, 1998). Modifying the plant rhizosphere microbiome is key to further advances and broader applications of bioaugmented phytoremediation (Werheni Ammeri et al., 2023). In endophytic phytoremediation, endophytic bacteria interact and exchange genes with bacterial communities in the rhizosphere and phyllosphere (Weyens et al., 2009). In this way, there is no need for the donor strain to survive, and the entire microbial community develops degradative capacity. Therefore, the use of endophyte augmentation in combination with phytoremediation or endophyte phytoaugmentation can provide an effective option for *in situ* treatment of wastewater and waste systems (Lauren & Claudia, 2016).

CONCLUSION

Plants have proven their effective roles in the remediation of pollutants in wastewater, which could be considered crucial in the near future. Scientists can assess the plants' role in removing contaminants from wastewater, which cannot be achieved by small organisms and microorganisms. New avenues of phytoremediation have been addressed in this review on several types of contaminants that exist in large quantities in wastewater and have harmful effects on the environment and human health. Scientists need to think about integrating microorganisms with plants to enhance their efficiency as microorganisms have a long history of pollutant biodegradation and remediation. Nevertheless, this depends on the types of microorganisms and plants used, their resistance, the level of contamination, and their soil physicochemical characteristics. However, these limitations could be surpassed by designing new microbial and plant species that express specific genes of interest or by combining microbial and/or phytoremediation with physicochemical strategies. Currently, there are no reports on the application of plants for radioactive element removal; these ideas might be beneficial for research.

FUNDING INFORMATION

Authors state no funding involved.

AUTHOR CONTRIBUTIONS

All authors contributed equally to writing and revising this manuscript.

CONFLICT OF INTEREST

Authors state no conflict of interest. Data availability statement: The datasets generated during and/or analyzed during the current study are available from the corresponding author on reasonable request.

REFERENCES

Ahmad, H. A., Ahmad, S., Cui, Q., Wang, Z., Wei, H., Chen, X., & Tawfik, A. (2022). The environmental distribution and removal of emerging pollutants, highlighting the importance of using microbes as a potential degrader: A review. Science of the Total Environment, 809, 151926.

Ahmed, T., Noman, M., Jiang, H., Shahid, M., Ma, C., Wu, Z., et al. (2022). Bioengineered chitosan-iron nanocomposite controls bacterial leaf blight disease by modulating plant defense response and nutritional status of rice (*Oryza sativa* L.). Nano Today, 45, 101547.

Ajibade, F. O., Adeniran, K. A., & Egbuna, C. K. (2013). Phytoremediation efficiencies of water hyacinth in removing heavy metals in domestic sewage (a case study of university of Ilorin, Nigeria). The International Journal of Engineering and Science, 2(12), 16–27.

Al-Baldawi, I. A., Abdullah, S. R. S., Almansoory, A. F., Ismail, N. I., Hasan, H. A., & Anuar, N. (2020). Role of Salvinia molesta in biodecolorization of methyl orange dye from water. Scientific Reports, 10(1), 13980.

Al-Rekab, E. A., & Al-Asadi, N. A. B. (2018). Histological effect of curcuma Longa and Zingiber officinale on the liver damage after treatment with hydrogen peroxide in rats (*Rattus norvegicus*). Journal of Thi-Qar University, 13(1), 113–126.

Al-Thani, R. F., & Yasseen, B. T. (2020). Phytoremediation of polluted soils and waters by native Qatari plants: Future perspectives. Environmental Pollution, 259, 113694.

Alalwan, H. A., Kadhom, M. A., & Alminshid, A. H. (2020). Removal of heavy metals from wastewater using agricultural byproducts. Journal of Water Supply: Research and Technology—AQUA, 69(2), 99–112.

Ali, S., Abbas, Z., Rizwan, M., Zaheer, I., Yavaş, İ., Ünay, A., et al. (2020). Application of floating aquatic plants in phytoremediation of heavy metals polluted water: A review. Sustainability, 12(5), 1927. https://doi.org/10.3390/su12051927

Aliko, V., Multisanti, C. R., Turani, B., & Faggio, C. (2022). Get rid of marine pollution: Bioremediation an innovative, attractive, and successful cleaning strategy. Sustainability, 14(18), 11784.

Alsghayer, R., Salmiaton, A., Mohammad, T., Idris, A., & Ishak, C. F. (2020). Removal efficiencies of constructed wetland planted with phragmites and vetiver in treating synthetic wastewater contaminated with high concentration of PAHs. Sustainability, 12(8), 3357.

Bala, M., Catena, F., Kashuk, J., De Simone, B., Gomes, C. A., Weber, D., & Moore, E. E. (2022). Acute mesenteric ischemia: Updated guidelines of the world society of emergency surgery. World Journal of Emergency Surgery, 17(1), 1–17.

Baldisserotto, C., Ferroni, L., Medici, V., Pagnoni, A., Pellizzari, M., Fasulo, M. P., Fagioli, F., Bonora, A., & Pancaldi, S. (2004). Specific intra-tissue responses to manganese in the floating lamina of *Trapa natans* L. Plant Biology, 6, 578–589.

Bell, T. H., Joly, S., Pitre, F. E., & Yergeau, E. (2014). Increasing phytoremediation efficiency and reliability using novel omics approaches. Trends in Biotechnology, 32(5), 271–280.

Bhargava, A., Carmona, F. F., Bhargava, M., & Srivastava, S. (2012). Approaches for enhanced phytoextraction of heavy metals. Journal of Environmental Management, 105, 103–120.

Boroş, M. N., Micle, V., & Avram, S. E. (2014). Study on the mechanisms of phytoremediation. ECOTERRA - Journal of Environmental Research and Protection, 11, 67–73.

Butler, E. A., Egloff, B., Wilhelm, F. H., Smith, N. C., Erickson, E. A., & Gross, J. J. (2003). The social consequences of expressive suppression. Emotion, 3, 48–67.

Bulter, T., Alcalde, M., Sieber, V., Meinhold, P., Schlachtbauer, C., & Arnold, F. H. (2003). Functional expression of a fungal laccase in *Saccharomyces cerevisiae* by directed evolution. Applied and Environmental Microbiology, 69(2), 987–995.

Calderón-Ospina, C. A., & Nava-Mesa, M. O. (2020). B vitamins in the nervous system: Current knowledge of the biochemical modes of action and synergies of thiamine, pyridoxine, and cobalamin. CNS Neuroscience & Therapeutics, 26(1), 5–13.

Chakraborty, A., Haskell, B. A., Keller, S., Speck, J. S., Denbaars, S. P., Nakamura, S., & Mishra, U. K. (2005). Demonstration of nonpolar m-plane InGaN/GaN light-emitting diodes on free-standing m-plane GaN substrates. Japanese Journal of Applied Physics, 44(1), L173.

Chang, Y. N., Zhu, C., Jiang, J., Zhang, H., Zhu, J. K., & Duan, C. G. (2020). Epigenetic regulation in plant abiotic stress responses. Journal of Integrative Plant Biology, 62(5), 563–580.

Chen, L., Liu, J., Zhang, W., Zhou, J., Luo, D., & Li, Z. (2021). Uranium (u) source, speciation, uptake, toxicity and bioremediation strategies in soil-plant system: A review. Journal of Hazardous Materials, 413, 125319.

Cherian, S., & Oliveira, M. M. (2005). Transgenic plants in phytoremediation: Recent advances and new possibilities. Environmental Science & Technology, 39(24), 9377–9390.

Cui, B., Liang, H., Li, J., Zhou, B., Chen, W., Liu, J., & Li, B. (2022). Development and characterization of edible plant-based fibers using a wet-spinning technique. Food Hydrocolloids, 133, 107965.

Czako, M., Feng, X., He, Y., Liang, D., & Marton, L. (2006). Transgenic spartina alterniflora for phytoremediation. Environmental Geochemistry and Health, 28, 103–110.

Dar, M. A. (2011). A review: Plant extracts and oils as corrosion inhibitors in aggressive media. Industrial Lubrication and Tribology, 63(4), 227–233.

Dixit, R., Wasiullah, D., Malaviya, D., Pandiyan, K., Singh, U., Sahu, A., et al. (2015). Bioremediation of heavy metals from soil and aquatic environment: An overview of principles and criteria of fundamental processes. Sustainability, 7(2), 2189–2212.

Dordio, A., Carvalho, A. P., Teixeira, D. M., Dias, C. B., & Pinto, A. P. (2010). Removal of pharmaceuticals in microcosm constructed wetlands using typha spp. And LECA. Bioresource Technology, 101(3), 886–892.

Doty, S. L. (2008). Enhancing phytoremediation through the use of transgenics and endophytes. New Phytologist, 179, 318–333.

Dsikowitzky, L., Nordhaus, I., Jennerjahn, T. C., Khrycheva, P., Sivatharshan, Y., Yuwono, E., et al. (2011). Anthropogenic organic contaminants in water, sediments and benthic organisms of the mangrove-fringed Segara Anakan Lagoon, Java, Indonesia. Marine Pollution Bulletin, 62(4), 851–862.

Eapen, S., Singh, S., & D'souza, S. (2007). Advances in development of transgenic plants for remediation of xenobiotic pollutants. Biotechnology Advances, 25, 442–451.

Germaine, K. J., Keogh, E., Garcia-Cabellos, G., Borremans, B., et al. (2004). Colonisation of poplar trees by GFP expressing bacterial endophytes. FEMS Microbiology Ecology, 48, 109–118.

Girma, A., Fischer, E., & Dumbo, B. (2015). Vascular plant diversity and community structure of Nandi forests, Western Kenya. Journal of East African Natural History, 103(2), 125–152.

Girma, S., Gong, Y., Görg, H., & Lancheros, S. (2015). Estimating direct and indirect effects of foreign direct investment on firm productivity in the presence of interactions between firms. Journal of International Economics, 95(1), 157–169.

Hassan, M. K., McInroy, J. A., & Kloepper, J. W. (2019). The interactions of rhizodeposits with plant growth-promoting rhizobacteria in the rhizosphere: A review. Agriculture, 9(7), 142.

Hegazy, A. K., Kabiel, H. F., & Fawzy, M. (2009). Duckweed as heavy metal accumulator and pollution indicator in industrial wastewater ponds. Desalination and Water Treatment, 12(1–3), 400–406.

Hossain, N., Bhuiyan, M. A., Pramanik, B. K., Nizamuddin, S., & Griffin, G. (2020). Waste materials for wastewater treatment and waste adsorbents for biofuel and cement supplement applications: A critical review. Journal of Cleaner Production, 255, 120261.

Hu, H. W., Chen, Q. L., & He, J. Z. (2022). The end of hunger: Fertilizers, microbes and plant productivity. Microbial Biotechnology, 15(4), 1050–1054.

Hu, X., Ji, Z., Gu, S., Ma, Z., Yan, Z., Liang, Y., & Liang, H. (2022). Mapping the research on desulfurization wastewater: Insights from a bibliometric review (1991–2021). Chemosphere, 314, 137678.

Huang, X. F., Ye, G. Y., Yi, N. K., Lu, L. J., Zhang, L., Yang, L. Y., & Liu, J. (2019). Effect of plant physiological characteristics on the removal of conventional and emerging pollutants from aquaculture wastewater by constructed wetlands. Ecological Engineering, 135, 45–53.

Ibarrolaza, A., Coppotelli, B. M., Del Panno, M. T., Donati, E. R., & Morelli, I. S. (2011). Application of the knowledge-based approach to strain selection for a bioaugmentation process of phenanthrene-and cr (VI)-contaminated soil. Journal of Applied Microbiology, 111(1), 26–35.

Jacobs, A., De Brabandere, L., Drouet, T., Sterckeman, T., & Noret, N. (2018). Phytoextraction of Cd and Zn with *Noccaea caerulescens* for urban soil remediation: Influence of nitrogen fertilization and planting density. Ecological Engineering, 116, 178–187.

Javed, H., Nagoor Meeran, M. F., Azimullah, S., Adem, A., Sadek, B., & Ojha, S. K. (2019). Plant extracts and phytochemicals targeting α-synuclein aggregation in Parkinson's disease models. Frontiers in Pharmacology, 9, 1555.

Jayaweera, M. W., & Kasturiarachchi, J. C. (2004). Removal of nitrogen and phosphorus from industrial wastewaters by phytoremediation using water hyacinth (Eichhornia crassipes (mart.) Solms). Water Science and Technology, 50(6), 217–225.

Jeevanantham, S., Hosimin, S., Vengatesan, S., & Sozhan, G. (2018). Quaternized poly (styrene-co-vinylbenzyl chloride) anion exchange membranes: Role of different ammonium cations on structural, morphological, thermal and physio-chemical properties. New Journal of Chemistry, 42(1), 380–387.

Jeevanantham, S., Saravanan, A., Hemavathy, R., Kumar, P. S., Yaashikaa, P., & Yuvaraj, D. (2019). Removal of toxic pollutants from water environment by phytoremediation: A survey on application and future prospects. Environmental Technology & Innovation, 13, 264–276.

Joseph, L., Jun, B.-M., Flora, J. R., Park, C. M., & Yoon, Y. (2019). Removal of heavy metals from water sources in the developing world using low-cost materials: A review. Chemosphere, 229, 142–159.

Kawahigashi, H. (2009). Transgenic plants for phytoremediation of herbicides. Current Opinion Biotechnology, 20, 225–230.

Khan, A. U., Khan, A. N., Waris, A., Ilyas, M., & Zamel, D. (2022). Phytoremediation of pollutants from wastewater: A concise review. Open Life Sciences, 17, 488–496.

Khan, M. I. R., Asgher, M., & Khan, N. A. (2014). Alleviation of salt-induced photosynthesis and growth inhibition by salicylic acid involves glycinebetaine and ethylene in mungbean (*Vigna radiata* L.). Plant Physiology and Biochemistry, 80, 67–74.

Khellaf, A., Garcia, N. M., Tajsic, T., Alam, A., Stovell, M. G., Killen, M. J., & Helmy, A. (2022). Focally administered succinate improves cerebral metabolism in traumatic brain injury patients with mitochondrial dysfunction. Journal of Cerebral Blood Flow & Metabolism, 42(1), 39–55.

Kidd, P., Prieto-Fernández, A., Monterroso, C., & Acea, M. (2008). Rhizosphere microbial community and hexachlorocyclohexane degradative potential in contrasting plant species. Plant Soil, 302, 233–247.

Klöppel, H., Kördel, W., & Stein, B.(1997). Herbicide transport by surface runoff and herbicide retention in a filter strip—Rainfall and runoff simulation studies. Chemosphere, 35(1–2), 129–141.

Kozminska, A., Wiszniewska, A., Hanus-Fajerska, E., & Muszyńska, E. (2018). Recent strategies of increasing metal tolerance and phytoremediation potential using genetic transformation of plants. Plant Biotechnology Report, 12, 1–14.

Krutz, L. J., Senseman, S. A., Zablotowicz, R. M., & Matocha, M. A. (2005). Reducing herbicide runoff from agricultural fields with vegetative filter strips: A review. Weed Science, 53(3).

Kumar, V., & Chopra, A. K. (2012). Monitoring of physico-chemical and microbiological characteristics of municipal wastewater at treatment plant, Haridwar (Uttarakhand) India. Journal of Environmental Science and Technology, 5(2), 109–118.

Kumar, V., & Chopra, A. K. (2012). Fertigation effect of distillery effluent on agronomical practices of *Trigonella foenum-graecum* L. (Fenugreek). Environmental Monitoring and Assessment, 184(3), 1207–1219.

Kurniawan, T. A., Othman, M. H. D., Hwang, G. H., & Gikas, P. (2022). Unlocking digital technologies for waste recycling in industry 4.0 era: A transformation towards a digitalization-based circular economy in Indonesia. Journal of Cleaner Production, 357, 131911.

Lauren, K. R., & Gunsch, C. K. (2016). Endophytic phytoaugmentation: Treating wastewater and runoff through augmented phytoremediation. Industrial Biotechnology, 12(2). https://doi.org/10.1089/ind.2015.0016

Li, J., Li, D., Xiong, C., & Hoi, S. (2022). Blip: Bootstrapping language-image pre-training for unified vision-language understanding and generation. In International Conference on Machine Learning (pp. 12888–12900). PMLR.

Liang, Y., Meggo, R., Hu, D., Schnoor, J. L., & Mattes, T. E. (2014). Enhanced polychlorinated biphenyl removal in a switchgrass rhizospheres by bioaugmentation with *Burkholderia xenovorans* LB400. Ecological Engineering, 71, 215–222.

Liu, M., Yu, H., Ouyang, B., Shi, C., Demidchik, V., Hao, Z., & Shabala, S. (2020). NADPH oxidases and the evolution of plant salinity tolerance. Plant, Cell & Environment, 43(12), 2957–2968.

Mahbub, K. R., Krishnan, K., Naidu, R., Andrews, S., & Megharaj, M. (2017). Mercury toxicity to terrestrial biota. Ecological Indicators, 74, 451–462.

Mastretta, C., Barac, T., Vangronsveld, J., Newman, L., Taghavi, S., & Lelie, D. V. D. (2006). Endophytic bacteria and their potential application to improve the phytoremediation of contaminated environments. Biotechnology and Genetic Engineering Reviews, 23(1), 175–188.

Mateo-Sagasta, J., Raschid-Sally, L., & Thebo, A. (2015). Global wastewater and sludge production, treatment and use. Wastewater: Economic Asset in an Urbanizing World, 15–38.

Miri, A., Dragovich, D., & Dong, Z. (2019). Wind-borne sand mass flux in vegetated surfaces–Wind tunnel experiments with live plants. Catena, 172, 421–434.

Mishra, P. K., Mishra, S., Selvakumar, G., Bisht, S. C., Bisht, J. K., Kundu, S., & Gupta, H. S. (2008). Characterisation of a psychrotolerant plant growth promoting pseudomonas sp. Strain PGERs17 (MTCC 9000) isolated from North Western Indian Himalayas. Annals of Microbiology, 58, 561–568.

Muthusaravanan, S., Sivarajasekar, N., Vivek, J. S., Paramasivan, T., Naushad, M., Prakashmaran, J., & Al-Duaij, O. K. (2018). Phytoremediation of heavy metals: Mechanisms, methods and enhancements. Environmental Chemistry Letters, 16, 1339–1359.

Naseer, H., Shaukat, K., Zahra, N., Hafeez, M. B., Raza, A., Nizar, M., & Ali, H. M. (2022). Appraisal of foliar spray of iron and salicylic acid under artificial magnetism on morpho-physiological attributes of pea (*Pisum sativum* L.) plants. PLoS ONE, 17(4), e0265654.

Nazir, M., Idrees, I., Idrees, P., Ahmad, S., Ali, Q., & Malik, A. (2020). Potential of water hyacinth (*Eichhornia crassipes* L.) for phytoremediation of heavy metals from waste water. Biological and Clinical Sciences Research Journal, 2020(1). https://doi.org/10.54112/bcsrj.v2020i1.6

Nwaehiri, U. L., Akwukwaegbu, P. I., & Nwoke, B. E. B. (2020). Bacterial remediation of heavy metal polluted soil and effluent from paper mill industry. Environmental Analysis, Health and Toxicology, 35(2), e2020009. https://doi.org/10.5620/eaht.e2020009

Olaniran, A. O., Balgobind, A., & Pillay, B. (2013). Bioavailability of heavy metals in soil: Impact on microbial biodegradation of organic compounds and possible improvement strategies. International Journal of Molecular Sciences, 14(5), 10197–10228. https://doi.org/10.3390/ijms140510197

Olette, R., Couderchet, M., Biagianti, S., & Eullaffroy, P. (2008). Toxicity and removal of pesticides by selected aquatic plants. Chemosphere, 70(8), 1414–1421. https://doi.org/10.1016/j.chemosphere.2007.09.016

Padmapriya, G., & Murugesan, A. G. (2012). Phytoremediation of various heavy metals (Cu, Pb and Hg) from aqueous solution using water hyacinth and its toxicity on plants. International Journal of Environmental Biology, 2(3), 97–103.

Partovinia, A., & Rasekh, B. (2018). Review of the immobilized microbial cell systems for bioremediation of petroleum hydrocarbons polluted environments. Critical Reviews in Environmental Science and Technology, 48(1), 1–38.

Pilon-Smits, E. (2005). Phytoremediation. Annual Review of Plant Biology, 56, 15–39.

Polti, M. A., Aparicio, J. D., Benimeli, C. S., & Amoroso, M. J. (2014). Simultaneous bioremediation of Cr (VI) and lindane in soil by actinobacteria. International Biodeterioration & Biodegradation, 88, 48–55.

Porra, R. J. (2002). The chequered history of the development and use of simultaneous equations for the accurate determination of chlorophylls a and b. Photosynthesis Research, 73, 149–156.

Purwanti, A., & Hardiyanti, M. (2018). Strategi Penyelesaian Tindak Kekerasan Seksual Terhadap Perempuan dan Anak Melalui RUU Kekerasan Seksual. Masalah-Masalah Hukum, 47(2), 138–148.

Purwanti, I. F., Obenu, A., Tangahu, B. V., Kurniawan, S. B., Imron, M. F., & Abdullah, S. R. S. (2020). Bioaugmentation of Vibrio alginolyticus in phytoremediation of aluminium-contaminated soil using Scirpus grossus and Thypa angustifolia. Heliyon, 6(9), e05004.

Purwanti, I. F., Titah, H. S., Tangahu, B. V., & Kurniawan, S. B. (2018). Design and application of wastewater treatment plant for "pempek" food industry, Surabaya, Indonesia. International Journal of Civil Engineering and Technology (IJCIET), 9, 1751–1765.

Raskin, I., Smith, R. D., & Salt, D. E. (1997). Phytoremediation of metals: Using plants to remove pollutants from the environment. Current Opinion in Biotechnology, 8(2), 221–226.

Rayu, S., Karpouzas, D. G., & Singh, B. K. (2012). Emerging technologies in bioremediation: Constraints and opportunities. Biodegradation, 23, 917–926.

Rustiana, Y., Kurniawan, S. B., Marlena, B., Hidayat, M. R., Kadier, A., Ma P. C., et al. (2023). Recent progress of phytoremediation-based technologies for industrial wastewater treatment. Journal of Ecological Engineering, 24(2), 208–220.

Ryan, R. P., Ryan, D. J., & Dowling, D. N. (2007). Plant protection by the recombinant, root-colonising Pseudomonas fluorescens F113rifPCB strain expressing arsenic resistance: Improving rhizoremediation. Letters in Applied Microbiology, 45(5), 668–674.

Salt, D. E., Blaylock, P. B. A., Nanda Kumar, V., Dushenkov, B. D., Ensley, I., & Raskin, I. (1995). Phytoremediation: A novel strategy for the removal of toxic metals from the environment using plants. Biotechnology, 13, 468–474.

Sas-Nowosielska, A., Kucharski, R., Małkowski, E., Pogrzeba, M., Kuperberg, J. M., & Kryński, K. J. E. P. (2004). Phytoextraction crop disposal—an unsolved problem. Environmental Pollution, 128(3), 373–379.

Shafiq, M., Alazba, A. A., & Amin, M. T. (2018). Removal of heavy metals from wastewater using date palm as a biosorbent: A comparative review. Sains Malaysiana, 47(1), 35–49.

Sharma, P., Pandey, A. K., Kim, S. H., Singh, S. P., Chaturvedi, P., & Varjani, S. (2021). Critical review on microbial community during in-situ bioremediation of heavy metals from industrial wastewater. Environmental Technology & Innovation, 24, 101826.

Shin, M.-N., Shim, J., You, Y., Myung, H., Bang, K.-S., Cho, M., Kamala-Kannan, S., & Oh, B.-T. (2012). Characterization of lead resistant endophytic bacillus sp. MN3-4 and its potential for promoting lead accumulation in metal hyperaccumulator Alnus firma. Journal of Hazardous Materials, 199, 314–320.

Singh, G., Singh, S., & Jindal, N. (2011). Environment friendly antibacterial activity of water chestnut fruits. Journal of Biodiversity and Environmental Sciences, 1(1), 26–34.

Solt, J. C., & Schlatter, J. C. (1998). Tests of catalytic-combustion technology show low emissions. Oil and Gas Journal, 96(14).

Song, Y. C., Kim, I. S., & Koh, S. C. (1998). Demulsification of oily wastewater through a synergistic effect of ozone and salt. Water Science and Technology, 38(4–5), 247–253.

Song, C., Qiu, Y., Li, S., Liu, Z., Chen, G., Sun, L., & Kitamura, Y. (2019). A novel concept of bicarbonate-carbon utilization via an absorption-microalgae hybrid process assisted with nutrient recycling from soybean wastewater. Journal of Cleaner Production, 237, 117864.

Sumiahadi, A., & Acar, R. (2018). A review of phytoremediation technology: Heavy metals uptake by plants. IOP Conference Series: Earth and Environmental Science, 142, 012023.

Sun, K., Liu, J., Jin, L., & Gao, Y. Z. (2014). Utilizing pyrene-degrading endophytic bacteria to reduce the risk of plant pyrene contamination. Plant Soil, 347, 251–263.

Tang, S., Liao, Y., Xu, Y., Dang, Z., Zhu, X., & Ji, G. (2020). Microbial coupling mechanisms of nitrogen removal in constructed wetlands: A review. Bioresource Technology, 314, 123759.

Thakur, S., Singh, L., Ab Wahid, Z., Siddiqui, M. F., Atnaw, S. M., & Din, M. F. M. (2016). Plant-driven removal of heavy metals from soil: Uptake, translocation, tolerance mechanism, challenges, and future perspectives. Environmental Monitoring and Assessment, 188(4), 206.

Toyama, T., Hanaoka, T., Tanaka, Y., Morikawa, M., & Mori, K. (2018). Comprehensive evaluation of nitrogen removal rate and biomass, ethanol, and methane production yields by combination of four major duckweeds and three types of wastewater effluent. Bioresource Technology, 250, 464–473.

Uchida, E., Ouchi, T., Suzuki, Y., Yoshida, T., Habe, H., Yamaguchi, I., & Nojiri, H. (2005). Secretion of bacterial xenobiotic-degrading enzymes from transgenic plants by an apoplastic expressional system: An applicability for phytoremediation. Environmental Science & Technology, 39, 7671–7677.

Usmani, Z., Sharma, M., Diwan, D., Tripathi, M., Whale, E., Jayakody, L. N., et al. (2022). Valorization of sugar beet pulp to value-added products: A review. Bioresource Technology, 346, 126580.

Van Aken, B. (2008). Transgenic plants for phytoremediation: Helping nature to clean up environmental pollution. Trends in Biotechnology, 26, 225–227.

Van Cauwenberghe, C., Van Broeckhoven, C., & Sleegers, K. (2016). The genetic landscape of Alzheimer disease: Clinical implications and perspectives. Genetics in Medicine, 18(5), 421–430.

Van Cauwenberghe, J., Visch, W., Michiels, J., & Honnay, O. (2016). Selection mosaics differentiate Rhizobium–host plant interactions across different nitrogen environments. Oikos, 125(12), 1755–1761.

Van Cauwenberghe, L., & Roote, D. S. (1998). In situ bioremediation. Ground-Water Remediation Technologies Analysis Center.

Wang, J., Wang, S., Chen, C., Hu, J., He, S., Zhou, Y., & Lin, J. (2022). Treatment of hospital wastewater by electron beam technology: Removal of COD, pathogenic bacteria and viruses. Chemosphere, 308, 136265.

Wang, J., Shen, J., Ye, D., Yan, X., Zhang, Y., Yang, W., & Pan, L. (2020). Disinfection technology of hospital wastes and wastewater: Suggestions for disinfection strategy during coronavirus disease 2019 (COVID-19) pandemic in China. Environmental Pollution, 262, 114665.

Werheni Ammeri, R., Eturki, S., Simeone, G. D. R., Ben Moussa, K., Hassen, W., Moussa, M., & Hassen, A. (2023). Effectiveness of combined tools: Adsorption, bioaugmentation and phytoremediation for pesticides removal from wastewater. International Journal of Phytoremediation, 25(11), 1474–1487.

Weyens, N., van der Lelie, D., Taghavi, S., & Vangronsveld, J. (2009). Phytoremediation: plant–endophyte partnerships take the challenge. Current Opinion in Biotechnology, 20(2), 248–254.

Yadav, K. K., Gupta, N., Kumar, A., Reece, L. M., Singh, N., Rezania, S., & Khan, S. A. (2018). Mechanistic understanding and holistic approach of phytoremediation: A review on application and future prospects. Ecological Engineering, 120, 274–298.

Yan, L., Yang, Y., Li, M., Zhang, Y., Zheng, L., Ge, J., & Lou, Z. (2021). Coupling of N7-methyltransferase and 3′-5′ exoribonuclease with SARS-CoV-2 polymerase reveals mechanisms for capping and proofreading. Cell, 184(13), 3474–3485.

15 Green Solutions for Heavy Metal Cleanup in Water through Phytoremediation

*Twinkle Rout, Bhagyashree Nanda,
Sunil Agrawala, and Dattatreya Kar*

INTRODUCTION

Heavy metal (HM) contamination is a pressing environmental issue that arises from the accumulation of toxic heavy metals in various ecosystems[1]. The consequences of HM contamination are far reaching, impacting not only the environment but also human health and biodiversity. Phytoremediation serves as a testament to nature's resilience and adaptability. Phytoremediation of HM-contaminated water has emerged as a sustainable and environmentally friendly solution to combat the environmental challenges[1,2]. The fundamental principle underlying phytoremediation involves selecting and cultivating specific plant species, known as hyperaccumulators, that have a natural affinity for heavy metal uptake. These plants, often native to contaminated areas, have evolved unique mechanisms to thrive in metal-rich soils and waters[3]. Through their roots, they absorb dissolved heavy metal ions, which are then transported and stored in the plant's tissues. This process not only reduces the metal concentration in the water but also concentrates it within the plant, effectively removing contaminants from the environment[2,3]. Consequently, phytoremediation not only purifies water but also offers the potential for metal recovery and recycling, making it an economically attractive option in addition to its environmental benefits.

By harnessing the unique capacities of plants, we can transform polluted water bodies into thriving ecosystems once again[1]. This approach not only restores water quality but also underscores the harmonious relationship between humans and the environment. As we navigate the challenges of pollution, phytoremediation stands as a beacon of hope, reminding us of the potential of eco-friendly solutions to create a cleaner, healthier world[2]. Heavy metals such as lead (Pb), cadmium (Cd), mercury (Hg), and arsenic (As), discharged from various industrial, agricultural, and urban activities, pose a grave threat to aquatic ecosystems and human health. Conventional methods of remediation often involve costly infrastructure and can generate secondary waste streams, making them less than ideal for long-term environmental sustainability[3,4]. In contrast, phytoremediation harnesses the remarkable natural abilities of certain plants to take up, sequester, and detoxify heavy metals,

DOI: 10.1201/9781003442295-15

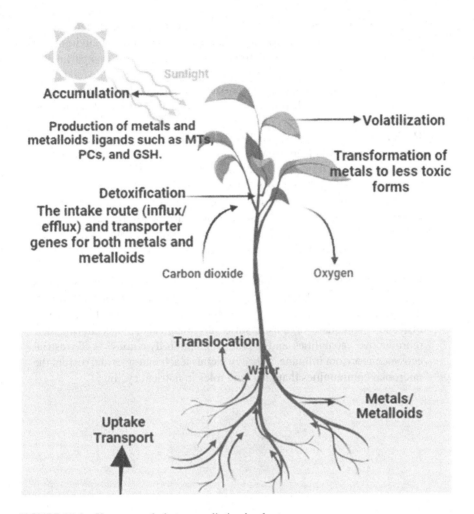

FIGURE 15.1   Heavy metal phytoremediation in plants.

offering a cost-effective and ecologically sound alternative for water purification[5] (Figure 15.1).

Background and Importance of Heavy Metal Contamination

1. **Industrialization and human activities:** The Industrial Revolution marked a turning point in human history, with rapid technological advancements leading to unprecedented economic growth. However, these advancements came at a price, as the surge in significant amounts of heavy metals were discharged into the environment as a result of industrial activity. Factories, power plants, and manufacturing units emitted heavy metal–laden effluents, polluting air, water bodies, and soil[3–5].

2. **Environmental consequences:** The ecological consequences of heavy metal contamination are alarming. These metals persist in the environment for extended periods, accumulating in various ecological compartments. Soil acts as a sink, absorbing heavy metals, which then enter the food chain through plant uptake and subsequent consumption by animals and humans. This bioaccumulation disrupts ecosystems, affecting species diversity and ecosystem stability[6].

3. **Human health implications:** Perhaps the most critical concern is the potential impact of heavy metal contamination on human health[5]. As heavy metals infiltrate the food chain, they accumulate in human tissues and organs, leading to various health issues. Lead exposure, for instance, can result in cognitive impairments, especially in children. Mercury accumulation in seafood can cause neurological disorders, while cadmium exposure has been linked to kidney damage and cancer. These health risks underscore the urgency of addressing heavy metal contamination[5,7].

4. **Biodiversity and ecosystem disruption:** Heavy metal contamination poses a significant threat to biodiversity. Aquatic ecosystems are particularly vulnerable, as heavy metals tend to accumulate in water bodies (Figure 15.2). Fish and other aquatic organisms suffer physiological damage, altering their reproductive capabilities and overall population dynamics[5-8]. Terrestrial ecosystems are not immune, as heavy metals leach into soils and disrupt the microbial communities that play vital roles in nutrient cycling[6].

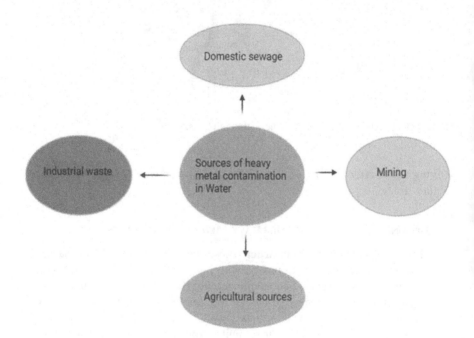

FIGURE 15.2   Sources of water contamination with heavy metal.

5. **Importance of remediation and prevention:** Addressing heavy metal contamination requires a multifaceted approach. Remediation strategies such as phytoremediation (using plants to extract metals from soil), bioremediation (microorganisms breaking down contaminants), and engineering solutions are crucial[6,8]. Equally important is prevention, including proper waste disposal, stringent regulations, and sustainable industrial practices. Public awareness and education play a pivotal role in advocating for responsible consumption and waste management[8].

NEED FOR SUSTAINABLE REMEDIATION APPROACHES

In an era defined by heightened environmental consciousness, the call for sustainable practices resonates across various industries, none more so than in the field of environmental remediation. Traditional remediation methods have often inadvertently exacerbated ecological challenges, prompting the need for a paradigm shift toward sustainable remediation approaches. The imperative to address contaminated sites while minimizing further ecological harm has led to the emergence of innovative strategies that prioritize both environmental health and human well-being[9,10].

The Current Landscape

Conventional remediation methods, such as excavation and landfilling, have been crucial in addressing contaminated sites. However, their limitations in terms of energy consumption, greenhouse gas emissions, and disruption of ecosystems have unveiled the necessity for more holistic and eco-friendly solutions. Sustainable remediation acknowledges the interconnectivity between human and ecological systems and seeks to rectify past environmental mistakes while preventing new ones[7,11].

Principles of Sustainable Remediation

1. **Minimization of environmental footprint:** Sustainable remediation focuses on reducing the overall impact on the environment. This includes minimizing the use of resources, energy, and water during the remediation process. Techniques like *in situ* treatment and bioremediation reduce the need for excessive excavation and transportation of contaminants, subsequently lowering energy consumption and emissions[12-14].
2. **Ecosystem integration:** Instead of treating contaminated sites as isolated entities, sustainable approaches consider the broader ecosystem. Methods like phytoremediation and ecological restoration leverage the natural processes of plants and microorganisms to restore balance to the ecosystem, improving soil and water quality in the process[15].
3. **Stakeholder engagement:** A sustainable approach recognizes the importance of involving local communities and stakeholders in the decision-making process. Transparent communication ensures that concerns are addressed and that the chosen remediation strategies align with the needs and values of the community[16].
4. **Life cycle assessment:** A crucial aspect of sustainable remediation involves evaluating the entire life cycle of a remediation project, from planning and

implementation to postremediation monitoring. This assessment considers the environmental, social, and economic impacts of each stage, aiding in the selection of the most viable and environmentally sound approach[16,17].

The Road Ahead

The shift toward sustainable remediation requires a collaborative effort from governments, industries, environmental consultants, and local communities. Research and development into innovative technologies and techniques, along with continuous education and awareness campaigns are vital to mainstreaming these eco-friendly practices[16]. Additionally, by enabling rules and financial incentives, government officials can be helpful in encouraging the adoption of sustainable techniques. The urgency to address environmental contamination while safeguarding the planet's well-being underscores the need for sustainable remediation approaches. By embracing principles that minimize ecological impact, harness natural processes, engage stakeholders, and assess the entire life cycle of remediation projects, we can pave the way for a healthier, more resilient future. The journey toward sustainable remediation is not only a scientific endeavor but a collective commitment to leave behind a legacy of environmental responsibility for generations to come[16].

ROLE OF PHYTOREMEDIATION IN WATER QUALITY RESTORATION

Water is a valuable resource that supports life on earth, but its quality is being threatened by a variety of contaminants, from organic compounds to heavy metals. Traditional methods of water treatment have clear disadvantages, sometimes requiring significant energy use and complex infrastructure. In this context, phytoremediation has come to be observed as a potential environmentally friendly approach that makes use of the plant's unique abilities to efficiently and naturally restore water quality[17,18].

Phytoremediation is a very sustainable and cost-effective approach which involves using plants to mitigate pollutants from water bodies due to their unique abilities to absorb, accumulate, transform, and degrade contaminants. Through mechanisms such as phytostabilization (reducing the mobility of contaminants), rhizofiltration (root filtration), phytoextraction (uptake and concentration in plant tissues), and phytodegradation (breaking down pollutants), phytoremediation offers a holistic solution to water pollution[19].

MECHANISMS OF PHYTOREMEDIATION

PHYTOEXTRACTION: PLANT–METAL INTERACTIONS AND UPTAKE

The complex interactions and absorption of different metals by plants are referred to as plant–metal interactions. The growth, development, and physiological activities of plants depend heavily on metals, some of which are essential and others nonessential. However, too much exposure to certain metals can harm plants and have a negative effect on their general health. For reducing metal-contaminated environments and improving plant development in varied conditions, understanding these interactions and absorption pathways is crucial[2,12,20].

Essential Metals and Nutrient Uptake

For the development and growth of plants, certain metals, sometimes referred to as micronutrients or trace elements, are essential. These include metals like copper (Cu), manganese (Mn), zinc (Zn), iron (Fe), molybdenum (Mo), and nickel (Ni)[12,20]. These essential metals are required in small amounts for various biochemical processes such as photosynthesis, enzyme activation, and nitrogen fixation. Plants have developed specific mechanisms to acquire these essential metals from the soil through their root systems. This process involves the secretion of metal-chelating compounds (phytochelatins), root morphology, and transporter proteins that facilitate the uptake and transport of these metals from the soil solution to the plant cells[21].

Nonessential Metals and Toxicity

Nonessential metals (referred to as heavy metals) include cadmium (Cd), mercury (Hg), lead (Pb), and arsenic (As)[21]. These metals don't play any role in plant growth, and they can be toxic even at low concentrations. They can also enter the plant's root system through various pathways, including direct uptake from the soil solution and absorption by foliar tissues. These metals can damage biomolecules, disrupt cellular processes, and interfere with nutrient uptake and transport. Plant responses to heavy metal stress include production of reactive oxygen species (ROS), altered gene expression, and synthesis of metal-binding proteins to sequester and detoxify the metals[22].

Phytoremediation and Hyperaccumulators

It is possible for certain plant species, referred to as hyperaccumulators, to store significant amounts of heavy metals in their tissues without experiencing harmful adverse consequences. These plants are possibilities for phytoremediation i.e., to remove, stabilize, or detoxify metal pollutants from soil and water, since they have specialized systems that allow them to withstand and collect metals. For restoring metal-contaminated areas, phytoremediation can be a viable and practical solution[23].

Implications for Agriculture and Environment

The management of the environment and agriculture is significantly affected by our understanding of plant–metal interactions and absorption. Maximizing the availability of vital micronutrients in agriculture is the key for increasing crop production and quality. A sufficient level of metal absorption by plants is ensured by using good soil management techniques, such as pH adjustments and the use of micronutrient fertilizers[17,18]. Plant–metal interactions and absorption mechanisms are intricate and multidimensional processes that have an impact on plant development, growth, and environmental health. Better management of metal-contaminated habitats, efficient phytoremediation techniques, and sustainable agriculture are all benefits of this understanding of these interactions[24].

PHYTOSTABILIZATION: PREVENTING METAL MIGRATION

A sustainable and environment friendly method to reduce the movement and dispersion of hazardous metals in polluted soils is phytostabilization. By immobilizing, sequestering, and stabilizing hazardous metals with plants, this technique prevents their mobility into the environment and lowers the danger of contaminating water sources, the air, and ecosystems. In places where typical cleanup approaches would be unfeasible or expensive, phytostabilization is very beneficial[25,26].

Mechanisms of Phytostabilization

Plant physiological and biochemical processes, as well as interactions between plant roots and the surrounding soil matrix, all contribute to phytostabilization. Infested soil causes plants to absorb metals through their roots. Some plants are capable of storing metals without significantly affecting their toxicity. These hyperaccumulator plants are essential for lowering metal concentrations in the soil solution. Plant roots release various organic compounds, such as organic acids and peptides that can form complexes with metals in the soil[26]. These complexes reduce the mobility and bioavailability of metals, making them less likely to leach into groundwater or be taken up by other plants. The rhizosphere, the zone of soil influenced by plant roots, experiences changes due to root exudates and microbial activity. These changes can alter the soil's physicochemical properties, promoting the formation of stable metal–mineral complexes that immobilize metals[8].

RHIZOFILTRATION: PLANT ROOTS AS NATURAL FILTERS

Rhizofiltration is a phytoremediation method that uses the plant root's inherent filters to remove contaminants from the soil and water[26]. With the growing concern about environmental pollution and its detrimental impact on ecosystems and human health, innovative and sustainable remediation strategies have gained popularity. By using plant roots' capacity to absorb and store pollutants, rhizofiltration aids in the detoxification of contaminated soil and water[27].

Mechanisms of Rhizofiltration

To filter and sequester pollutants, plant roots use a variety of methods. Adsorption onto root surfaces, absorption into root tissues, volatilization, and microbial decomposition supported by the rhizosphere are the main mechanisms in rhizofiltration. Exudates from plant roots alter the chemical makeup of the soil, facilitating interactions between the root zone and contaminants. These interactions cause pollutants to become immobile and change, which lowers their toxicity and mobility[28].

PHYTOVOLATILIZATION: CONVERSION OF METALS TO VOLATILE FORMS

An advanced phytoremediation method called phytovolatilization uses certain plants' inherent capacity to change metals from their solid, nonvolatile condition to volatile forms. In this process, metal ions are taken up by plant roots, transported to the aboveground plant components, and then changed into gaseous compounds and discharged into the environment. By successfully lowering metal concentrations and minimizing their environmental impact, phytovolatilization offers a fresh means of managing metal pollution in soils and water[27–29].

Mechanisms of Phytovolatilization

Metalloids including selenium (Se), arsenic (As), and mercury (Hg), as well as some nonmetal elements like boron (B), predominantly enhance the phytovolatilization process. Arsenate reductases and methyltransferases, two specialized enzymes found in plants, are essential for converting these metals to their volatile forms. Through active or passive transport processes, the plant roots absorb metal ions from the soil or water[30]. The plant's vascular system carries the metal ions to the sections above ground[30]. Within the aboveground plant tissues, the metals are enzymatically transformed into gaseous compounds. For example, arsenic can be converted into volatile arsines (e.g., trimethylarsine, arsine gas) and mercury into dimethylmercury or elemental mercury. Then the volatile metal compounds are released into the atmosphere through stomata. From there, they can disperse and be diluted within the surrounding air[29,30].

Suitable Plant Species and Environmental Factors

Certain plant species are particularly adept at facilitating phytovolatilization due to their ability to accumulate specific metals and their capacity to produce the necessary enzymes. For instance, ferns, willows, and members of the Brassicaceae family (e.g., Indian mustard) have been studied for their phytovolatilization potential.

Environmental factors, including soil pH, moisture content, and metal concentrations, have a significant impact on how well phytovolatilization works. The process may be optimized by changing these variables, improving the intake and conversion of metals into volatile forms[31] (Figure 15.3).

PLANT SELECTION FOR PHYTOREMEDIATION

Selection of the right plant species is intimately related to the overall effectiveness of phytoremediation[31,32]. The optimal performance of the chosen plants is ensured by careful evaluation of variables such contaminant specificity, metal tolerance, growth characteristics, and environment compatibility. Phytoremediation can serve as a potent tool for restoring polluted ecosystems while boosting sustainability and environmental health by choosing the proper plants[32].

- **Hyperaccumulator plants:** An intriguing group of plants known as hyperaccumulator plants has the incredible ability to collect extremely high concentrations of certain elements, notably heavy metals, in their tissues without experiencing deleterious consequences. These plants have a remarkable ability to take in and store large amounts of pollutants. Because of this distinctive characteristic, phytoremediation has attracted a lot of attention as a potential method of addressing soil pollution, promoting environmental restoration, and maybe even recovering important metals[33]. An interesting approach to addressing heavy metal pollution and environmental deterioration is through the use of hyperaccumulator plants. We may pave the way for more efficient and ecologically friendly solutions to pollution-related problems by further examining and using the potential of hyperaccumulator plants[33]. Examples include *Thlaspi*

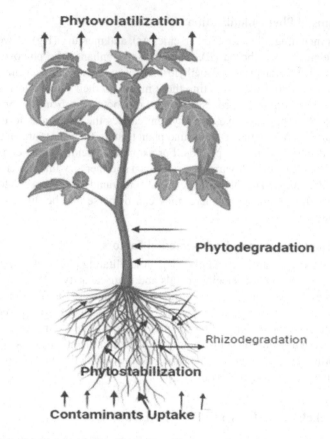

FIGURE 15.3 Multiple methods of phytoremediation.

caerulescens (cadmium and zinc hyperaccumulator) and *Alyssum* species (nickel hyperaccumulator)[34].

• **Non-hyperaccumulator plants and their remediation roles:** Although hyperaccumulator plants have drawn a lot of attention for their unique ability to accumulate enormous amounts of specific elements, non-hyperaccumulator plants also play crucial roles in various phytoremediation scenarios. These plants might not accumulate metals at extraordinary levels, but their distinct characteristics and functions contribute to the overall success of environmental remediation efforts[34,35]. Non-hyperaccumulator plants, although not specialized in accumulating metals, can still participate in phytoextraction by gradually taking up and translocating contaminants to their aerial parts. These plants complement hyperaccumulators and contribute to overall metal reduction in the soil. Some non-hyperaccumulator plants release root exudates that stimulate microbial activity, promoting the degradation of organic contaminants in the rhizosphere. They also play a pivotal role in restoring degraded landscapes by enhancing biodiversity, stabilizing ecosystems, and providing habitat for wildlife and preventing soil erosion[35].

Examples of non-hyperaccumulator plants in phytoremediation[36]:

Grasses (Poaceae family):
- Grasses like ryegrass (*Lolium* spp.) are often used for phytostabilization due to their fibrous root systems and ability to hold soil in place.

Legumes (Fabaceae family):
- Legumes like clover (*Trifolium* spp.) contribute nitrogen fixation, enriching soil fertility and enhancing overall ecosystem health.

Trees and shrubs:
- Woody plants such as willows (*Salix* spp.) contribute to phytostabilization, erosion control, and habitat creation.

CASE STUDIES IN PHYTOREMEDIATION

Phytoremediation, a sustainable and eco-friendly approach to addressing environmental contamination, has yielded several successful case studies across the globe. Phytoextraction, a subset of phytoremediation, focuses on using plants to extract pollutants, particularly heavy metals, from contaminated soils. Examining these success stories not only highlights the effectiveness of the technique but also offers valuable insights and lessons for future phytoremediation projects[37].

- **Case Study 1. *Thlaspi caerulescens* at a zinc and cadmium polluted site:** In the metal-rich soils of Sudbury, Canada, *Thlaspi caerulescens*, a zinc and cadmium hyperaccumulator, was cultivated. Over the years, this plant successfully extracted significant amounts of zinc and cadmium from the soil[34]. The key lessons from this case study include:
  - Plant selection matters: Choosing hyperaccumulator species that can thrive in the specific contaminated environment is crucial for success.
  - Phytoextraction efficiency: The amount of metals extracted can be substantial, but it might require long-term cultivation and monitoring.
- **Case Study 2. Indian mustard (*Brassica juncea*) for cadmium remediation:** Indian mustard, a well-known accumulator of cadmium, was used to remediate cadmium-contaminated farmland in China. The results demonstrated that Indian mustard effectively reduced soil cadmium levels. Lessons from this case study include:
  - Soil amendments: Incorporating soil amendments can enhance metal availability and uptake by plants.
  - Metal recovery: After phytoextraction, harvested plant biomass can potentially be processed for metal recovery[38].
- **Case Study 3. Sunflowers (*Helianthus annuus*) for lead contamination:** Sunflowers were planted in a former lead battery recycling site in Spain, effectively reducing lead concentrations in the soil[39].
  - Rhizosphere influence: The release of root exudates can influence metal mobility and availability in the rhizosphere.
  - Economic and aesthetic benefits: Sunflowers can provide additional benefits like beautifying the area and attracting pollinators.

APPLICATIONS IN CONTAMINATED AQUATIC ENVIRONMENTS

While often applied to terrestrial environments, phytostabilization can also be effectively employed in contaminated aquatic environments to mitigate the spread of pollutants and restore ecological balance. This approach capitalizes on the unique properties of aquatic plants to stabilize sediments and water bodies contaminated with heavy metals and other pollutants[40].

1. **Root uptake and sediment binding:**
   * Aquatic plants take up pollutants through their roots, reducing metal concentrations in the water column. Root systems stabilize sediments by binding soil particles and contaminants.
2. **Rhizosphere interactions:**
   * Root exudates enhance microbial activity in the sediment, promoting the breakdown and immobilization of contaminants.
3. **Biofilm formation:**
   * Plant roots and submerged surfaces encourage the growth of biofilms, which can trap and immobilize pollutants.

* **Case Study 1. Water hyacinth (*Eichhornia crassipes*) in eutrophic water bodies:**
   * Water hyacinth effectively removes excess nutrients from water bodies, reducing eutrophication and improving water quality[41] (Table 15.1).
* **Case Study 2. *Phragmites australis* for heavy metal stabilization:**
   * *Phragmites australis*, commonly known as common reed, has been utilized to stabilize heavy metals in aquatic sediments[42].

RHIZOFILTRATION IMPLEMENTATION FOR POINT SOURCE POLLUTION

Point source pollution, originating from discrete and identifiable sources, poses significant threats to water bodies and ecosystems. Addressing this type of pollution requires targeted and efficient remediation strategies. Rhizofiltration, a specialized phytoremediation technique, offers a promising approach to tackle point source pollution by utilizing the natural filtration abilities of plant roots to remove pollutants from contaminated water[43]. Roots act as a natural filter in capturing pollutants as water flows through them. Plant roots selectively accumulate pollutants like heavy metals, nutrients, and organic compounds, depending on their characteristics[44]. Rhizofiltration offers a targeted and efficient solution for addressing point source pollution in water bodies. By harnessing the natural filtration capabilities of plant roots, this technique holds great potential to reduce pollutant loads and improve water quality. With careful site selection, appropriate plant species, and well-designed systems, rhizofiltration can contribute significantly to the restoration and preservation of aquatic ecosystems impacted by point source pollution[43] (Table 15.1).

TABLE 15.1
Plants Responsible for Removal of Pollutants in Aquatic Environment

Plants	Contaminants	Phytoremediation	References
Lemna minor (Duckweed)	Azo-dyes (acid blue dye), C30-C40 hydrocarbons and petroleum hydrocarbons	Dye transformation into by-products enhances phytoremediation, while higher carbon chain hydrocarbons are converted into lower carbon chain by plants, with decreasing removal rate over 120 days.	61, 62
Elodea sp. (American waterweed)	Dichlorodiphenyl trichloroethane (DDT)	The process involves the bioaccumulation and phytotransformation of DDT into DDD.	62
E. crassipes (Water hyacinth)	Bisphenol A, ethion, di-n-hexyl phthalate, and pentabromo-diphenyl ether, along with ethion, dicofol, and pentachlorophenol, found in pesticide-contaminated water	Phytodegradation of organic contaminants, which are then absorbed by plants.	63, 64
B. juncea (Indian mustard)	Phorate	Plant's absorption and uptake of phorate led to a 68.28% decrease in phorate concentration within five days.	65
Eleocharis ochrostachys (Spike rush)	Water contaminated with PAHs, which is a type of diesel	*E. ochrostachys* plants demonstrated superior results in phytoremediation of PAHs from 1% diesel-contaminated wastewater.	66
E. crassipes (Water hyacinth)	Palladium (Pd)	Phytoremediation of palladium involves the use of rhizofiltration.	64, 67
Zinnia elegans	Chromium (Cr)	The plant's roots have been found to be more prone to chromium bioaccumulation due to improved phytostabilization.	68
Phragmites australis (Common reed)	(Co, Ni, Mo, Cd, Pb, Cr, Cu, Fe, Mn, Zn, and Hg) and trace elements (As, Se, Ba)	Phytoremediation of estuarine sediments polluted with heavy metals.	42

INNOVATIVE APPROACHES AND MULTI-PLANT STRATEGIES

Innovative techniques and multi-plant strategies have increased the effectiveness and breadth of phytoremediation, enabling it to deal with a wider range of pollutants and challenging environments. Plant–microbe interactions, genetic engineering, and nanotechnology approaches hold promise for further enhancing the efficacy of phytoremediation and contributing to sustainable environmental management[45].

Multi-Plant Strategies

1. Polyculture and mixed planting:

- Combining different plant species in the same area creates a diverse ecosystem that enhances pollutant removal efficiency.
- Each plant species contributes unique abilities, such as hyperaccumulation, nutrient uptake, and root exudates[46].

2. Succession planting:

- Sequentially planting different species based on their growth rates and pollutant accumulation abilities optimizes contaminant removal over time.

3. Companion planting:

- Companion plants release chemicals that enhance pollutant solubility or mobilize nutrients, facilitating their uptake by other plants.

4. Microbe–plant partnerships:

- Establishing mycorrhizal associations and other symbiotic relationships between plants and microbes can boost pollutant degradation and nutrient cycling.

Innovative Approaches

1. Nanotechnology and phytonanoremediation:

- Engineered nanoparticles can be used to enhance pollutant uptake, transport, and transformation in plants.
- Phytonanoremediation combines nanotechnology with phytoremediation, enabling efficient removal of contaminants like heavy metals.

2. Genetic engineering and transgenic plants:

- Genetic modifications can enhance plant metal uptake, tolerance, and pollutant degradation abilities.
- Transgenic plants engineered for specific phytoremediation traits are being explored for improved performance.

3. Constructed wetlands and floating treatment islands:

- Engineered wetlands and floating islands populated with plants offer efficient pollutant removal from water bodies.
- These systems provide controlled environments for enhanced nutrient and metal uptake.

4. Hydroponic phytoremediation:

- Growing plants in nutrient-rich water solutions allows for efficient nutrient and metal uptake without soil limitations.
- Hydroponic systems are adaptable to various pollutant scenarios[47].

OVERCOMING CHALLENGES AND FUTURE DIRECTIONS

To overcome the obstacles associated with phytoremediation requires interdisciplinary cooperation, innovative tactics, and a patient approach. By taking into consideration the unique factors of each site, exploring novel technologies, and integrating phytoremediation with complementary methods, we can surmount the challenges and establish a more sustainable and efficient approach to environmental remediation. A deep comprehension of site-specific conditions is essential for customizing phytoremediation strategies to maximize their effectiveness. Encouraging beneficial partnerships between plants and microbes can expedite the degradation of contaminants and the cycling of nutrients. Integrating phytoremediation with other techniques such as bioremediation or physical interventions can fine-tune the outcomes. The development of plants with enhanced capabilities for metal uptake, tolerance, and pollutant breakdown can significantly enhance efficacy. Furthermore, educating communities and stakeholders about the advantages and limitations of phytoremediation serves to garner support and foster cooperation[2].

- **Transgenic plants:** Advances in genetic engineering could lead to the development of more efficient phytoremediation-specific plant traits.
- **Nanotechnology integration:** Continued research into nanotechnology's role in enhancing plant–metal interactions may yield breakthroughs.
- **Ecosystem-level approaches:** Scaling up phytoremediation to ecosystem-level applications can have broader environmental impacts.

GENETIC ENGINEERING FOR ENHANCED METAL UPTAKE AND TOLERANCE

In the field of phytoremediation, genetic engineering has become a potent technique that allows researchers to improve plant features like metal absorption and tolerance. Researchers are attempting to design organisms that are more effective in removing heavy metals and other pollutants from the environment by modifying the genetic composition of plants[48]. This innovative approach holds great promise for revolutionizing the field of phytoremediation. Accessing the full potential of genetic engineering in phytoremediation requires careful consideration of ecological effects, regulatory frameworks, and ethical concerns as this technology develops[49]. In order to increase the absorption of desired metals from the soil, scientists can design plants to overexpress metal transporters in their roots. It is possible to genetically modify plants to create chelators, which are organic compounds that bind to metals and increase their solubility and absorption. The ability of plants to securely store accumulated metals can be improved by manipulating the genes responsible for metal sequestration in vacuoles. A plant's antioxidant defense system can be improved genetically to help it better survive oxidative stress brought on by metals[48,49].

CASE STUDY: *ARABIDOPSIS THALIANA* FOR CADMIUM PHYTOREMEDIATION

Arabidopsis thaliana has been successfully engineered by researchers, a model plant to express genes responsible for enhanced metal uptake and accumulation. The engineered plants exhibited improved cadmium uptake and tolerance, demonstrating the potential of genetic engineering in phytoremediation[50].

- **Precision editing techniques:** Advances in genome editing, such as CRISPR-Cas9, enable precise modifications in plant genomes, enhancing the efficiency of genetic engineering[51].

- **Field trials and scaling up:** Conducting field trials with genetically engineered plants will provide insights into their performance and environmental impact on a larger scale[51,52].

INTEGRATING PHYTOREMEDIATION WITH OTHER REMEDIATION TECHNIQUES

Environmental contamination is a complex challenge that often requires multifaceted solutions. Integrating phytoremediation with other remediation techniques offers a synergistic approach that capitalizes on the strengths of each method to address a broader range of contaminants, enhance efficiency, and accelerate the restoration of polluted environments. Different techniques target different pollutants, allowing for a more comprehensive remediation approach[2,29]. Combined methods can work in tandem to accelerate pollutant removal, reducing the overall time required for cleanup. Customized integration can be tailored to the unique characteristics of each contaminated site. It can mitigate the risks associated with relying solely on one technique. Combining plants that stabilize contaminants (phytostabilization) with microbial communities that degrade pollutants (bioremediation) can synergistically enhance the overall effectiveness of remediation[53]. Enhancing soil conditions through amendments like chelating agents can improve the availability of contaminants for uptake by hyperaccumulator plants (phytoextraction). Planting native species for phytoremediation in conjunction with erosion-control techniques like vegetation cover prevents soil disturbance and contaminant spread. Integrating constructed wetlands with phytoremediation techniques like phytofiltration enhances water quality by combining natural filtration mechanisms[29,53].

CASE STUDY: INTEGRATING PHYTOEXTRACTION AND MYCOREMEDIATION

Researchers have successfully combined phytoextraction with mycoremediation, using fungi to break down complex organic contaminants. This integration demonstrated enhanced removal of pollutants, showcasing the potential of combining plant and fungal strategies[54].

NANOTECHNOLOGY IN PHYTOREMEDIATION

The evolving landscape of environmental contamination demands innovative solutions that go beyond conventional methods. Nano- and biotechnological advancements have emerged as powerful tools in phytoremediation, offering unprecedented potential to enhance contaminant uptake, transformation, and plant–microbe interactions. These cutting-edge approaches are poised to reshape the field of phytoremediation, making it more efficient, targeted, and sustainable[55]. Nano- and biotechnological advancements are pushing the boundaries of phytoremediation, offering solutions that are faster, more targeted, and environmentally friendly. By using these advanced technologies, we can revolutionize environmental cleanup, making phytoremediation a more potent and versatile tool in the fight against pollution[56]. Engineered nanoparticles can enhance metal uptake by plants through increased surface area and reactivity by facilitating the penetration of plant root cells and improving contaminant absorption. Combining nanotechnology with phytoremediation creates new opportunities for precise pollutant targeting. Phytonanoremediation strategies improve metal solubility and help in plant uptake.

FUTURE OUTLOOK

Water is a finite and essential resource, vital for human well-being, ecosystems, and economic development. As population growth, industrialization, and climate change place increasing stress on water availability, the need for sustainable water resource management has become paramount. Achieving sustainable water management involves a holistic approach that balances human needs, ecological health, and long-term viability. Ecological restoration, sustainable technology, and environmental protection all come together in phytoremediation. It is positioned to play a significant part in determining a cleaner, healthier future for our world thanks to its wide range of applications, cutting-edge methodologies, and ever-expanding potential. Phytoremediation can mend our ecosystems and leave a beneficial legacy for future generations as we continue to research, develop, and incorporate its processes. Our understanding of the mechanisms and uses of various phytoremediation techniques has come a long way over the years. This field has had achievements, advancements, and the incorporation of cutting-edge technology in everything from phytoextraction to rhizofiltration[57]. The field of phytoremediation is at the forefront of cutting-edge remedies as environmental problems continue to worsen. The potential for further developing and broadening phytoremediation approaches is enormous given continuing study and advances in several scientific fields. The potential for advances in phytoremediation research is intriguing, providing a route to more effective, adaptable, and long-lasting heavy metal cleaning techniques. The creation of specialized microbial communities that enhance phytoremediation processes can result from a deeper knowledge of plant–microbe interactions[31]. Designing microbial consortia to promote pollution breakdown and nutrient cycling in the rhizosphere might open up new remediation prospects[58]. The exact alteration of plant features via the use of genetic engineering techniques will enhance metal absorption, tolerance, and pollution degradation[58]. Integrating genomics, transcriptomics, and metabolomics can provide insights into plant responses to pollutants and guide the selection of more effective species[59]. Tailoring nanoparticles for optimal pollutant interaction and plant uptake could revolutionize phytonanoremediation strategies. Integrating nanoparticles with beneficial microorganisms can create dynamic synergies for pollutant degradation and immobilization[60]. Utilizing wetlands for phytoremediation can simultaneously restore ecosystems, improve water quality, and mitigate flood risks. Incorporating remote sensing, drones, and sensor networks can provide real-time data for more efficient and adaptive phytoremediation strategies. Advanced computational models can predict plant performance, pollutant behavior, and system responses, aiding in decision-making.

CONCLUSION

The future of phytoremediation research is full of potential breakthroughs that could reshape the landscape of heavy metal removal from water. The convergence of genetic engineering, biotechnology, nanotechnology, and digital innovations offers unprecedented opportunities to enhance the efficiency, scope, and effectiveness of phytoremediation techniques. By integrating scientific advancements with ethical considerations and public engagement, we can pave the way for a cleaner, healthier, and more sustainable planet. Sustainable water resource management is a complex but vital endeavor that requires the collaboration of governments, industries, communities, and individuals. By adopting holistic approaches, embracing technological innovations, and prioritizing the needs of both people and ecosystems, we can work toward a future where water resources are managed sustainably, ensuring a resilient and prosperous planet for generations to come.

REFERENCES

1. Briffa J, Sinagra E, Blundell R. Heavy metal pollution in the environment and their toxicological effects on humans. Heliyon. 2020;6(9):e04691.
2. Yan A, Wang Y, Tan SN, MohdYusof ML, Ghosh S, Chen Z. Phytoremediation: a promising approach for revegetation of heavy metal-polluted land. Frontiers in Plant Science. 2020;11:359.
3. Yadav R, Singh G, Santal AR, Singh NP. Omics approaches in effective selection and generation of potential plants for phytoremediation of heavy metal from contaminated resources. Journal of Environmental Management. 2023;336:117730.
4. Agarwal P, Rani R. Strategic management of contaminated water bodies: Omics, genome-editing and other recent advances in phytoremediation. Environmental Technology & Innovation. 2022;27:102463.
5. Sahota NK, Sharma R. Insight into pharmaceutical waste management by employing bioremediation techniques to restore environment. In Handbook of Solid Waste Management: Sustainability through Circular Economy (pp. 1795–1826). Singapore: Springer Nature Singapore; 2022.
6. Naz M, Dai Z, Hussain S, Tariq M, Danish S, Khan IU, Qi S, Du D. The soil pH and heavy metals revealed their impact on soil microbial community. Journal of Environmental Management. 2022;321:115770.
7. Padhye LP, Srivastava P, Jasemizad T, Bolan S, Hou D, Sabry S, Rinklebe J, O'Connor D, Lamb D, Wang H, Siddique KH. Contaminant containment for sustainable remediation of persistent contaminants in soil and groundwater. Journal of Hazardous Materials. 2023;6:131575.
8. Ferreira PA, Brunetto G, Giachini AJ, Soares CR. Heavy metal uptake and the effect on plant growth. In Heavy Metal Remediation: Transport and Accumulation in Plants (pp. 127–154). : New York: Nova Science Publishers, Inc.; 2014.
9. Akpor O, Muchie B. Environmental and public health implications of wastewater quality. African Journal of Biotechnology. 2011;10(13):2379–2387.
10. Zhang S, Wang J, Zhang Y, Ma J, Huang L, Yu S, et al. Applications of water-stable metal-organic frameworks in the removal of water pollutants: a review. Environmental Pollution. 2021;291:118076.
11. Aghili S, Golzary A. Greening the earth, healing the soil: A comprehensive life cycle assessment of phytoremediation for heavy metal contamination. Environmental Technology & Innovation. 2023;32:103241.

12. Purwadi I, Erskine PD, Casey LW, van der Ent A. Recognition of trace element hyper-accumulation based on empirical datasets derived from XRF scanning of herbarium specimens. Plant and Soil. 2023;492:429–438.

13. Balali-Mood M, Naseri K, Tahergorabi Z, Khazdair MR, Sadeghi M. Toxic mechanisms of five heavy metals: Mercury, lead, chromium, cadmium, and arsenic. Frontiers in Pharmacology. 2021;12:643972.

14. Pandey VC, editor. Assisted Phytoremediation. Elsevier; 2021.

15. Kurade MB, Ha YH, Xiong JQ, Govindwar SP, Jang M, Jeon BH. Phytoremediation as a green biotechnology tool for emerging environmental pollution: a step forward towards sustainable rehabilitation of the environment. Chemical Engineering Journal. 2021;415:129040.

16. Rizzo E, Bardos P, Pizzol L, Critto A, Giubilato E, Marcomini A, Albano C, Darmendrail D, Döberl G, Harclerode M, Harries N. Comparison of international approaches to sustainable remediation. Journal of Environmental Management. 2016;184:4–17.

17. Vardhan KH, Kumar PS, Panda RC. A review on heavy metal pollution, toxicity and remedial measures: Current trends and future perspectives. Journal of Molecular Liquids. 2019;290:111197.

18. Obaideen K, Shehata N, Sayed ET, Abdelkareem MA, Mahmoud MS, Olabi AG. The role of wastewater treatment in achieving sustainable development goals (SDGs) and sustainability guideline. Energy Nexus. 2022;7:100112.

19. Khan AU, Khan AN, Waris A, Ilyas M, Zamel D. Phytoremediation of pollutants from wastewater: A concise review. Open Life Sciences. 2022;17(1):488–96.

20. REDDY K, KUMAR G. Green and sustainable remediation of polluted sites: New concept, assessment tools, and challenges. Ce/papers. 2018;2(2–3):83–92.

21. Arif N, Yadav V, Singh S, Singh S, Ahmad P, Mishra RK, Sharma S, Tripathi DK, Dubey NK, Chauhan DK. Influence of high and low levels of plant-beneficial heavy metal ions on plant growth and development. Frontiers in Environmental Science. 2016;4:69.

22. Kumar V, Singh J, Kumar P. Heavy metals accumulation in crop plants: Sources, response mechanisms, stress tolerance and their effects. In Contaminants in Agriculture and Environment: Health Risks and Remediation (pp. 38–57). Haridwar, India: Agro Environ Media; 2019.

23. Ali H, Khan E, Sajad MA. Phytoremediation of heavy metals—concepts and applications. Chemosphere. 2013;91(7):869–881.

24. Babu SO, Hossain MB, Rahman MS, Rahman M, Ahmed AS, Hasan MM, Rakib A, Emran TB, Xiao J, Simal-Gandara J. Phytoremediation of toxic metals: A sustainable green solution for clean environment. Applied Sciences. 2021;11(21):10348.

25. Bolan NS, Park JH, Robinson B, Naidu R, Huh KY. Phytostabilization: a green approach to contaminant containment. Advances in Agronomy. 2011;112:145–204.

26. Kristanti RA, Tirtalistyani R, Tang YY, Thao NT, Kasongo J, Wijayanti Y. Phytoremediation Mechanism for Emerging Pollutants: A Review. Tropical Aquatic and Soil Pollution. 2023;3(1):88–108.

27. Boulkhessaim S, Gacem A, Khan SH, Amari A, Yadav VK, Harharah HN, Elkhaleefa AM, Yadav KK, Rather SU, Ahn HJ, Jeon BH. Emerging trends in the remediation of persistent organic pollutants using nanomaterials and related processes: A review. Nanomaterials. 2022;12(13):2148.

28. Bilgaiyan P, Shivhare N, Rao NG. Phytoremediation of Wastewater through Implemented Wetland–A Review. In E3S Web of Conferences 2023 (Vol. 405, p. 04026). EDP Sciences.

29. Sharma JK, Kumar N, Singh NP, Santal AR. Phytoremediation technologies and their mechanism for removal of heavy metal from contaminated soil: An approach for a sustainable environment. Frontiers in Plant Science. 2023;14:1076876.

30. Jabeen R, Ahmad A, Iqbal M. Phytoremediation of heavy metals: physiological and molecular mechanisms. The Botanical Review. 2009;75:339–364.
31. Mocek-Płóciniak A, Mencel J, Zakrzewski W, Roszkowski S. Phytoremediation as an Effective Remedy for Removing Trace Elements from Ecosystems. Plants. 2023;12(8):1653.
32. Khan S, Masoodi TH, Pala NA, Murtaza S, Mugloo JA, Sofi PA, Zaman MU, Kumar R, Kumar A. Phytoremediation prospects for restoration of contamination in the natural ecosystems. Water. 2023;15(8):1498.
33. Rascio N, Navari-Izzo F. Heavy metal hyperaccumulating plants: how and why do they do it? And what makes them so interesting?. Plant Science. 2011 Feb 1;180(2):169–181.
34. Cosio C, Martinoia E, Keller C. Hyperaccumulation of cadmium and zinc in *Thlaspi caerulescens* and *Arabidopsis halleri* at the leaf cellular level. Plant Physiology. 2004;134(2):716–725.
35. Chandra R, Kumar V, Singh K. Hyperaccumulator versus nonhyperaccumulator plants for environmental waste management. In Phytoremediation of Environmental Pollutants (pp. 43–80). CRC Press; 2018.
36. Hall J, Soole K, Bentham R. Hydrocarbon phytoremediation in the family Fabacea—A review. International Journal of Phytoremediation. 2011;13(4):317–332.
37. Tan HW, Pang YL, Lim S, Chong WC. A state-of-the-art of phytoremediation approach for sustainable management of heavy metals recovery. Environmental Technology & Innovation. 2023;30:103043.
38. Bhadkariya RK, Jain VK, Chak GP, Gupta SK. Remediation of cadmium by Indian mustard (*Brassica juncea* L.) from cadmium contaminated soil: A phytoextraction study. International Journal of Environment. 2014;3(2):229–237.
39. Al-Jobori KM, Kadhim AK. Evaluation of sunflower (*Helianthus annuus* L.) for phytoremediation of lead contaminated soil. Journal of Pharmaceutical Sciences and Research. 2019;11(3):847–854
40. Ihsanullah A, Abbas A, Al-Amer AM, Laoui T, Al-Marri MJ,Nasser MS, et al. Heavy metal removal from aqueous solutionby advanced carbon nanotubes: critical review of adsorptionapplications. Separation and Purification Technology. 2016;157:141–161.
41. Auchterlonie J, Eden CL, Sheridan C. The phytoremediation potential of water hyacinth: A case study from Hartbeespoort Dam, South Africa. South African Journal of Chemical Engineering. 2021;37:31–36.
42. Cicero-Fernández D, Peña-Fernández M, Expósito-Camargo JA, Antizar-Ladislao B. Role of *Phragmites australis* (common reed) for heavy metals phytoremediation of estuarine sediments. International Journal of Phytoremediation. 2016;18(6):575–582.
43. Kristanti RA, Ngu WJ, Yuniarto A, Hadibarata T. Rhizofiltration for removal of inorganic and organic pollutants in groundwater: A review. Biointerafce Research in Applied Chemistry. 2021;4:1232647.
44. Yadav R, Singh S, Kumar A, Singh AN. Phytoremediation: A wonderful cost-effective tool. In Cost Effective Technologies for Solid Waste and Wastewater Treatment (pp. 179–208). Elsevier; 2022.
45. Ijaz A, Imran A, Anwar ulHaq M, Khan QM, Afzal M. Phytoremediation: recent advances in plant-endophytic synergistic interactions. Plant and Soil. 2016;405:179–195.
46. Wu FY, Chung AK, Tam NF, Wong MH. Root exudates of wetland plants influenced by nutrient status and types of plant cultivation. International Journal of Phytoremediation. 2012;14(6):543–553.
47. Pivetz BE. Phytoremediation of contaminated soil and ground water at hazardous waste sites. US Environmental Protection Agency, Office of Research and Development, Office of Solid Waste and Emergency Response; 2001.
48. Kumar K, Shinde A, Aeron V, Verma A, Arif NS. Genetic engineering of plants for phytoremediation: Advances and challenges. Journal of Plant Biochemistry and Biotechnology. 2023;32(1):12–30.

49. Fulekar MH, Singh A, Bhaduri AM. Genetic engineering strategies for enhancing phytoremediation of heavy metals. African Journal of Biotechnology. 2009;8(4).

50. Ben Saad R, Ben Romdhane W, Baazaoui N, Bouteraa MT, Ben Hsouna A, Mishra A, Ćavar Zeljković S. Assessment of the cadmium and copper phytoremediation potential of the *Lobularia maritima Thioredoxin* 2 gene using genetically engineered tobacco. Agronomy. 2023;13(2):399.

51. Arora L, Narula A. Gene editing and crop improvement using CRISPR-Cas9 system. Frontiers in Plant Science. 2017;8:1932.

52. Naing AH, Park DY, Park HC, Kim CK. Removal of heavy metals using Iris species: A potential approach for reclamation of heavy metal-polluted sites and environmental beautification. Environmental Science and Pollution Research. 2023;12:1–3.

53. Bittencourt GA, de Souza Vandenberghe LP, Martínez-Burgos WJ, Valladares-Diestra KK, de Mello AF, Maske BL, Brar SK, Varjani S, de Melo Pereira GV, Soccol CR. Emerging contaminants bioremediation by enzyme and nanozyme-based processes–A review. iScience. 2023;26(6):106785.

54. Akpasi SO, Anekwe IM, Tetteh EK, Amune UO, Shoyiga HO, Mahlangu TP, Kiambi SL. Mycoremediation as a Potentially Promising Technology: Current Status and Prospects—A Review. Applied Sciences. 2023;13(8):4978.

55. Gill R, Naeem M, Ansari AA, Gill SS. Phytoremediation and Management of Environmental Contaminants: Conclusion and Future Perspectives. In Phytoremediation: Management of Environmental Contaminants (Volume 7, pp. 599–603). Cham: Springer International Publishing; 2023.

56. Biswal T. Nano-phytoremediation and Its Applications. In Phytoremediation: Management of Environmental Contaminants (Volume 7, pp. 335–364). Cham: Springer International Publishing; 2023.

57. Sharma M, Agarwal S, Agarwal Malik R, Kumar G, Pal DB, Mandal M, Sarkar A, Bantun F, Haque S, Singh P, Srivastava N. Recent advances in microbial engineering approaches for wastewater treatment: A review. Bioengineered. 2023;14(1):2184518.

58. Phang LY, Mohammadi M, Mingyuan L. Underutilised plants as potential phytoremediators for inorganic pollutants decontamination. Water, Air, & Soil Pollution. 2023;234(5):306.

59. Qi S, Wang J, Zhang Y, Naz M, Afzal MR, Du D, Dai Z. Omics approaches in invasion biology: Understanding mechanisms and impacts on ecological health. Plants. 2023;12(9):1860.

60. Chaudhary P, Ahamad L, Chaudhary A, Kumar G, Chen WJ, Chen S. Nanoparticle-mediated bioremediation as a powerful weapon in the removal of environmental pollutants. Journal of Environmental Chemical Engineering. 2023;11(2):109591.

61. Ansari AA, Naeem M, Gill SS, Al Zuaibr FM. Phytoremediation of contaminated waters: An eco-friendly technology based on aquatic macrophytes application. The Egyptian Journal of Aquatic Research. 2020;46(4):371–376.

62. Ekperusi AO, Nwachukwu EO, Sikoki FD. Assessing and modelling the efficacy of Lemnapaucicostata for the phytoremediation of petroleum hydrocarbons in crude oil-contaminated wetlands. Scientific Reports. 2020;10(1):8489.

63. De Laet C, Matringe T, Petit E, Grison C. *Eichhornia crassipes*: A powerful bioindicator for water pollution by emerging pollutants. Scientific Reports. 2019;9(1):7326.

64. Dhir B. Phytoremediation: Role of Aquatic Plants in Environmental Clean-Up. New Delhi: Springer; 2013.

65. Rani R, Padole P, Juwarkar A, Chakrabarti T. Phytotransformation of phorate by *Brassica juncea* (Indian Mustard). Water, Air, & Soil Pollution. 2012;223:1383–1392.

66. Sbani NH, Abdullah SR, Idris M, Hasan HA, Al-Baldawi IA, Jehawi OH, Ismail NI. Remediation of PAHs-contaminated water and sand by tropical plant (*Eleocharis*

ochrostachys) through sub-surface flow system. Environmental Technology & Innovation. 2020;20:101044.

67. Deyris PA, Grison C. Nature, ecology and chemistry: An unusual combination for a new green catalysis, ecocatalysis. Current Opinion in Green and Sustainable Chemistry. 2018;10:6–10.

68. Panda A, Patra DK, Acharya S, Pradhan C, Patra HK. Assessment of the phytoremediation potential of *Zinnia elegans* L. plant species for hexavalent chromium through pot experiment. Environmental Technology & Innovation. 2020;20:101042.

16 Application of Metal Hyperaccumulator Plants in Phytoremediation

Swapnashree Satapathy, Dattatreya Kar, and Ananya Kuanar

INTRODUCTION

The phenomenon addressed in this chapter can be referred to as "anthropogenic metal ion pollution" or "anthropogenic metal ion contamination." This term refers to the growing amount of dangerous metal ions in the environment as a result of human activity. This pollution can harm ecosystems in a number of ways, such as by interfering with biological processes, poisoning living things, and creating long-term ecological imbalances. High levels of metals in soil, such as cadmium (Cd), zinc (Zn), lead (Pb), and chromium (Cr), can have negative impacts on plant growth and metabolic processes (Agrawal and Sharma, 2006). Health hazards for humans as well as animals are related to the accumulation of these hazardous metals in plants (Wang et al., 2003). High concentrations of heavy metals, especially those with hazardous traits, inhibit respiratory and photosynthetic activity, which may cause plant death. They also adversely impact several physiological and metabolic processes (Garbisu and Alkorta, 2001; Schmidt, 2003). The generation of reactive oxygen species (ROS) is more frequent when a metal is hazardous (Pagliano et al., 2006). It causes DNA damage, decreases membrane integrity, and induces oxidative stress (Quartacci et al., 2001). Due to anthropogenic influence, certain atypical plant species exhibit the capacity to flourish and thrive in environments characterized by both naturally occurring metal-rich soils and heavy metal–contaminated soils. Moreover, the presence of metal contaminants in the soil alters its composition and structural attributes, concurrently exerting adverse impacts on the microbial community (Kozdroj and van Elsas, 2001; Kurek and Bollag, 2004).

Excessive metal ions from contaminated areas are remediated through the application of chemical and biological techniques. Chemical remediation involves the application of specific chemicals to restore the environment. However, it is important to note that this method is not universally applicable, as different types of metal ions may require distinct chemical treatments for effective removal and remediation (Chaney et al., 1997). Phytoremediation is considered the most effective biological method for remedying contaminated soil and water. It is viewed as an advanced and promising technology for the restoration of polluted sites, and it often proves to be a cost-effective alternative compared to traditional methods like physicochemical approaches (Raskin, 1996; McGrath et al., 2001). According to Brooks et al. (1977), "hyperaccumulator" plants, in contrast to "excluder" plants, can actively absorb exceptionally large amounts of one or more heavy metals from the soil. Furthermore, due to the fact that heavy metals do not accumulate in the roots but are instead transported to the aerial parts of the plant, concentrations in aboveground organs, especially leaves, can be 100–1000 times greater in hyperaccumulating species compared to non-hyperaccumulating species. According to Rascio (1977) and Reeves (2006), they exhibit no signs of phytotoxicity. Although hyperaccumulation is a distinct trait, it also depends on hypertolerance, an essential trait that shields plants from heavy metal toxicity, to which hyperaccumulator plants are just as susceptible as non-hyperaccumulators. Numerous studies conducted over the past few decades, as exemplified by Barman et al. (2000), have observed the phenomenon of heavy metal hyperaccumulation in certain plants. This underscores the importance of conducting advanced research into the molecular underpinnings of phytoremediation technology.

It is important to remember that the types of heavy metals, the specific plant species involved, and various soil properties like pH, organic matter content, and cation exchange capacity, among others, can all influence the hyperaccumulation of heavy metals, as indicated by Xian and Shokohifard (1989). Experimental data in the field of metal biology reveal that, depending on factors like the metal's oxidation state, chemical form, dosage, and method of exposure, even essential metals for a plant's normal growth, like iron and copper, may become hazardous. Metals and their compounds are enriched to harmful levels, in contrast to the organic waste produced by some living plants. Many studies, including those conducted by Rai et al. (1995) and Cardwell et al. (2002), have demonstrated that various aquatic plants can absorb hazardous metals such as lead, chromium, mercury (Hg), and others through their root systems. These metals are then transported to the stems and leaves of the plants without causing toxicity symptoms. This process contributes to the reduction of hazardous metal levels in contaminated environments. This metal intake is heavily influenced by metal type and chemical speciation, as well as plant environment factors such as terrestrial, aquatic, and so on. As a result, plant selection is critical for polluted site restoration.

Many research labs throughout the world have made remediating soils with metal contamination one of their main objectives. Environmental management faces the challenge of developing low-cost treatment systems that incorporate plants. The practice of phytoremediation is a current, environmentally friendly technique. The process involves utilizing metal hyperaccumulator plants for the purpose of remediating contaminated soil, water, and the surrounding environment (Prasad and Freitas, 2003). Studies by Baker and Walker (1989) and Baker et al. (1991) have noted that these plants demonstrate flowering without displaying any signs of toxicity. Hyperaccumulator plants possess the ability to extract various metals and accumulate them in their upper regions, demonstrating a notable level of metal tolerance. This characteristic renders them a valuable resource for remediating landscapes contaminated with metals. According to Baker and Walker (1989), in specific plant species, the concentration of metals or metalloids in the aboveground biomass can exceed that of neighboring plants by more than an order of magnitude, and in some cases, up to four times higher. Krämer (2010) reported that a total of 500 plant species exhibit this distinctive level of heavy metal accumulation. Baker et al. (1991) conducted a notable field trial on the phytoextraction of zinc and cadmium, which has garnered significant attention for the cleanup of soils tainted with heavy metals (USEPA, 2000). Numerous comprehensive analyses summarizing various pivotal aspects of this distinctive green technology have been disseminated (Padmavathiamma and Li, 2007; Krämer, 2010). Recent reviews, with an emphasis on the molecular and physiological origins of phytoremediation, provide overarching insights into prevailing research trends. The third search round concluded in June 2022, encompassing a thorough examination of databases including Biological Abstracts (BIOSIS), CAB Abstracts (CAB International), Web of Science (ISI), Biological Sciences (CSA), Plant Science (CSA), and Biosis Previews (BIOSIS). Given the potential necessity for multiple crop cycles spanning several years to reduce metal levels to meet approved regulatory standards, it is prudent to regard phytoremediation as a protracted remediation strategy. This approach is particularly applicable in locations where the conditions favor the implementation of phytoremediation techniques. This new remediation approach is competitive and might even be better than the current conventional technologies. It is really intriguing that the chapter also emphasizes the application of heavy metal accumulator plants or model plants in phytoremediation.

PHYTOREMEDIATION

In simple terms, phytoremediation is the process of removing or reducing inorganic and organic toxins from the environment using plants and associated microbes (Visioli, 2013). Pollutants can be eliminated from soil or water through various techniques, including physical removal, transfer to another medium, biological degradation, or encapsulation (Arifin et al., 2012). Phytoremediation is a novel approach that uses the ability of plant roots to initially absorb pollutants before eventually collecting them on the shoot tissue through translocation along the stem. Phytoremediation, which is more recent than other traditional remediation techniques, has great promise as a green technology to provide our decomposing environment with the required treatment. To date, many plant species have been considered as potential phytoremediation agents (Da Conceição Gomes et al., 2016).

During phytoremediation, plants that grow in soil or water containing heavy or substantial metals can absorb or tolerate these elements in various ways, depending on the physiological processes at play and the specific types of metals involved (Pilon-Smits and Freeman, 2006). As outlined by Halder and Ghosh (2014), there are five distinct phytoremediation techniques: phytoextraction, phytostabilization, phytofiltration, phytovolatilization, and phytotransformation in Figure 16.1.

PHYTOEXTRACTION

Metals are translocated by plants and accumulated in a form that may be removed from their tissue during the phytoextraction process (Nwoko, 2010). The terms *phytoextraction*, *phytoabsorption*, and *phytoaccumulation* are often used interchangeably in regard to this kind of phytoremediation (Ali et al., 2013). This appears to be the most significant phytoremediation technique for removing metals from contaminated

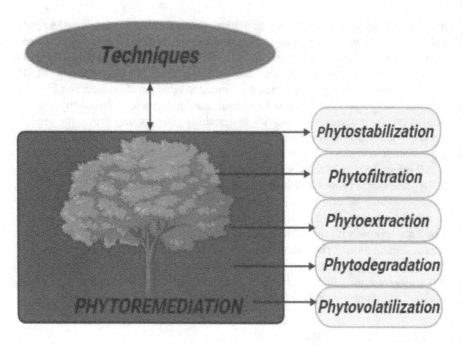

FIGURE 16.1   Phytoremediation techniques. (Ref: Durairaj et al. 2021.)

water, sediment, and soil. Several factors, including soil properties, metal bioavailability, speciation, and plant species, have an impact on this remediation procedure's effectiveness. However, the plant's cultivated shoot biomass frequently contains high amounts of absorbed metals (Walliwalagedara et al., 2010). Continuous phytoextraction involves the utilization of hyperaccumulator plants, which naturally accumulate extremely high levels of hazardous pollutants during the course of their existence. This approach differs from induced phytoextraction methods, where accelerants or chelators are introduced to the soil to boost the formation of toxins at a specific time. Chelators like EDTA are employed to extract heavy metal pollutants like chromium, lead, nickel (Ni), cadmium, copper (Cu), and zinc from the soil. These chelators bind to the metals, making them more available for uptake by plants like *Brassica juncea* (Indian mustard) and *Helianthus annuus* (sunflower), which subsequently accumulate the metals from the soil (Turgut et al., 2004).

Due to their advantages of safe disposal, cost-effectiveness, and ease of handling, hyperaccumulators are highly desirable. In phytoremediation, the hyperaccumulation of contaminants by plants is often considered more significant than the sheer volume of biomass produced. Hyperaccumulation refers to the unique potential of some plants to amass remarkably high concentrations of specific contaminants in their tissues, even if the overall biomass or plant size may be relatively modest. This elevated concentration of contaminants in the plant tissues is crucial for the efficient elimination of environmental contaminants, enabling the effective cleanup of contaminated sites. Therefore, a plant's capacity to hyperaccumulate contaminants is a pivotal factor in the success of phytoremediation processes (Walliwalagedara et al., 2010).

PHYTOFILTRATION

When toxins from surface wastewater are taken up or deposited by plant roots, it is referred to as phytofiltration or rhizofiltration. This process prevents the contaminants from releasing into underground water (Sarma, 2011). This method, known as phytofiltration, is a type of phytoremediation that has the potential to be implemented directly in polluted water bodies by cultivating plants (Da Conceição Gomes et al., 2016). While it is typically associated with aquatic plant species, there have been suggestions to apply it to terrestrial plants as well. In this approach, terrestrial plants facilitate the precipitation of metals with the assistance of bacteria in the root biofilter (Shah et al., 2024). Numerous terrestrial plants, such as hydroponically cultivated grasses, have demonstrated effectiveness in successfully removing metals through phytofiltration (Khilji and Sajid, 2020). Studies have revealed that a variety of aquatic macrophyte species are capable of phytofiltration. In their study on phytoremediation in differently polluted water conditions containing heavy metals, Liao and Chang (2004) discovered that *Eichhornia crassipes*, commonly known as water hyacinth, not only absorbs and accumulates metal contaminants but also exhibits a high growth rate and enhanced biomass production. These characteristics collectively make it a highly effective agent for phytofiltration. Potential phytofiltration agents need to be capable of withstanding high metal concentrations, develop rapidly and thoroughly, and generate a significant amount of aboveground biomass. One advantage of efficient phytofiltration compared to phytoextraction is that plants utilized for phytofiltration are generally anticipated to have a restricted capacity to transport absorbed metals from their roots to their aboveground shoot tissues. This restricted translocation of contaminants is advantageous as it leads to reduced contamination in other parts of the plant, thereby lowering the risk of the plant itself becoming toxic. In contrast, phytoextraction endeavors to amass contaminants in the aboveground portions of the plant, potentially posing a higher risk of toxicity and necessitating more meticulous management. Phytofiltration's focus on retaining contaminants in the roots minimizes this risk (Lee et al., 2003).

PHYTOSTABILIZATION

In this process, pollutants undergo transformation through continuous precipitation in the plant's rhizosphere, converting them into forms that are less hazardous or more readily taken up by organisms. To achieve this, methods such as leaching, erosion control, and preventing surface overflow are employed (Nwoko, 2010). It can be used to stabilize metals in polluted soil, sediment, or water, preventing them from migrating to other sections of food crops or beneath the surface and entering the food chain. Indeed, in phytostabilization, the process involves metal absorption by plant roots, followed by precipitation and the subsequent reduction of metals within the plant's rhizosphere. For instance, the highly toxic Cr^{6+} may be converted into the less harmful Cr^{3+} (Wu et al., 2012). It's crucial to remember that phytostabilization primarily restricts the movement of metals and prevents their dispersion but does not completely eliminate them from the contaminated environmental compartment. Consequently, phytostabilization is often considered an interim or temporary approach for managing pollutant contamination. While it offers certain advantages over other phytoremediation techniques, such as reduced risk of contaminant spreading, it may not provide a permanent solution for completely removing contaminants from the environment (Vangronsveld et al., 2009).

PHYTODEGRADATION

Phytotransformation, also known as phytodegradation, represents another part of phytoremediation. This approach involves the chemical conversion of pollutants and other substances by plant metabolism, effectively rendering the contaminants inert within both the root and shoot tissues of the plant (Tangahu et al., 2011). This process involves not only the action of rhizosphere microorganisms associated with plants but also the metabolic enzymes within the plants themselves, which work on the surrounding pollutants to transform them into less hazardous forms (Dobos and Puia, 2009). It was also suggested that this process works against inorganic pollutants, such as metals, in which case a method equivalent to phytostabilization would be used to change hazardous metals into less toxic forms (Abd El et al., 2012). Phytotransformation, while a valuable phytoremediation method, does have certain limitations. It tends to be less precise and efficient than other techniques. It requires more time for the transformation process, demands specific soil conditions and the presence of underground water, and often necessitates the application of soil additives to enhance its effectiveness. These factors collectively contribute to its relative complexity and potentially lower its efficacy compared to alternative phytoremediation approaches.

PHYTOVOLATILIZATION

Contaminants are converted into a volatile state during the phytovolatilization process and removed into the environment through the stomata on plant leaves (Nwoko, 2010). However, as this mechanism only carries toxins from one environmental compartment to another, precipitation could eventually restore them to their source in the soil, so it may not be as popular as other phytoremediation methods like phytoextraction and phytofiltration (Sarma, 2011). According to Wang et al., the process of phytovolatilization, elemental arsenic (As) is converted into selenoaminoacids, such as selenomethionine (Wang et al., 2012). Subsequently, through methylation, selenomethionine is transformed into a volatile and less harmful form known as dimethylselenide. This conversion process has a vital role in reducing the toxicity of the element and facilitating its release into the atmosphere.

HYPERACCUMULATOR PLANTS FOR HEAVY METALS: WHAT ARE THEY?

The term *hyperaccumulator* is employed to characterize plants that possess the remarkable ability to accumulate exceptionally high quantities of one or more soil-based heavy metals, distinguishing them from excluder plants (Brooks et al., 1977). Furthermore, heavy metals are not sequestered in the roots but rather transported to the shoot and accumulate in aboveground parts, particularly the leaves. As a result, the levels of heavy metals in hyperaccumulating species can be 100 to 1000 times higher than in non-hyperaccumulating species. These hyperaccumulators exhibit no signs of phytotoxicity, as reported by Reeves (2006). Hypertolerance, a crucial property that enables plants to withstand heavy metal toxicity, is a trait to which hyperaccumulator plants are just as sensitive as non-hyperaccumulator plants (Chaney et al., 1997). Hyperaccumulation is indeed a distinctive characteristic, but it also relies on hypertolerance. Only around 450 angiosperm species, less than 0.2% of all species that are known to exist, have been confirmed as hyperaccumulators of the heavy metals arsenic, cobalt (Co), cadmium, copper, manganese (Mn), nickel, selenium (Se), lead, antimony (Sb), thallium (Tl), and zinc. Despite an increasing number of reported cases, it is plausible that numerous hyperaccumulators remain undiscovered in natural habitats (Karimi, 2009). The presence of hyperaccumulator species across a diverse range of closely related plant families indicates that this trait has independently evolved multiple times in response to specific ecological conditions.

The term *hyperaccumulator* was initially coined to characterize plants that could accumulate over 1 mg of nickel per kilogram of dry weight in their shoots, as nickel toxicity typically starts at concentrations of 10 to 15 µg/g in most plant vegetative organs. Over time, specific threshold values emerged for each heavy metal, based on their unique phytotoxicity. According to these criteria, hyperaccumulator plants, when grown in their native soils, accumulate more than 10 mg/g of manganese or zinc, over 1 mg/g of arsenic, cobalt, copper, chromium, nickel, selenium, lead, antimony, or thallium, and over 0.1 mg/g of cadmium in their aerial organs without experiencing phytotoxic effects (Table 16.1) (Verbruggen et al., 2009). While some "facultative metallophytes" are capable of growing on nonmetalliferous soils, they

TABLE 16.1
Examples of Hyperaccumulator Plants Showing Different Parts of Plant

Hyperaccumulator Plant	Metal(s)	Metal-Accumulating Plant Portion	References
Alyssum bertolonii	Ni	Shoot	(Lee et al., 2003)
Alyssum corsicum	Ni	Shoot	(Lee et al., 2003)
Alyssum murale	Ni	Leaves	(Bani et al., 2009)
Corrigiola telephiifolia	As	Aboveground plant parts	(García-Salgado et al., 2012)
Eleocharis acicularis	Cu, Zn	Shoots	(Ha et al., 2011)
Pteris cretica	As	Root	(Zhao et al., 2002)
Solanum photeinocarpum	Cd	Root	(Zhang et al., 2011)
Thlaspi caerulescens	Cd, Zn	Shoots	(Banasova et al., 2008)
Schima superba	Mn	Leaves	(Yang, 2008)
Azolla pinnata	Cd	Root	(Rai, 2008)
Isatis pinnatiloba	Ni	Aboveground plant parts	(Altinözlü et al., 2012)
Euphorbia cheiradenia	Pb	Shoots	(Chehregani et al., 2009)
Pteris vittata	As	Root	(Kalve et al., 2011)

Source: Ullah et al., 2014.

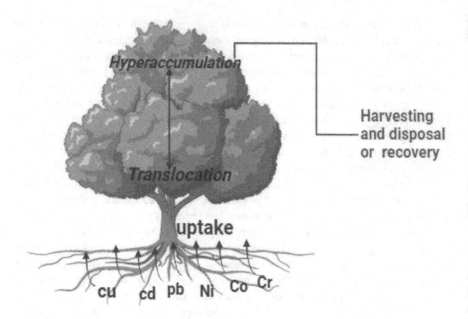

FIGURE 16.2   Process of heavy metal hyperaccumulating plants. (Ref: Rascio and Navari-Izzo, 2011.)

are more commonly found in habitats enriched with metals. Most hyperaccumulators are exclusive to soils rich in metalliferous materials and are classified as "strict metallophytes." However, certain "facultative metallophytes" can also be found in such environments (Assunção et al., 2003). It's noteworthy to mention that there are populations of several species reported to exhibit both metallicolous and nonmetallicolous traits. Among these species are the zinc hyperaccumulators *A. halleri* and *T. caerulescens*, which demonstrate hyperaccumulation as a species-specific characteristic present across all populations (Bert, 2000). In contrast, in some species such as the Cd hyperaccumulators and the Zn hyperaccumulator *S. alfredii*, this feature is not unique to the species as a whole and is instead only found in metallicolous populations in Figure 16.2 (Verbruggen et al., 2009).

HOW CAN HEAVY METALS HYPERACCUMULATE IN PLANTS?

According to Deng et al., 2007, the degree of hyperaccumulation of the levels of one or more heavy metals can differ greatly between different plant species, as well as between populations and ecotypes of the same species. There are three distinct features that set hyperaccumulators apart from similar non-hyperaccumulator species. These attributes encompass a notably heightened ability to detoxify and store significant heavy metal concentrations in their leaves, a substantially enhanced ability to accumulate heavy metals from the soil, and a swifter and more efficient transport of metals from the roots to the shoots. Much research has been done on model plants like *Thlaspi caerulescens* and *Arabidopsis halleri* (Krämer, 2010). While the

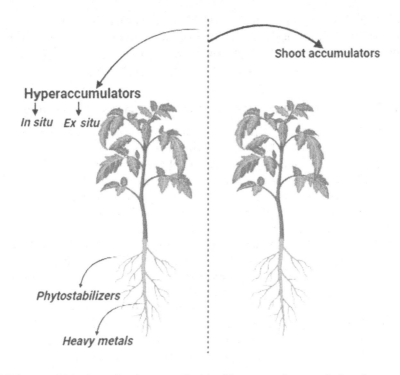

Shoot accumulators

Hyperaccumulators

In situ   Ex situ

Phytostabilizers

Heavy metals

FIGURE 16.3   Halophytes for phytoremediation of heavy metal accumulating plants.

foundational processes of hyperaccumulation are rooted in genes shared by both hyperaccumulator and non-hyperaccumulator plants, it is crucial to note that these genes undergo distinct patterns of expression and regulation in the two plant categories. This distinction is highly significant as it underscores the substantial differences between these two plant types in Figure 16.3 (Verbruggen et al., 2009).

POTENTIAL APPLICATION OF HYPERACCUMULATION IN PHYTOREMEDIATION

In the last two decades, the adoption of plant species in sustainable soil remediation methods, collectively referred to as phytoremediation, has been on the rise. Utilizing plants for the purpose of remediating polluted soils has emerged as an approach offering significant advantages when compared to conventional civil engineering techniques (Tiwari, 2021). There are a number of recent comprehensive evaluations examining the most significant elements of soil metal phytoremediation. Chaney et al. (1995) were pioneers in suggesting the utilization of heavy metal hyperaccumulator plants for remediating polluted areas. However, it's worth noting that these plants often exhibit small biomasses, slow growth rates, and shallow root systems, factors which may constrain the pace at which they can effectively extract metals (Ebbs et al., 1997). It has been found that the ability to hyperaccumulate metals is more important for phytoremediation than the annual biomass yield (Chaney et al., 1995).

In conclusion, although extensive research has been conducted on this subject, there have been relatively few field experiments or commercial operations showcasing hyperaccumulators' effectiveness for phytoextraction. Nonetheless, it is crucial to regard this approach as a viable long-term remediation strategy for heavily polluted soils.

QATAR'S HALOPHYTES ARE PROMISING HEAVY METAL HYPERACCUMULATORS

Several studies suggest that certain plants native to Qatar, many of which are heavy metal hyperaccumulators, have the potential to serve as efficient options for phytoremediation. Prominent instances include plants pertaining to the *Zygophyllum* genus, which have demonstrated either metal tolerance or accumulation characteristics when assessed in polluted soil and wastewater environments (Rejeb et al., 2013). Other plants like *Phragmites australis* and *Typha domingensis* have also shown promise for phytoremediation (Yasseen, 2022). *Typha domingensis* exhibits the ability to detoxify heavy metals from solution cultures and industrial wastewater. Additionally, the tree species *Prosopis juliflora* has demonstrated effectiveness in removing heavy metals from the environment (Lone et al., 2008). Furthermore, *Medicago* species such as *Glycine max* and *Medicago sativa,* which have shown substantial phytoremediation abilities for petroleum-polluted soils, are plants with comparable potential (Njoku et al., 2009). Over the years, plants like *Phragmites australis* have shown promising results in remediating oil-polluted soils. They also offer viable solutions for the phytoremediation of wastewater generated from conventional oil and gas production.

COMMON HEAVY METAL CONTAMINANTS

The processes by which plants tolerate and/or accumulate common heavy metal pollutants are explained in the section that follows.

PHYTOREMEDIATION OF ARSENIC (AS)

Many investigations suggested the use of *P. vittata* for soil and water phytoremediation. This may remediate damaged soil areas in 10 years or less and lower water arsenic levels to <10 ppb (Salido et al., 2003). The ideal phytoremediator would store arsenic in a less hazardous form while accumulating it at quantities comparable to *P. vittata.* The two main forms of arsenic present in fish, arsenobetaine and arsenocholine, are not hazardous to humans and are easily eliminated in urine. One potential phytoremediation approach involves the use of transgenic organisms capable of converting inorganic forms of arsenic into compounds that resemble them. This transformation can be achieved through the expression of just two specific genes that significantly improved *Arabidopsis'* arsenic tolerance and accumulation (Dhankher et al., 2002).

PHYTOREMEDIATION OF CD

In preliminary experiments conducted by Brown et al., 1994, agricultural soil from the UK contaminated with sewage sludge, as well as soil samples from closed mines in Silver Bow Creek, Montana, Palmerton, Pennsylvania, and France, were used to investigate cadmium and zinc phytoremediation. *Thlaspi caerulescens* demonstrated the capacity to decrease levels of cadmium and zinc in the soils in each instance, although only to a specific depth and at rates necessitating over 15 years to eliminate the majority of the metals. Nevertheless, it is crucial to acknowledge that *T. caerulescens* faced challenges in growing in agricultural soil in the UK due to copper toxicity, underscoring the necessity for phytoremediator plants capable of withstanding elevated levels of multiple toxic substances (Lombi et al., 2001). The plant's restricted height and limited biomass further constrain its remediation potential. In a separate investigation, it was noted that *Thlaspi caerulescens* exhibited the highest accumulation of cadmium and zinc in its leaves, while *Brassica juncea* eliminated a higher Zn and an equivalent amount of Cd, attributable to its larger size. This implies that if *T. caerulescens* were subjected to genetic modifications to promote larger growth, or if *Brassica juncea* were engineered to accumulate greater levels of Cd and Zn, phytoextraction could evolve into a more efficacious remediation strategy. These findings underscore the potential for genetic and engineering interventions to amplify the efficiency of phytoremediation processes.

PHYTOREMEDIATION OF CR

Chromium, existing in both its hexavalent (Cr[VI]) and trivalent (Cr[III]) states, can pose significant risks to human health and exert detrimental impacts on ecosystems when found in estuaries and groundwater. It's worth noting that while chromium in soil is often transformed into the more hazardous Cr(III), the presence of Cr(VI) is a significant concern (Kamnev, 2003). In terms of remediation efforts, certain natural materials like coconut husks and biogases have been identified as effective absorbers of Cr(VI) in estuaries (Parimala et al., 2004). Additionally, studies have shown that *Betula* and *Salix* trees have the capacity to take up chromium, making them potentially valuable tools for phytoremediation of groundwater contaminated with Cr(VI) (Pulford et al., 2001). Recent research has also highlighted that *Salsola kali*, commonly known as tumbleweed, possesses the ability to absorb Cr(VI), suggesting its potential utility in phytoremediation efforts (Gardea-Torresdey et al., 2005). This finding may open up new avenues for the effective cleanup of environments contaminated with this form of chromium.

PHYTOREMEDIATION OF CU

Copper is recognized for its potential harm to plants when present in elevated concentrations, leading plants to employ various ligands as defense mechanisms such as citrate, PC2, PC3, and metallothioneins to bind and regulate Cu in their tissues (Murphy et al., 1999). As a result, most Cu-tolerant plants are excluders, which means they prevent excess Cu from entering their tissues, and large Cu accumulators are rare. For instance, *Elsholtzia splendens* was initially thought to be a Cu hyperaccumulator, but further research revealed that it is actually a tolerant excluder, similar to plants like *Silene vulgaris, Mimulus guttatus,* and *Elsholtzia argyi* (Jiang et al., 2004). Although 37 species of copper hyperaccumulators have been identified, primarily from the Congo's Shaban Copper Arc, further investigation is required to ascertain whether the elevated Cu levels observed are attributable to hyperaccumulation or the deposition of Cu dust on leaves (Song et al., 2004). Additionally, copper can be absorbed by soil amendments such as phosphate, which can aid ongoing phytoremediation efforts (Wu et al., 2004). In the context of wastewater treatment, the water hyacinth (*Eichhornia crassipes*) exhibits potential for phytoremediation of trace amounts of Cu pollution, as it can absorb significant quantities of Cu from aqueous environments, rendering it a valuable tool for environmental cleanup (Liao and Chang, 2004).

PHYTOREMEDIATION OF HG

Numerous studies have primarily focused on the conversion of organomercurials into volatile elemental mercury (Hg^0), which can be released into the atmosphere, rather than employing plants for mercury phytoextraction. The University of Georgia's Meagher research group achieved this transformation by introducing bacterial genes, specifically *MerA* (which encodes organomercuric lyase) and *MerB* (which encodes mercuric reductase), into plants (Rugh et al., 1996). *MerB* effectively disrupts the mercury–carbon bond, enabling *MerA* to convert ionic mercury into its elemental form. Cottonwood and poplar transgenic trees that express *MerA* and/or *MerB* serve as phytoremediators that do not require removal or replacement each season (Rugh et al., 1998). The development of *MerB*, which was immediately targeted at the ER and cell wall, served as an outstanding example of the significance of actual subcellular targeting. Due to the predominant intracellular presence of hydrophobic organomercurials, *MerB* activity seems to have been channeled toward the secretory pathway. Interestingly, even though plants generated targeted *MerB* at a rate tenfold or lower than untargeted *MerB*, they demonstrated the capacity to convert comparable quantities of elemental mercury (Hg^0). In a study conducted by Ruiz et al. (2003), a noteworthy quantity of protein was successfully produced by expressing both *MerA* and *MerB* in chloroplasts, potentially offering additional advantages.

PHYTOREMEDIATION OF NI

In Oregon and Washington, *Alyssum* hybrids with suitable traits for phytomining nickel have been developed (Chaney et al., 1999), and the technique has been utilized (Lee et al., 2003). Numerous species of *Alyssum*, known as nickel hyperaccumulators, have undergone extensive study for their potential application in the phytoremediation of mine sites (McGrath and Zhao, 2003). Additionally, alternative metal hyperaccumulators are under investigation for their potential in phytomining gold, nickel, and thallium from soils (Anderson et al., 1999; Boominathan et al., 2004). The augmentation of shoot-bound nickel accumulation can be achieved through the activation of genes sourced from various hyperaccumulator species, including the metal tolerance proteins (MTPs). Furthermore, the simultaneous overproduction of many MTP genes in combination with genes conferring tolerance to metal stress, such as the SAT genes, holds promise for enhancing nickel phytomining and phytoremediation technologies.

PHYTOREMEDIATION OF PB

The most significant obstacle to efficient Pb phytoremediation is that only 0.1% of the Pb in soil can be removed (Huang et al., 1997). Using soil components like EDTA has served as the primary objective of phytoremediation attempts to increase the amount of Pb that may be absorbed (Blaylock et al., 1997). Chelators can enhance solubility and absorption. Nevertheless, if the quantity of lead transported to the shoots remains substantially lower than the Pb content in the soil, there is an increased risk of mobilized Pb-EDTA spreading within the soil and potentially contaminating groundwater. Additional methods for solubilizing Pb and then transferring it to the leaves must be discovered to optimize Pb phytoremediation. The introduction in *Arabidopsis* of the glutathione-Cd vacuolar transporter YCF-1 has been observed to heighten tolerance levels and significantly augment lead accumulation (Song et al., 2003).

PHYTOREMEDIATION OF SE

B. juncea's overexpression of ATP sulfurylase confers the capacity for effective phytoremediation of selenium-contaminated sites, resulting in a twofold increase in Se accumulation in shoots and elevated biomass production. This enhancement stems from the heightened activity of ATP sulfurylase, an enzyme in charge of transforming selenium into selenite (Van Huysen et al., 2004). *B. juncea* plants, which overexpress ATP sulfurylase and currently have a bioconcentration factor of approximately 10, might be produced by any improvement in accumulation or volatilization and are ideal for effective phytoremediation. Further study has focused on Se metabolism's overproduced enzymes. The quantity of selenium accumulated and the amount of selenium absorbed by proteins both significantly decrease in *Arabidopsis* when the mammalian enzyme that converts selenium to elemental selenium and alanine is highly expressed (Pilon et al., 2003). Noteworthy studies highlight the crucial role of strategically localizing novel proteins within cells of transgenic plants. These investigations have demonstrated that cytosolic variants of selenocysteine lyase enhance tolerance to selenium, while chloroplastic variants diminish tolerance.

CONCLUSION AND FUTURE OUTLOOK

The role of hyperaccumulators in modern environmental biotechnology has developed into a new, commercially important area of study within the science of phytoextraction. A significant portion of this research has taken place in meticulously controlled laboratory environments, albeit in a comparatively short amount of time (Bizily et al., 2003). It is imperative that further investigations be undertaken to optimize both ecological and economic efficiencies in this domain. These studies have explored the contributions of essential and facultative accumulator plant species to metal accumulation and detoxification, aligning with the latest advancements in the field of phytoremediation. Recent advances in genetic and molecular studies of metal transporters have elucidated the intricate processes governing metal absorption, accumulation, and detoxification in hyperaccumulator plants. These revelations have also shed light on their pivotal role in the phytoremediation process. Phytoremediation, an environmentally sustainable and green approach, holds substantial promise for effectively cleansing contaminated environments. Green plants' incredible capacity to consume elements from their surroundings and undergo biological transformation has given rise to a new field related to plant biology. Due to their great potential for phytoextraction of metals from the environment, metal hyperaccumulating plants have attracted increasing interest. Heavy metals can accumulate in species that are hyperaccumulators such as nickel, copper, lead, cobalt, zinc, and manganese at concentrations 100 to 1000 times higher than non-accumulator plant species. In addition to cleaning up contaminated areas, they can also extract valuable metals from soils which are rich in metals and produce profits.

Phytoremediation, especially when using metal hyperaccumulator plants, exhibits substantial potential as a widespread approach for the remediation of hazardous environmental pollutants. To optimize its efficacy as a remediation and purification strategy, phytoremediation involving metal hyperaccumulator plants necessitates interdisciplinary cooperation across domains such as soil chemistry, microbiology, plant biology, and environmental engineering. By working together, experts from these disciplines can develop and implement phytoremediation solutions that address pollution and promote environmental health and sustainability.

REFERENCES

Abd El-Kader, A.A., Hussein, M.M., & Alva, A.K. (2012). Response of Jatropha on a clay soil to different concentrations of micronutrients. *Am J Plant Sci*, 3, 1376–1381.

Agrawal, V., & Sharma, K. (2006). Phytotoxic effects of Cu, Zn, Cd and Pb on in vitro regeneration and concomitant protein changes in *Holarrhena antidysenterica*. *Biologia Plantarum*, 50, 307–310.

Ali, H., Khan, E., & Sajad, M.A. (2013). Phytoremediation of heavy metals—concepts and applications. *Chemosphere*, 91(7), 869–881.

Altinözlü, H., Karagöz, A., Polat, T., & Ünver, İ. (2012). Nickel hyperaccumulation by natural plants in Turkish serpentine soils. *Turk J Bot*, 36(3), 269–280.

Anderson, C.W.N., Brooks, R.R., Chiarucci, A., Lacoste, C.J., Leblanc, M., Robinson, B.H., Simcock, R., & Stewart, R.B. (1999). Phytomining for nickel, thallium and gold. *J Geochem Explore*, 67, 407–415.

Arifin, A., Parisa, A., Hazandy, A.H., Mahmud, T.M., Junejo, N., Fatemeh, A., ... Majid, N.M. (2012). Evaluation of cadmium bioaccumulation and translocation by *Hopea odorata* grown in a contaminated soil. *Afr J Biotechnol*, 11(29).

Assunção, A.G., Bookum, W.M., Nelissen, H.J., Vooijs, R., Schat, H., & Ernst, W.H. (2003). Differential metal-specific tolerance and accumulation patterns among *Thlaspi caerulescens* populations originating from different soil types. *New Phytol*, 159(2), 411–419.

Baker, A.J.M., & Walker, P.L. (1989). Ecophysiology of Metal Uptake by Tolerant Plants. In: Heavy Metal Tolerance in Plants: Evolutionary Aspects, Shaw, A.J. (Ed.), CRC Publ, 155–177.

Baker, A.J.M., Reeves, R.D., & McGrath, S.P. (1991). In situ Decontamination of Heavy Metal Polluted Soils Using Crops of Metal Accumulating Plants: A Feasibility Study. In: In situ Bioremediation, Hinchee, R.E. and R.F. Olfenbuttel (Eds.). Butterworth-Heinemann, Stoneham, MA, 539–544.

Banasova, V., Horak, O., Nadubinska, M., Ciamporova, M., & Lichtscheidl, I. (2008). Heavy metal content in *Thlaspi caerulescens* J. et C. Presl growing on metalliferous and non-metalliferous soils in Central Slovakia. *Int J Environ Pollut*, 33(2–3), 133–145.

Bani, A., Echevarria, G., Mullaj, A., Reeves, R., Morel, J.L., & Sulçe, S. (2009). Nickel hyperaccumulation by Brassicaceae in serpentine soils of Albania and northwestern Greece. *Northeastern Nat*, 16(sp5), 385–404.

Barman, S.C., Sahu, R.K., Bhargava, S.K., & Chaterjee, C. (2000). Distribution of heavy metals in wheat, mustard, and weed grown in field irrigated with industrial effluents. *Bull Environ Contam Toxicol*, *64*, 489–496.

Bert, V., Macnair, M.R., De Laguerie, P., Saumitou-Laprade, P., & Petit, D. (2000). Zinc tolerance and accumulation in metallicolous and nonmetallicolous populations of *Arabidopsis halleri* (Brassicaceae). *New Phytol*, 146(2), 225–233.

Bizily, S.P., Kim, T., Kandasamy, M.K., & Meagher, R.B. (2003). Subcellular targeting of methyl mercury lyase enhances its specific activity for organic mercury detoxification in plants. *Plant Physiol*, 131, 463–471.

Blaylock, M.J., Salt, D.E., Dushenkov, S., Zakharova, O., Gussman, C., Kapulnik, Y., … Raskin, I. (1997). Enhanced accumulation of Pb in Indian mustard by soil-applied chelating agents. *Environ Sci Technol*, 31(3), 860–865.

Boominathan, R.R., Saha-Chaudhury, N.M., Sahajwalla, V., & Doran, P.M. (2004). Production of nickel bio-ore from hyperaccumulator plant biomass: Applications in phytomining. *Biotechnol Bioeng*, 86, 243–250.

Brooks, R.R., Lee, J., Reeves, R.D., & Jaffré, T. (1977). Detection of nickeliferous rocks by analysis of herbarium specimens of indicator plants. *J Geochem Explor*, 7, 49–57.

Brown, S., Chaney, R., Angle, J.S., & Baker, A.J.M. (1994). Phytoremediation potential of *Thlaspi caerulescens* and bladder campion for zinc- and cadmium-contaminated soil. *J Environ Qual*, 23, 1151–1157.

Cardwell, A.J., Hawker, D.W., & Greenway, M. (2002). Metal accumulation in aquatic macrophytes from southeast Queensland, Australia. *Chemosphere*, 48(7), 653–663.

Chaney, R.L., Angle, J.S., Li, Y.M., & Baker, A.J.M. (1999). Method for phytomining of nickel, cobalt and other metals from soil. U.S. Patent No, 5944872, (continuation in-part of US Patent 5711784).

Chaney, R.L., Li, Y.M., Brown, S.L., Angle, J.S., & Baker, A.J. (1995). Hyperaccumulator based phytoremediation of metal-rich soils. Current topics in plant biochemistry, physiology and molecular biology. Will plants have a role in bioremediation, 10–12.

Chaney, R.L., Malik, M., Li, Y.M., Brown, S.L., Brewer, E.P., Angle, J.S., & Baker, A.J.M. (1997). Phytoremediation of soil metals. *Curr Opin Biotechnol*, 8, 279–284.

Chehregani, A., Noori, M., & Yazdi, H.L. (2009). Phytoremediation of heavy-metal-polluted soils: Screening for new accumulator plants in Angouran mine (Iran) and evaluation of removal ability. *Ecotoxicol Environ Saf*, 72(5), 1349–1353.

Da Conceição Gomes, M.A., Hauser-Davis, R.A., de Souza, A.N., & Vitória, A.P. (2016). Metal phytoremediation: General strategies, genetically modified plants and applications in metal nanoparticle contamination. *Ecotoxicol Environ Saf*, 134, 133–147.

Deng, D.M., Shu, W.S., Zhang, J., Zou, H.L., Lin, Z., Ye, Z.H., & Wong, M.H. (2007). Zinc and cadmium accumulation and tolerance in populations of *Sedum alfredii*. *Environ Pollut*, 147(2), 381–386.

Dhankher, O.P., Li, Y., Rosen, B.P., Shi, J., Salt, D., Senecoff, J.F., Sashti, N.A., & Meagher, R.B. (2002). Engineering tolerance and hyperaccumulation of arsenic in plants by combining arsenate reductase and gamma-glutamylcysteine synthetase expression. *Nat Biotechnol*, 20, 1140–1145.

Dobos, L., & Puia, C. (2009). The most important methods for depollution of hydrocarbons polluted soils. *Bull Uni Agricul Sci Veter Med Cluj Nap Agr*, 66(1).

Durairaj, K., Senthilkumar, P., Velmurugan, P., Dhamodaran, K., Kadirvelu, K., & Kumaran, S. (2021). Sol-gel mediated synthesis of silica nanoparticle from Bambusa vulgaris leaves and its environmental applications: Kinetics and isotherms studies. *J Sol Gel Sci Technol*, 90, 653–664.

Ebbs, S.D., Lasat, M.M., Brady, D.J., Cornish, J., Gordon, R., & Kochian, L.V. (1997). Phytoextraction of Cadmium and Zinc from a Contaminated Soil. *J Environ Qual*, 26(5), 1424–1430.

Garbisu, C., & Alkorta, I. (2001). Phytoextraction: A cost-effective plant-based technology for the removal of metals from the environment. *Bioresour Technol*, 77(3), 229–236.

García-Salgado, S., García-Casillas, D., Quijano-Nieto, M.A., & Bonilla-Simón, M.M. (2012). Arsenic and heavy metal uptake and accumulation in native plant species from soils polluted by mining activities. *Wat Air Soil Poll*, 223, 559–572.

Gardea-Torresdey, J.L., la, de, Rosa, G., Peralta-Videa, J.R., Montes, M., Cruz-Jimenez, G., & CanoAguilera, I. (2005). Differential uptake and transport of trivalent and hexavalent chromium by tumbleweed (*Salsola kali*). *Arch Environ Contam Toxicol*, 48, 225–232.

Ha, N.T.H., Sakakibara, M., & Sano, S. (2011). Accumulation of Indium and other heavy metals by *Eleocharis acicularis*: An option for phytoremediation and phytomining. *Bioresour Technol*, 102(3), 2228–2234.

Halder, S., & Ghosh, S. (2014). Wetland macrophytes in purification of water. *Int J Environ Sci*, 5(2), 432–437.

Huang, J., Chen, J., Berti, W., & Cunningham, S. (1997). Phytoremediation of lead-contaminated soils: Role of synthetic chelates in lead phytoextraction. *Environ Sci Technol*, 31, 800–805.

Jiang, L.Y., Yang, X.E., & He, Z.L. (2004). Growth response and phytoextraction of copper at different levels in soils by *Elsholtzia splendens*. *Chemosphere*, 55, 1179–1187.

Kalve, S., Sarangi, B.K., Pandey, R.A., & Chakrabarti, T. (2011). Arsenic and chromium hyperaccumulation by an ecotype of Pteris vittata–prospective for phytoextraction from contaminated water and soil. *Cur Sci*, 888–894.

Kamnev, A.A. (2003). Phytoremediation of heavy metals: An overview. *Recent advan marine biotechnol*, 8, 269–318.

Karimi, N., Ghaderian, S.M., Maroofi, H., & Schat, H. (2009). Analysis of arsenic in soil and vegetation of a contaminated area in Zarshuran. *Iran Int J Phytoremed*, 12(2), 159–173.

Khilji, S.A., & Sajid, Z.A. (2020). Phytoremediation potential of lemongrass (Cymbopogon flexuosus Stapf.) grown on tannery sludge contaminated soil. *Appl Ecol Env Res*, 18(6).

Kozdroj, J., & van Elsas, J.D. (2001). Structural diversity of microorganisms in chemically perturbed soil assessed by molecular and cytochemical approaches. *J Microbiol Methods*, 43(3), 197–212.

Krämer, U. (2010). Metal hyperaccumulation in plants. *Ann Rev Plant Biol*, 61, 517–534.

Kurek, E., & Bollag, J.M. (2004). Microbial immobilization of cadmium released from CdO in the soil. *Biogeochemistry*, 69, 227–239.

Lee, S., Moon, J., Ko, T.S., Petros, D., Goldsbrough, P.B., & Korban, S.S. (2003). Overexpression of Arabidopsis phytochelatin synthase paradoxically leads to hypersensitivity to cadmium stress. *Plant Physiol*, 131, 656–663.

Liao, S., & Chang, N. (2004). Heavy metal phytoremediation by water hyacinth at constructed wetlands in Taiwan. *J Aquatic Plant Manag*, 42, 60–68.

Lombi, E., Zhao, F., McGrath, S., Young, S., & Sacchi, G. (2001). Physiological evidence for a high-affinity cadmium transporter highly expressed in a *Thlaspi caerulescens* ecotype. *New Phytol*, 149, 53–60.

Lone, M.I., He, Z.L., Stoffella, P.J., & Yang, X.E. (2008). Phytoremediation of heavy metal polluted soils and water: Progresses and perspectives. *J Zhejiang Univ Sci*, 9(3), 210–220.

McGrath, S.P., Zhao, F.J., & Lombi, E. (2001). Plant and rhizosphere processes involved in phytoremediation of metal-contaminated soils. *Plant Soil*, 232, 207–214.

McGrath, S.P., & Zhao, F.J. (2003). Phytoextraction of metals and metalloids from contaminated soils. *Curr Opin Biotechnol*, 14, 277–282.

Murphy, A.S., Eisinger, W.R., Shaff, J.E., Kochian, L.V., & Taiz, L. (1999). Early copper-induced leakage of K(+) from Arabidopsis seedlings is mediated by ion channels and coupled to citrate efflux. *Plant Physiol*, 121, 1375–1382.

Njoku, K.L., Akinola, M.O., & Oboh, B.O. (2009). Phytoremediation of crude oil contaminated soil: The effect of growth of Glycine max on the physico-chemistry and crude oil contents of soil.

Nwoko, C.O. (2010). Trends in phytoremediation of toxic elemental and organic pollutants. *Afr J Biotechnol*, 9(37), 6010–6016.

Padmavathiamma, P.K., & Li, L.Y. (2007). Phytoremediation technology: Hyper-accumulation metals in plants. *Water Air Soil Pollut*, 184, 105–126.

Pagliano, C., Raviolo, M., Dalla Vecchia, F., Gabbrielli, R., Gonnelli, C., Rascio, N., ... La Rocca, N. (2006). Evidence for PSII donor-side damage and photoinhibition induced by cadmium treatment on rice (*Oryza sativa* L.). *J Photochem Photobiol B*, 84(1), 70–78.

Parimala, V., Krishnani, K.K., Gupta, B.P., Jayanthi, M., & Abraham, M. (2004). Phytoremediation of chromium from seawater using five different products from coconut husk. *Bull Environ Contam Toxicol*, 73(1), 31–37.

Pilon, M., Owen, J.D., Garifullina, G.F., Kurihara, T., Mihara, H., Esaki, N., & Pilon-Smits, E.A. (2003). Enhanced selenium tolerance and accumulation in transgenic Arabidopsis expressing a mouse selenocysteine lyase. *Plant Physiol*, 131, 1250–1257.

Pilon-Smits, E.A., & Freeman, J.L. (2006). Environmental cleanup using plants: Biotechnological advances and ecological considerations. *Fron Ecol Environ*, 4(4), 203–210.

Prasad, M.N.V., & Freitas, H.M.D.O. (2003). Metal hyperaccumulation in plants-biodiversity prospecting for phytoremediation technology. *Electron J Biotechnol*, 6, 275–321.

Pulford, I.D., Watson, C., & McGregor, S.D. (2001). Uptake of chromium by trees: Prospects for phytoremediation. *Environ Geochem Health*, 23, 307–311.

Quartacci, M.F., Cosi, E., & Navari-Izzo, F. (2001). Lipids and NADPH-dependent superoxide production in plasma membrane vesicles from roots of wheat grown under copper deficiency or excess. *J Exp Bot*, 52(354), 77–84.

Rai, P.K. (2008). Phytoremediation of Hg and Cd from industrial effluents using an aquatic free floating macrophyte *Azolla pinnata*. *Int J Phytoremed*, 10(5), 430–439.

Rai, U.N., Sinha, S., Tripathi, R.D., & Chandra, P. (1995). Wastewater treatability potential of some aquatic macrophytes: Removal of heavy metals. *Ecol Eng*, 5(1), 5–12.

Rascio, N. (1977). Metal accumulation by some plants growing on zinc-mine deposits. *Oikos*, 29(2), 250–253.

Rascio, N., & Navari-Izzo, F. (2011). Heavy metal hyperaccumulating plants: How and why do they do it? And what makes them so interesting? *Plant Sci*, 180(2), 169–181.

Raskin, I. (1996). Plant genetic engineering may help with environmental cleanup. *Proc Natl Acad Sci USA*, 93, 3164–3166.

Reeves, R.D. (2006). Hyperaccumulation of Trace Elements by Plants. In: Phytoremediation of Metal-Contaminated Soils. Earth and Environmental Sciences, J.L. Morel, G. Echevarria, N. Goncharova (Eds.). Springer, 1–25.

Rejeb, K.B., Ghnaya, T., Zaier, H., Benzarti, M., Baioui, R., Ghabriche, R., ... Abdelly, C. (2013). Evaluation of the Cd^{2+} phytoextraction potential in the xerohalophyte Salsola kali L. and the impact of EDTA on this process. *Ecol Engineer*, 60, 309–315.

Rugh, C.L., Senecoff, J.F., Meagher, R.B., & Merkle, S.A. (1998). Development of transgenic yellow poplar for mercury phytoremediation. *Nat Biotechnol*, 16, 925–928.

Rugh, C.L., Wilde, H.D., Stack, N.M., Thompson, D.M., Summers, A.O., & Meagher, R.B. (1996). Mercuric ion reduction and resistance in transgenic Arabidopsis thaliana plants expressing a modified bacterial merA gene. *Proc Natl Acad Sci USA*, 93, 3182–3187.

Ruiz, O.N., Hussein, H.S., Terry, N., & Daniell, H. (2003). Phytoremediation of organomercurial compounds via chloroplast genetic engineering. *Plant Physiol*, 132, 1344–1352.

Salido, A.L., Hasty, K.L., Lim, J.M., & Butcher, D.J. (2003). Phytoremediation of arsenic and lead in contaminated soil using Chinese brake ferns (*Pteris vittata*) and Indian mustard (*Brassica juncea*). *Int J Phytoremediation*, 5, 89–103.

Sarma, H. (2011). Metal hyperaccumulation in plants: A review focusing on phytoremediation technology. *J Environ Sci Tech*, 4(2), 118–138.

Schmidt, U. (2003). Enhancing phytoextraction: The effect of chemical soil manipulation on mobility, plant accumulation, and leaching of heavy metals. *J Environ Qual*, 32(6), 1939–1954.

Shah, D., Kamili, A., Sajjad, N., Tyub, S., Majeed, G., Hafiz, S., ... Maqbool, I. (2024). Phytoremediation of pesticides and heavy metals in contaminated environs. *Aqua Cont: Tol Bioremed*, 189–206.

Song, J., Zhao, F.J., Luo, Y.M., McGrath, S.P., & Zhang, H. (2004). Copper uptake by *Elsholtzia splendens* and *Silene vulgaris* and assessment of copper phytoavailability in contaminated soils. *Environ Pollut*, 128, 307–315.

Song, W.Y., Sohn, E.J., Martinoia, E., Lee, Y.J., Yang, Y.Y., Jasinski, M., Forestier, C., Hwang, I., & Lee, Y. (2003). Engineering tolerance and accumulation of lead and cadmium in transgenic plants. *Nat Biotechnol*, 21, 914–919.

Tangahu, B.V., Sheikh Abdullah, S.R., Basri, H., Idris, M., Anuar, N., & Mukhlisin, M. (2011). A review on heavy metals (As, Pb, and Hg) uptake by plants through phytoremediation. *Int J Chem Eng*, 2011(1), 939161.

Tiwari, S. (2021). Mechanism of heavy metal hyperaccumulation in plants. *Bioremed Sci*, 103–120.

Turgut, C., Pepe, M.K., & Cutright, T.J. (2004). The effect of EDTA and citric acid on phytoremediation of Cd, Cr, and Ni from soil using *Helianthus annuus*. *Environ Poll*, 131(1), 147–154.

Ullah, A., Mushtaq, H., Ali, H., Munis, M., Javed, M., & Chaudhary, H. (2014). Diazotrophs-assisted phytoremediation of heavy metals: A novel approach. *Env Sci Poll R*, 22(4), 2505–2514.

USEPA. (2000). Introduction to Phytoremediation. United States Environmental Protection Agency, Washington, DC., USA.

Van Huysen, T., Terry, N., & Pilon-Smits, E.A.H. (2004). Exploring the selenium phytoremediation potential of transgenic Indian mustard overexpressing ATP sulfurylase or cystathionine-γ-synthase. *Int J Phytoremediation*, 6, 1–8.

Vangronsveld, J., Herzig, R., Weyens, N., Boulet, J., Adriaensen, K., Ruttens, A., ... Mench, M. (2009). Phytoremediation of contaminated soils and groundwater: Lessons from the field. *Environ Sci Poll R*, 16, 765–794.

Verbruggen, N., Hermans, C., & Schat, H. (2009). Molecular mechanisms of metal hyperaccumulation in plants. *New Phytol*, 181(4), 759–776.

Visioli, G., & Marmiroli, N. (2013). The proteomics of heavy metal hyperaccumulation by plants. *J Proteomics*, 79, 133–145.

Walliwalagedara, C., Atkinson, I., van Keulen, H., Cutright, T., & Wei, R. (2010). Differential expression of proteins induced by lead in the Dwarf Sunflower *Helianthus annuus*. *Phytochem*, 71(13), 1460–1465.

Wang, D., Wen, F., Xu, C., Tang, Y., & Luo, X. (2012). The uptake of Cs and Sr from soil to radish (*Raphanus sativus* L.)-potential for phytoextraction and remediation of contaminated soils. *J Env Radioact*, 110, 78–83.

Wang, Q.R., Cui, Y.S., Liu, X.M., Dong, Y.T., & Christie, P. (2003). Soil contamination and plant uptake of heavy metals at polluted sites in China. *J Environ Sci Health Part A*, 38(5), 823–838.

Wu, L., Li, Z., Akahane, I., Liu, L., Han, C., Makino, T., ... Christie, P. (2012). Effects of organic amendments on Cd, Zn and Cu bioavailability in soil with repeated phytoremediation by *Sedum plumbizincicola*. *Int J Phytoremediation*, 14(10), 1024–1038.

Wu, L.H., Li, H., Luo, Y.M., & Christie, P. (2004). Nutrients can enhance phytoremediation of copper-polluted soil by Indian mustard. *Environ Geochem Health*, *26*, 331–335.

Xian, X., & In Shokohifard, G. (1989). Effect of pH on chemical forms and plant availability of cadmium, zinc, and lead in polluted soils. *Water Air Soil Pollut*, *45*, 265–273.

Yang, L. (2008). Phytoremediation: An ecotechnology for treating contaminated sites. *Prac Period Hazard Toxic Radioact Waste Manag*, 12(4), 290–298.

Yasseen, B.T., & Al-Thani, R.F. (2022). Endophytes and halophytes to remediate industrial wastewater and saline soils: Perspectives from Qatar. *Plants*, 11(11), 1497.

Zhang, X., Liu, X., Liu, S., Liu, F., Chen, L., Xu, G., ... Cao, Z. (2011). Responses of Scirpus triqueter, soil enzymes and microbial community during phytoremediation of pyrene contaminated soil in simulated wetland. *J Hazard Mat*, 193, 45–51.

Zhao, F.J., Dunham, S.J., & McGrath, S.P. (2002). Arsenic hyperaccumulation by different fern species. *New Phytol*, 156(1), 27–31.

Index

Printed in the United States
by Baker & Taylor Publisher Services